Handbook of Electrical Construction Tools and Materials

Handbook of Electrical Construction Tools and Materials

Gene Whitson

McGraw-Hill
New York San Francisco Washington, D.C. Auckland Bogotá
Caracas Lisbon London Madrid Mexico City Milan
Montreal New Delhi San Juan Singapore
Sydney Tokyo Toronto

Library of Congress Cataloging-in-Publication Data

Whitson, Gene.
    Handbook of electrical construction tools and materials / Gene Whitson.
        p.   cm.
    Includes index.
    ISBN 0-07-069920-8
    1. Electric apparatus and appliances.   2. Electric engineering—Equipment and supplies.   I. Title.
TK452.W55   1996
621.319'3—dc20
                                                                                   95-42985
                                                                                                  CIP

McGraw-Hill
A Division of The McGraw-Hill Companies

Copyright © 1996 by the McGraw-Hill Companies, Inc. All rights reserved. Printed in the United States of America. Except as permitted under the United States Copyright Act of 1976, no part of this publication may be reproduced or distributed in any form or by any means, or stored in a data base or retrieval system, without the prior written permission of the publisher.

1 2 3 4 5 6 7 8 9 0   AGM/AGM   9 0 1 0 9 8 7 6

ISBN 0-07-069920-8

National Electrical Code® and NEC® are registered trademarks of the National Fire Protection Association, Inc., Quincy, MA 02269

Printed and bound by Quebecor/Martinsburg.

This book is printed on acid-free paper.

Information contained in this work has been obtained by the McGraw-Hill Companies, Inc. ("McGraw-Hill") from sources believed to be reliable. However, neither McGraw-Hill nor its authors guarantee the accuracy or completeness of any information published herein, and neither McGraw-Hill nor its authors shall be responsible for any errors, omissions, or damages arising out of use of this information. This work is published with the understanding that McGraw-Hill and its authors are supplying information but are not attempting to render engineering or other professional services. If such services are required, the assistance of an appropriate professional should be sought.

Contents

	Introduction	1
Chapter 1	Conduit Benders	3
Chapter 2	Pipe Threaders and Equipment	11
Chapter 3	Fish Tapes and Wire Pulling Systems	19
Chapter 4	Drills, Drill Bits and Screw Guns	37
Chapter 5	Saws and Blades	49
Chapter 6	Cable Tools and Equipment	57
Chapter 7	Welders and Torches	63
Chapter 8	Generators, Power Distribution, Lighting	69
Chapter 9	Digging and Compacting Equipment	75
Chapter 10	Material Handling Equipment	81
Chapter 11	Scaffolding	91
Chapter 12	Test Equipment	109
Chapter 13	Miscellaneous Tools	121
Chapter 14	Electrical Materials Introduction	139
Chapter 15	Conduit	143
Chapter 16	Conduit Fasteners	153

Chapter 17	Conduit Fittings	165
Chapter 18	Conduit Bodies	177
Chapter 19	Outlet Boxes	185
Chapter 20	Pull and Junction Boxes	221
Chapter 21	Wire and Cable	233
Chapter 22	Electrical Connectors and Terminators	259
Chapter 23	Introduction to Wiring Devices	293
Chapter 24	Plugs and Receptacles	297
Chapter 25	Switches	347
Chapter 26	Contactors and Relays	365
Chapter 27	Motor Controls	375
Chapter 28	Electric Lighting	393
Chapter 29	Overcurrent and Disconnecting Devices	437
Chapter 30	Electric Motors	455
Index		471

Handbook of Electrical Construction Tools and Materials

Introduction

Electrical contracting is a business requiring astute professionals in every phase of the operation. Due to the highly competitive nature of the business, there is no time or money to waste on incompetent people. Employees in the electrical construction industry will be expected to perform, know their job, take pride in their work, and do it well. Without the ability and willingness to adhere to these basic standards they cannot hope to succeed.

Long before the dawn of recorded history, men were inventing and using tools for prying, cutting, hammering and twisting. These tools greatly increased the quantity and quality of work produced, but at the same time, because they were heavy or sharp or provided great leverage, their capacity to injure the user also increased. All tools must be handled carefully to avoid injury. In some industries, more than half the bodily injury accidents involve the use of hand tools.

Fortunately, the safe way to use a hand tool is usually the best way, not only to avoid accidents but also to prevent damage to the tool. Take good care of your tools. Use them carefully and you will have less need for a first aid kit. Keep your tools clean, protect them against damage from corrosion, wipe off accumulated grease and if possible, occasionally dip dirty tools in cleaning fluids and wipe them clean. Lubricate adjustable and other moving parts to prevent wear and misalignment.

Keep the edge on cutting tools sharp. Sharp tools make work easier, improve the accuracy of your work, save time, and are safer to use than dull tools. Tools, like good friends, should be treated right and not abused. You can prevent many accidents by using the right tool for the work instead of a tool that is too heavy or too light. Use the most suitable tool and avoid misuse of tools, such as prying with a screwdriver or file instead of with a pinch bar.

Every successful electrical contractor has large sums invested in tools, equipment, materials, and employee wages. The purpose of this book is to identify the tools and materials used in the electrical construction industry and illustrate methods to preserve the value of the assets and properly use and control the materials.

Quality people using quality tools and materials in an efficient manner are the only means available to the contractor to fight inflation and maintain a competitive edge in the electrical construction industry.

This handbook has been prepared to enable anyone involved in the electrical construction industry to readily identify and determine the use of the principal tools and materials being used in the industry. By use of the handbook, many years of on-the-job training can be accomplished almost instantly.

Much time, money, and effort are expended in the electrical construction industry on tools, equipment, and material. It is not unusual in a contracting establishment to find $500,000 to $1,000,000 invested in tools and equipment, and the material inventory may run to several million dollars.

Proper tool and material flow to the various jobs is of vital importance in the electrical construction industry. Many studies have been made relative to the satisfactory profitable completion of jobs, and invariably, all of the studies show that one of the primary reasons for cost overruns is the lack of proper coordination of delivery of tools and materials to the project.

The persons responsible for the delivery of tools and materials to the job site must have a comprehensive knowledge of the tools and materials they will be handling. It is to these persons that this handbook is dedicated.

Chapter 1

Conduit Benders

The Greenlee No. 555 Bender is a high-speed bender built especially for bending $\frac{1}{2}$- through 2-in rigid steel and aluminum conduit. Bending shoes are also available for electrical metallic tubing (EMT) in sizes $\frac{1}{2}$ in through 2 in. Additional accessories include "shotgun shoes" which can bend four pieces of $\frac{1}{2}$ in or $\frac{3}{4}$ in rigid steel or aluminum conduit at one time, or three pieces of 1 in conduit.

This bender is electrically operated. It has one shoe in place on the front of the machine and one shoe in storage on the back. There are support rollers for each of the bending shoes and it is necessary to use the correct support roller for each bending shoe.

The bending shoes and supporting rollers are marked in bending stations. For example, a $\frac{1}{2}$ in through $1\frac{1}{4}$ in rigid shoe would be a three station shoe and require a three station supporting roller. In every case, the number of stations is dependent on the number of available conduit sizes covered by the bending shoe. The shoe for bending $\frac{1}{2}$ in through $1\frac{1}{4}$ in EMT would cover sizes $\frac{1}{2}$ in through $1\frac{1}{4}$ in, requiring a four station shoe. The supporting rollers for EMT bending shoes will not interchange with the rigid rollers and therefore cannot be used with the rigid shoes.

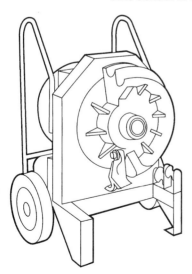

Greenlee No. 555 Electric bender

This bender is standard equipment for bending rigid conduit. It is recommended for making numerous bends of the same type and is therefore especially useful in performing prefab work. The EMT shoes and shotgun shoes are optional equipment.

Greenlee No. 880 Bender

The No. 880 bender is a hydraulic bender basically equipped with a hand pump and capable of bending ½ in through 2 in rigid conduit. Greenlee electric pumps Nos. 915, 940 and 960 are available and will operate with this bender. The bender is supplied with sizes ½-, ¾-, 1-, 1¼- 1½- and 2-in bending shoes. A large metal storage box is standard equipment.

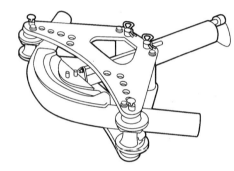

Greenlee No. 880 bender

Any bender that will bend rigid conduit will also bend aluminum conduit. However, rigid benders cannot be used to bend EMT unless special EMT shoes are used.

Oil protecting caps for the pump, hoses, and hydraulic rams are supplied with this unit. These caps should always be in place when the unit is stored, and removed only when the bender is being used. Dirt and water are the principal enemies of hydraulic systems and these caps prevent such contaminants from entering the hydraulic oil supply of the pumps.

All Greenlee Benders have a decal attached to the bender to indicate degrees of bend, radius, and length of travel. These decals are very important to the person operating the bender and should be treated with care. They should not be subjected to metal scrapings, piles of equipment placed on them or anything that might distort or destroy the decal.

As with all hydraulic benders, this one has no value unless there is enough hydraulic oil in the pump to obtain full performance and bending stroke. The oil level in the Greenlee hand pump can be checked by removing the back cap on the cylinder of the pump. There is a screw located under the cap which can be removed to check the oil level. It is difficult to add oil to the pump through the screw opening. The pump can be placed on end with the hose coupling facing upward. The pump head can then be removed with hand pressure. Removing the pump head will expose the body of oil in the pump. Place a clean stick or metal object carefully into the center of the pump reservoir and push the plunger to the bottom of the reservoir. Oil may then be added to fill the reservoir to the top.

Once any bender hand pump is refilled with oil, it should be connected to the ram and the ram pumped to its full extension. Place the pump on end with the hose coupling or release device on the pump facing up. Open the release valve fully, letting the ram retract completely. This procedure should be performed at least three times and will allow air to escape from the system.

Greenlee No. 882 Flip Top Bender

Bending capabilities of the No. 882 bender are $1\frac{1}{4}$ in through 2 in EMT conduit. This bender is also available with either a manual hand pump or an electric hydraulic pump. The No. 882 is designed, with use of proper shoes, to bend EMT, IMC, and rigid conduit. The flip top hinged frame top plate opens for easy and fast loading and unloading.

Cleanliness is essential to the proper operation of hydraulic equipment, especially with the various hoses and connectors. A speck of dirt inside a hydraulic valve can cause a malfunction in the hydraulic system.

The electric pumps have a vent which must be open when the pump is operating. This vent should be closed to prevent oil leakage and contamination when the pump is in storage. The pump should not be stored on its side, but should remain standing upright on the base, as designed for use in operation.

The electric pumps operate at 10,000 pounds pressure per square inch, the pressure being controlled by relief valves. Hydraulic oil blowing out of a broken hose could injure persons working around the pump. Therefore, any worn or leaky hoses should be replaced prior to operating the pump.

Greenlee No. 882 Flip Top bender

Greenlee No. 885 Rigid Bender

This bender is designed to bend heavy rigid (GRC) pipe and will bend $1\frac{1}{4}$ in to 4 in in one pass or 5 in pipe in a segmental mode. It is supplied with a large metal carrying container which should always be used to store

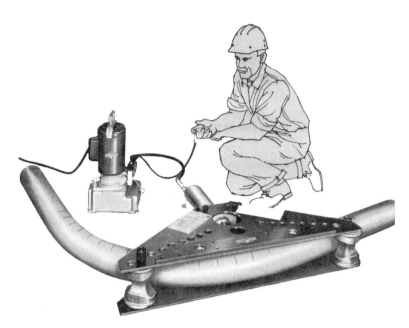

Greenlee No. 885 rigid conduit bender

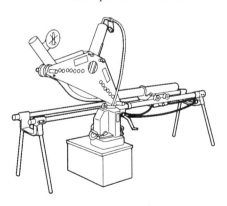

Greenlee No. 1802 bending table

the bender and applicable bending shoes. No other parts should ever be stored in this container as the parts required for operation of this tool are very heavy duty and not to be confused with lighter weight parts.

When using the hydraulic benders, it is often advantageous to establish a level surface working area for bending the conduit. This method enables the operator to obtain accurate bends. Greenlee supplies the No. 1802 Bending Table which is used with the No. 777, No. 880, No. 883, No. 883-4, No. 884, and No. 885 hydraulic pipe benders. The two ends of the table are connected with two ten foot lengths of standard 2 in rigid conduit which form the side shafts of the table.

Conduit is secured on the table by means of the chain pipe vise. A chain is used to slide conduit back and forward on the table. The table end components are secured to the side shafts by means of large hairpin-type clips.

Mechanical Benders

Greenlee No. 1800 ratchet bender

Much electrical work requires conduit in $\frac{1}{2}$-, $\frac{3}{4}$-, or 1-in sizes. Bending these smaller sizes of conduit does not usually require the use of the electric or hydraulic benders. Consequently, mechanical and Chicago-type benders have been developed.

The mechanical benders are mounted on a frame with wheels to support them when being used. A four or five foot length of 1 in conduit is used for a handle. The bender is capable of bending up to $1\frac{1}{2}$ in rigid conduit.

Greenlee No. 1818 Mechanical Bender

This mechanical bender is capable of bending $\frac{3}{4}$ in through 2 in EMT, $\frac{1}{2}$ in through $1\frac{1}{2}$ in rigid, and $\frac{1}{2}$ in through 2 in aluminum conduit. This bender is supplied with a set of six bending shoes, two follower bars, a roller pin, a cushion roller unit, a tie bar, and hairpin clips.

Greenlee No. 1818 mechanical bender

Hand Benders

EMT bender

EMT hand benders are capable of bending ½ in, ¾ in, 1 in, and 1¼ in EMT conduit. These are hand-operated, manual benders and consist of a shoe, with a hook on one end for holding the conduit in the shoe while bending, and a piece of rigid steel conduit for a handle. They are always marked to indicate the size conduit they will bend. Some of them list a size for bending rigid conduit also.

The type of bender designed for bending rigid steel conduit is often referred to as a "hickey." The hickey has a very short shoe with a holding hook. A pipe handle is always required to use either of these benders.

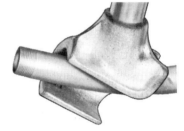
Hickey bender

Offset Benders

Greenlee "Little Kicker" offset bender

Many times, when the conduit is adjacent to the destination box, it is necessary to make a short S-shaped offset in the conduit to allow it to be connected to the box. The offset bender is designed specifically for this purpose.

The Greenlee No. 1810 is capable of making an offset bend in either ½ in EMT, while the No. 1811 is designed for ¾ in conduit. These are one shot benders and the ½ in can be used as supplied by the manufacturer. The ¾ in model, however, usually requires that it be mounted on a board base and secured by screws through adapters built into the bender.

PVC Benders

Greenlee No. 850 PVC bender

Greenlee manufactures a portable PVC conduit bender called the No. 850. This bender is heated by propane and uses liquid trimethylene glycol as a bending agent to transfer the heat to the conduit. A tank of butane gas with a pressure regulator is required to use this bender.

The liquid trimethylene glycol is supplied in a reusable five gallon bucket and the bender has a capacity of $7\frac{1}{2}$ gallons of the agent. When the conduit is inserted in this bender, it becomes very hot, so it is necessary to wear gloves.

The No. 850 bender is used to bend $\frac{1}{2}$ in through 5 in PVC conduit. Once the desired bend is completed it must be cooled with water to return the conduit to a rigid state.

The heat for the No. 850 is controlled by an internal thermostat, and the heat-up time is 20 minutes. Large 5 in conduit will be heated to bending temperature in less than two minutes, while $\frac{1}{2}$ in may be bent after about 30 seconds.

Electric PVC Benders

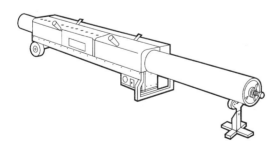

Typical electric PVC bender

The Electric PVC Benders are available in 110 V and 220 V sizes. The 110 V bender will handle $\frac{1}{2}$ in to $1\frac{1}{2}$ in PVC and the 220 V will handle up to 6 in conduit. The 220 V bender is mounted on wheels to allow it to be easily moved from one location to another. Both the 110 V and the 220 V units are equipped with wooden handles to protect the operator from the hot metal sides of the heater.

PVC Heating Blankets

The PVC heating blankets are available in three sizes for $\frac{1}{2}$ in to $1\frac{1}{2}$ in PVC conduit, 2 in to 4 in PVC conduit and for 5 in to 6 in PVC conduit. The heating blanket will permit the bending of PVC to any angle or radius and will also heat and realign conduit already installed.

These heaters are versatile enough to be used on stub-ups for alignment into gutters or panels, on conduit emerging from walls or floors, or for overhead or trench applications. The thermostat control provides even heating for uniform bending and the narrow internal wire spacing heats PVC conduit 20 to 40% faster than older-style blankets.

Multiple circuit wiring allows blankets to generate heat even if an internal wire is damaged. The deeply-grooved internal hook surface helps prevent conduit from slipping and the pre-curled design allows easy wrapping around conduit.

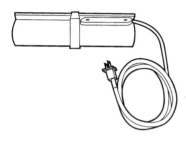

PVC heating blanket

Chapter 2

Pipe Threaders and Equipment

In the electrical construction industry, it often becomes necessary to cut pipe, and in many cases, to thread or rethread both pipe and bolts. To accomplish either cutting or rethreading operations, various tools are used, such as pipe cutters, reamers, dies, power drives, ratchets, and wrenches. Cutting of pipes and bolts by use of various types of saws will be illustrated in a later chapter.

Ratchet Pipe Threaders

Ratchet pipe threader

These are manually operated. The threader handle has a built-in ratchet with the capability of operating both forward and reverse. The units are available in sizes 00, lllR and 12R ratchet-type drop-head threaders and will accommodate dies for threading $\frac{1}{8}$ in to 2 in pipe. The dies can be detached from the handle and the $\frac{1}{8}$ in, $\frac{1}{2}$ in, $\frac{3}{4}$ in, 1 in, $1\frac{1}{4}$ in, $1\frac{1}{2}$ in, and 2 in dies will all fit in the ratchet head of the same handle.

For efficient operation it is necessary to remove excessive thread filings from the threading teeth in the dies and there should not be any broken teeth in the die. The teeth can be replaced by removing the four screws on the top and inserting the dies according to numbers to match the numbers on the die head.

The die teeth can be taken out and reversed to obtain a flush thread or to thread a short piece of pipe sticking out of a slab or through a wall, etc.

Three-way pipe dies are also available. These are basically the same tool except that they are not reversible and three die heads are built onto one tool.

The ratchet threaders may be used in conjunction with a power head by mounting the pipe in the power head and either holding the threader in position or locking it in some way.

Power Drives and Threaders

Ridgid No. 141 threader

The Ridgid Tool Company supplies the No. 141 Geared Pipe Threader. The No. 141 is an adjustable die that handles $2\frac{1}{2}$ in, 3 in, $3\frac{1}{2}$ in, and 4 in conduit with one set of dies. There are two handles on the side of the die head that are pulled directly out and turned to the properly-marked size for the pipe to be threaded. The No. 141 is usually driven by either a power head or the 700 style portable threader. The dies in this unit can be replaced. In replacing them, there is no need to remove the top of the threader. There is a lock screw at the back of the adjustable section where the die is turned toward the 4 in cutting size. Remove one screw, turn the dies a little past the 4 mark and the dies will pull out of the head. When replacing the dies it is necessary to match the identifying numbers with the numbers on the head.

Cutting oil must be used to lubricate the pipe and dies when using this threader.

Ridgid 4 PJ Series Pipe Threader

This threader is supplied with $2\frac{1}{2}$ in through 4 in dies. The main difference between this threader and No. 141 is that it takes a die change to change pipe size, as these dies are not adjustable. Each die of the set of five must be removed completely from the threader. The dies are marked $2\frac{1}{2}$ in, 3 in, $3\frac{1}{2}$ in, and 4 in.

Most dies have automatic stops which operate a clutch release and stop the threading action if the machine is not shut off at the proper time. This will prevent damage to the dies and possibly to the machine or the operator. The start and stop position of the die is marked on the drive shaft. There is a ring labeled Stop and the machine should be stopped when this position is reached.

Two types of drive shaft may be used to adapt the 4PJ to a power head: the Universal shaft and the Ridgid No. 844 drive shaft. The Universal Shaft is a flexible shaft, adjustable in length and used to connect the threader to the power head. This shaft is a necessity when threading a piece of pipe that is already bent to any degree. By inserting the hex end of the shaft into the power head, clamping the pipe in a stationary chain vise and setting the threader on the end of the pipe, the die will revolve on the pipe and cut the desired threads.

The Ridgid No. 844 drive shaft is shorter and more compact than the Universal shaft. It has a female receiver that adapts to the drive end of the die and is secured with an Allen wrench. The drive shaft is placed into the throat of the power head and the die hangs from the power head. A Ridgid 758 loop is available with this set-up, which is secured to the die and loops around one of the support arms on the power head.

When using this short-shaft method, a pipe support roller is needed if the pipe is 24 ft or longer. When threading pipe in this manner, the pipe is inserted into the die and the pipe does the turning. When using the Universal shaft, the die does the turning.

Pipe Cutters

Pipe cutter

Pipe cutters are available in sizes $\frac{1}{8}$ in, $\frac{1}{2}$ in through $1\frac{1}{4}$ in, $\frac{1}{2}$ in through 2 in, 1 in through 3 in, 2 in through 4 in and 4 in through 6 in.

The smaller sizes have one cutting blade and two rollers, while the large sizes may have two cutting blades and two rollers. All of the cutters have a handle which is threaded and used to adjust the pressure on the cutting blade. For ease of operation and a smooth cut, care should be taken to assure that all rollers are clean and turning properly, that the threading adjustment arm on the unit is clean and turning properly, and that cutting wheels are sharp, free from nicks, and are turning freely.

Ridgid No. 700 Portable Power Drive

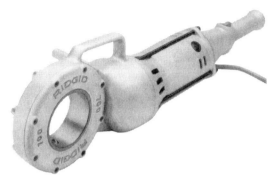

This tool is often referred to as a pony threader. It will accept dies such as the Ridgid 12R die set, from $\frac{1}{2}$ in through 2 in and it is used to thread conduit. There are attachments available to convert it to a puller and an adapter which enables it to be used to drive the large $2\frac{1}{2}$ in to 4 in pipe dies. Though the No. 700 is classed as a heavy duty electric power tool, it is comparably light in weight and easily portable.

Ridgid No. 700 Power Drive

Ridgid No. 535 Pipe and Bolt Threading Machine

This machine has the capacity of threading $\frac{1}{8}$ in through 2 in conduit and $\frac{1}{4}$ in through 2 in steel rod.

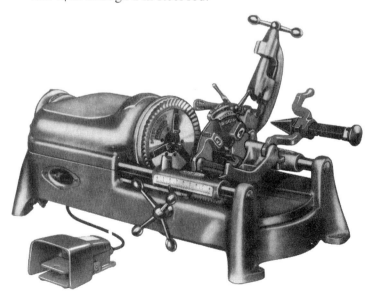

Ridgid No. 535 pipe and bolt threading machine

The 535 has a built-in oiler, pipe cutter and a pipe reamer. The die heads are interchangeable and adjustable. One die head will cut $\frac{1}{2}$ in and $\frac{3}{4}$ in pipe by merely adjusting the die to proper size. The 1 in to 2 in die head will cut by adjusting the die in the same manner. The die heads are simple to remove or replace as they snap in and out of the machine.

The machine has forward and reverse operating positions and the dies have a quick-release feature which allows the dies to be raised off of the work and the work removed without having to reverse the machine.

A nipple chuck is available for the No. 535 that enables the threading of nipples up to 2 in.

The power-driven threaders can be extremely dangerous because of the very powerful gear drives. Anything such as clothing, hair, fingers, or hands which are caught in the machine are apt to be torn off, as the machine may make six or more revolutions even after the power is turned off.

Ridgid No. 200 and No. 300 Power Heads

The No. 200 and No. 300 heads handle pipe dies ranging from $\frac{1}{8}$ in through 2 in and bar and bolt thread $\frac{1}{4}$ in through 2 in.

The two power heads are basically the same, with the difference being in the weight of the two units. The No. 300 weighs approximately 77 lbs and the No. 200 has a net weight of almost 120 lbs.

Each of the power heads has a forward and reverse switch which will not move directly from forward to reverse without a pause in the Off position. This pause is designed to prevent the motor from being switched from one direction to the other while it is still running.

The No. 200 and No. 300 may be utilized with geared threading heads with $2\frac{1}{2}$ in to 4 in dies and also geared threading heads for $2\frac{1}{2}$ in to 6 in dies.

Ridgid No. 300 power head

The power heads are also used to wrap pipe with plastic tape. The pipe is inserted into the front of the machine and is supported at the other end by a roller stand. When being used in this manner, the front and rear jaws of the machine should be locked onto the pipe for proper support of the pipe.

A tripod stand is generally used to support the power heads, but they may also be mounted on any firm flat surface which will support not only their weight but also the torque generated when they are in use.

Ridgid No. 450 Tri-Stand Chain Vise

The Tri-Stand Chain Vise is used to secure a section of pipe in a stationary position so that it can be cut or threaded. The vise has a roller-type chain similar to a bicycle chain, only much heavier, and a screw-down handle. The handle and chain must be kept lubricated to prevent rust from making them inoperable.

Ridgid No. 450 Tri-Stand vise

Pipe Chain and Strap Wrenches

Various sizes of chain wrenches are available which will handle pipe from $\frac{1}{2}$ in to 6 in in diameter. The chains have ear grips on them which lock the chain onto the handle. The chains are replaceable, which allows the handles to be used with a new chain. The strap-type wrenches are designed to prevent marring the outside finish of the pipe.

Pipe wrench

Oil Bucket and Pump

Ridgid No. 318 oiler

The oil bucket and hand pump are used to apply cutting oil to the dies and pipe during the cutting or threading operation. The hand pump pumps oil onto the work through a rubber hose connected to the oil storage bucket. The bucket is designed with a screen and pan on top which screens out the filings when the oil is returned to the bucket placed under the work. This procedure saves the oil and prevents an oil spill on the floor. The oil in the bucket should be checked often to ascertain that there is no water in the oil. Water pumped onto the dies during operation will destroy the dies.

Pipe Taps

Pipe tap

Pipe Taps are commonly available from $\frac{1}{8}$ in to 2 in and are available in larger sizes for special applications. There are two basic types of pipe taps: tapered and straight thread. The straight thread can be used for all electrical conduit applications but not for pipe that has to hold water pressure, etc. The tapered tap can be used for electrical applications and is a necessity for anything that holds pressure.

Reasonable care is required in storing the taps to prevent dulling of the threads. Cutting oil should be applied when they are used.

In addition to cutting new threads, the taps may also be used to chase or clean out existing threads.

Machine Taps

The machine thread taps are used on all fixture boxes and wiring devices, etc. The popular sizes are $\frac{6}{32}$, $\frac{8}{32}$, $\frac{10}{32}$, $\frac{10}{24}$, and $\frac{12}{24}$. The first number indicates the screw size and the second number indicates the number of threads per inch. A "T" handle wrench is used to hold the tap centered in the work and to turn it. Also available are plug taps and bottoming taps in the same sizes. The plug taps have a tapered point for easy starting, but will

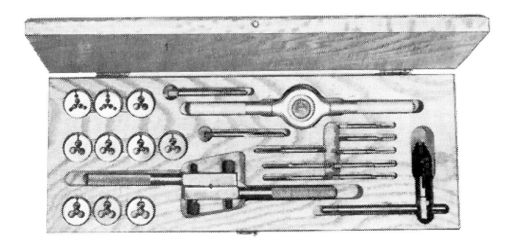

Machine tap and die set

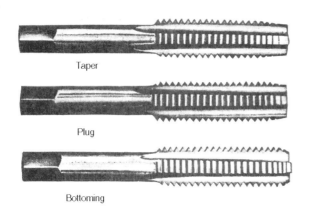

The three types of common taps

not thread to the bottom of a blind hole. The bottoming taps are used to run thread to the bottom of a drilled blind hole.

Chapter 3

Fish Tapes and Wire Pulling Systems

Jet Line Fish Tapes

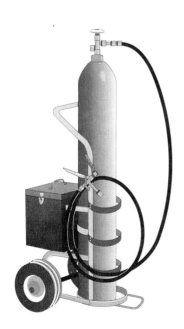

Jet Line fish tape system

The Jet Line system is a portable CO_2 system used to blow a string through the smaller sizes of conduit. This string, in turn, is used to pull in heavier cord or rope that will be attached to conductors or cable that will be pulled into and through the conduit system.

Cartons of string for use with ½, ¾ and 1 in conduit are supplied with the unit. The string is inserted and secured into a round piece of sponge rubber (often called a *piston* or *mouse*) which is sized to fit the inside diameter of the conduit.

The Jet Line CO_2 bottle is attached to the conduit by means of a tapered rubber cone that fits over the end of the conduit. Then the Jet Line bottle is triggered, releasing enough CO_2 to push the piston-like string container through the conduit to the opening on the opposite end.

The CO_2 in the Jet Line bottle is under high pressure and should be handled with care to prevent injury. The gas bottle, nor any of the other Jet Line attachments, should never be used as a blower for any purpose other than for wire-pulling operations.

Steel Fish Tapes

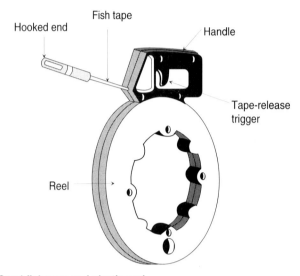

Steel fish tape and plastic reel

Steel fish tapes are available $\frac{1}{8}$- and $\frac{1}{4}$-in widths and in lengths of 50 – 100 ft. They are normally shipped and stored in a circular container called a *reel*.

In use, the tape is manually pushed through the conduit between pull points. Conductors are then attached to the end of the tape as shown in the illustrations. After being lubricated, the conductors are pulled back into the conduit to complete the conductor run between pull points or outlet boxes.

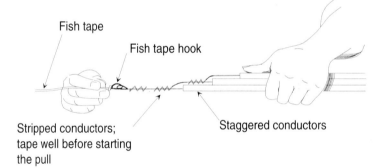

Conductors are attached to the end of the tape and then pulled through the conduit system

Nylon Fish Tapes

Nylon fish tapes are made of nonconductive material and are especially valuable when pushing into an energized switch gear or service box. The nylon tapes are available in 25-, 75- and 100-ft lengths. Due to a tendency to bind or bunch-up in conduit, the 100-ft lengths are seldom used.

Fiberglass Fish Tapes

Fiberglass fish tape

This relatively new type of nonconductive fish tape pushes stiff and straight, yet it is flexible enough to go easily through multiple 90° bends in different planes. Fiberglass fish tapes are rapidly replacing nylon types.

Greenlee No. 690 Vacuum Fish Tape

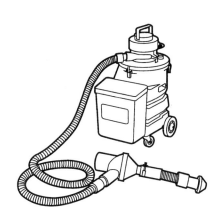

Greenlee No. 690 vacuum fish tape

The vacuum fish tape is used in the reverse order of the jet line fish tape. Instead of blowing a rubber piston through the conduit, the vacuum pulls the piston through. The machine creates a strong vacuum and also may be used as a power blower.

The sponge rubber pistons used to vacuum through the conduit are available in sizes $\frac{1}{2}$, $\frac{3}{4}$, 1, $1\frac{1}{4}$, $1\frac{1}{2}$, 2, $2\frac{1}{2}$, 3, $3\frac{1}{2}$, and 4 in. Three sizes of pulling string available are the 45, 72, and 90 lb test. The vacuum hose supplied with the machine is similar to a vacuum sweeper hose and is flexible, wire-lined and plastic-coated. The hose is connected to the applicable conduit by a metal tube with a rubber adapter to fit the various sizes of conduit.

A string dispenser is supplied with the machine which consists of a metal canister called a play out gun. The ball of string is placed in the canister and a trigger release allows the string to flow freely.

Crevice tools and dust-brush attachments are available for use with the No. 690. The unit may be used as a vacuum cleaner to clean panels, etc. Before use as a vacuum cleaner all of the various tools and attachments should be removed from the vacuum tank.

Hand-Powered Wire Puller

The hand-powered wire pullers consist of a cable on a reel, a ratchet device and a handle. The cable has a looped eye on the end to facilitate fastening the wire or cable to be pulled to it. The ratchet has a release and reverse lever that allows the pulling cable to be dispensed freely to the beginning position. The two shafts on the handle side of this tool are for developing low and high torque and speed.

Hand-powered wire puller

Motor Driven Wire Puller

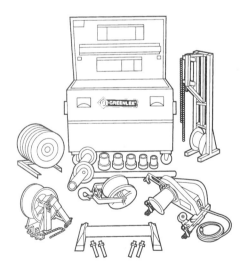

Greenlee No. 686 wire pulling system

The Greenlee No. 686 is a good example of a power driven wire pulling machine. The heart of the Greenlee No. 686 is the Pullins motor and capstand. The machine is supplied with vise chains and hand screws that allow the unit to be mounted on a pipe or solid beam.

The motor is equipped with an anti-reverse lock which prevents a back lash if the pulling rope or cable snaps under the strain of being pulled. A 300-ft length of $3/4$ in Polypro rope with a rope dispensing stand is standard with the machine. An extension T-boom, pipe adapter and sheaves are also available.

Easy Tugger Cable Puller System

Greenlee Easy-Tugger wire-pulling system

The easy tugger puller system features easy, one man set-up. The system is fast, pulls up to 22 ft per minute (fpm), and light weight.

The system includes a cable puller power unit, pipe adapter sheave, a $5/8$ in × 300-ft multiples rope, a rope stand and storage box.

Rated pulling capacity is 2000 lbs maximum and 1400 lbs continuous pulling speed. The unit will attain a pulling speed of 11 fpm at 2000 lbs or 22 fpm at 1400 pounds.

Tugger Cable Puller Package

This complete system includes the Tugger power unit; pipe adapter sheave; flexible pipe adapter; 12-in hook-type cable sheave; mobile extension boom; 2, $2\frac{1}{2}$, 3, $3\frac{1}{2}$, and 4 in extension bushings; $9/16$-in × 300 ft double-braided polyester rope; rope reel stand; floor mount and steel storage box with casters. The unit will pull with 4000 lbs of force with less than 36 lbs of operator effort. The right angle sheave allows the operator to stand out of the direct line of force. An audio alarm high-force warning and circuit breaker shut-off at maximum force is included for added safety. There is a tapered capstan and patented rope guidance system for better operator control.

Mounting of the unit is accomplished by captivated mounting chains and serrated grippers for positive mounting. Vise chains are included.

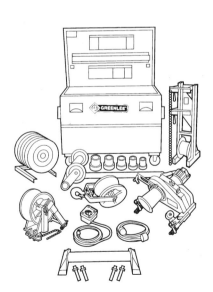

Greenlee Tugger cable-pulling package

The capacity of the unit is 4,000 lbs maximum and 2,500 lbs continuous. The pulling speed is 6 ft per minute at 4,000 lbs and 16 fpm at 2,000 lbs.

Self-Propelled Cable Puller

For heavy pulling jobs the Kebi self-propelled cable puller offers a 5000-lb line pull with a boom extension kit that is 28 in wide.

The puller drive and steering is battery operated which permits easy portability of the unit on the job site. The heavy-duty capstan on the top of the machine and built-in sheave on the front permit the handling of large cable.

The machine is equipped with sturdy crank-up jacks to stabilize it during operation. The large pneumatic balloon tires on the wheels facilitate use over rough terrain and also offers protection for floors.

The Kebi self-propelled cable puller

The Kebi Self-Propelled Reel Carrier

The Kebi Company also manufactures a self propelled reel carrier. The machine is driven and steered with battery power to permit one man to handle four 36 in reels or three 42 in reels.

Kebi self-propelled reel carrier

Built-in jacks and stabilizers lock the carrier in position when in operation. The permanent sheave on the front end guides the cable while it is being pulled.

Cable Paralleling & Coiling Machine

Many times on the larger type projects, it is necessary to pull parallel runs of cable. Chances are the supplier will impose a fee for paralleling and cutting to length the cable ordered. Other times the supplier is not equipped to perform this function. In these cases it is better if the electrical contractor has the equipment for measuring and paralleling the cable.

The model No. 2132B Cart (take up model) is ideal for use in the contractor's yard or on the job site. It has a 2500 lb capacity and will handle reels up to 54 in diameter by 32 in wide.

The complete unit includes turning axle, bushings, replaceable bearing pads, swivel meter bracket, and is illustrated with power unit and meter accessories.

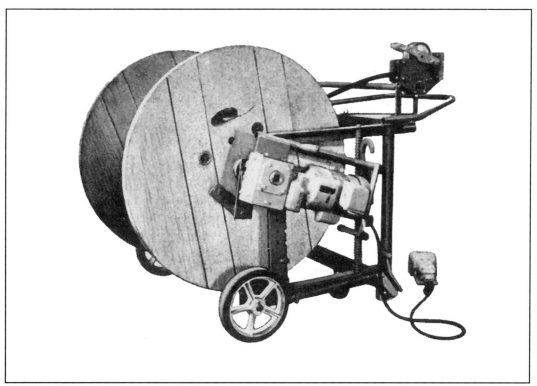

Cable paralleling and coiling machine

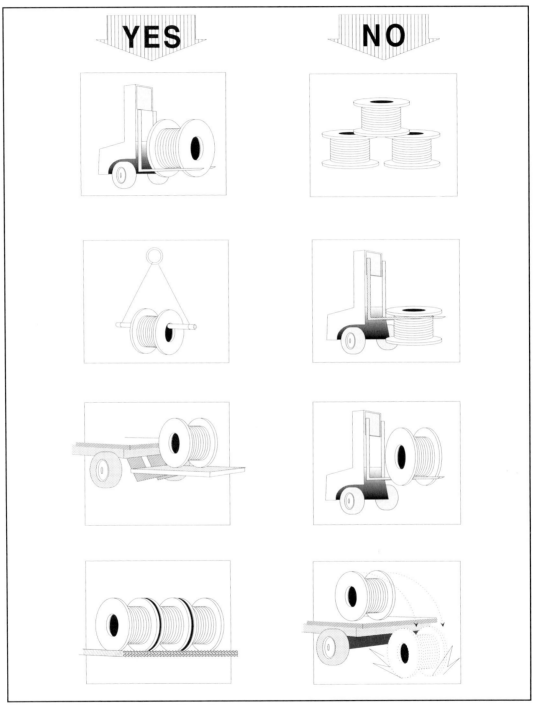

Do's and don'ts of transporting cable reels

Wire Pulling Lubricants

When pulling cable into a conduit, it is necessary to lubricate the insulation to reduce friction on the cable making it much easier to pull and reduce the strain on the insulation. The Ideal Company produces six types of lubricating compounds, each with a different composition making them suitable for use with different cable or for specific pulling applications.

Yellow 77 Wire Pulling Lubricant

Ideal Yellow 77 wire pulling lubricant

The "Yellow 77" is a general purpose construction and maintenance wax compound that allows a smooth pull without damaging the insulation. The wax-based formulation is exceptionally slippery and protects the cable during the pull. A special additive causes the lubricant to coat the cable and conduit walls. The smooth creamy texture and paste viscosity makes this type easy to handle and apply uniformly to the cable.

The "Yellow 77" will not wash off and the pull can be made through standing water without losing coverage or lubricity. The substance is non-toxic and safe to use almost anywhere as it meets all OSHA and Toxic Substances Control Act standards. This pulling compound is non-staining and allows easy clean-up with mild soap and water.

The lubricant is available in 15-oz aerosol, 1-qt squeeze bottle, 1-qt container, 1-gal pail, 5-gal pail, and 55-gal drum.

Yellow 190 Wire Pulling Lubricant

The "Yellow 190" lubricant may be stored and used in high temperatures without breakdown as it is stable from 30°F up to 190°F and has excellent freeze-thaw recovery characteristics. The lubricant may be used with all types of cable jackets since the compatibility range is extended to include low density polyethylene insulations that is common to applications such as communications cable.

Ideal Yellow 190 pulling lubricant

This type of lubricant can also be safely used on cable with a semi-conducting jacket.

Use of the 190 lubricant assures a smooth, easy pull as the outstanding lubricity is provided by the wax emulsion base. The average coefficient of friction is 0.055. The lubricant stays on the cable even if water is present in the conduit.

The lubricant is available in the same quantities as the 77.

Aqua-Gel 11 Cable Pulling Lubricant

The Aqua-Gel 11 cable pulling lubricant is a clean, slow drying, water-based compound especially formulated to provide maximum tension reduction in high-stress electrical and communications cable pulling operations.

The lubricant has a superior lubricity with an average friction coefficient of just 0.052. It dries to a semi-liquid, waxy film that will not clog and retains high lubricity characteristics for months after application. The lubricant is safe to use in all job environments with all types of cable jackets including low density polyethylene and semi-conducting jackets.

Aqua-Gel 11 comes in 1-qt squeeze bottles, 1-gal pails, 5-gal drums, 55-gal drums and also in convenient bags to cover the full range of applications. The bags can be easily adapted to various pulling situations. They can be used to pre-lubricate the conduit or inserted into conduit at access points, such as vaults or manholes to help relubricate. Multiple packs can be used at various intervals for long runs.

Bags are packed in a 5-gal pail to protect against breakage or leakage. The bags are available in half gal size (designed for conduit with a 3 in or larger I.D.) and qt size (for conduit with a 2 in or larger I.D.).

Aqua-Gel CW Cable Pulling Lubricant

Aqua-Gel CW cable pulling lubricant is a versatile, all weather compound for use in pulling all types of cable even at sub-freezing temperatures. It can be stored between jobs without loss of performance due to freezing. The lubricant will not freeze until -28° F, yet contains no alcohol, cosolvents, or other toxic ingredients. The lubricant is compatible with all cable types and maintains superior lubricity even at low temperatures with a coefficient of friction of 0.055 providing additional protection to cold, stiff cables. Ease of application allows pouring, hand or pumping methods.

Aqua-Gel CW is packaged in 1-qt squeeze bottles, 1-gal jugs, or 5-gal pails.

Aqua-Gel 11P Cable Pulling Lubricant

Aqua-Gel llP cable pulling lubricant offers easy and simple application. Once the pull rope has been brought back through the conduit, pour or pump approximately $\frac{1}{3}$ to $\frac{1}{2}$ of the pail into the conduit (for normal pulls up to 200 ft, one $3\frac{1}{2}$-gal container can be used initially. For pulls greater than 200 ft, a 5-gal pail is recommended.) Begin the pull. With approximately $\frac{1}{3}$ of the pull complete, pour or pump the balance of the lubricant into the conduit and complete your pull. Based upon the length of the pull and conditions, additional lubrication may be needed.

Pourable, non-staining formulation clings to the cable eliminating hand application which provides a cleaner and safer application. The superior lubricity offers a coefficient of friction of .050. The pour or pump application options save time and money and the non-flammable, non-combustible formula assures safer use.

The lubricant has a wide usable temperature range from 28° to 180° F. Universal application ensures compatibility with all cable jackets including semi-conducting material and low density polyethylene.

The product is available in $3\frac{1}{2}$- and 5-gal pail, 55 gal drum and $\frac{1}{2}$-gal bags (packed 6 bags per pail).

Optic-Lube Cable Pulling Lubricant

Optic-Lube is specially formulated for utilities, engineers and contractors who install delicate fiber optic cable. Use anywhere, it is non-corrosive on aluminum, copper, PVC, galvanized steel, or FRE conduit.

This lubricant has superior adherence characteristics and may be used with confidence on all types of cable jacket materials. It is non-staining, easy to work with and cleans with water. The compound is stable under temperature extremes from 32° to 200° F, non-flammable and environmentally safe.

Convenient sizes include 16 oz. aerosol, l-qt tub, l-qt squeeze bottle, 1-gal pail, 5-gal pail and 55-gal drum.

Wire-Lube Wire Pulling Lubricant

Standard wire pulling lubricant is soap based and suitable for shorter pulls. The lubricant dries to a slippery powder that will not harden or build up.

Cable Gripping Gloves

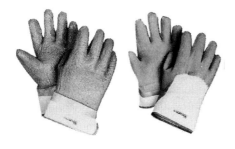

The cable gripping gloves are designed for handling cable that is covered with various lubricants. The non-slippery surface ensures a firmer grip on lubricated cable without wiping off the lubricant.

The gloves are available either rubber-or vinyl-coated in only one size.

Cable gripping gloves

Fiber Optic Cable Puller

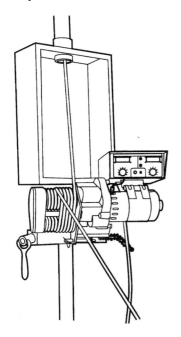

The fiber optic cable puller is designed specifically to pull fiber optic cable without damaging the delicate cable. The pulling speed and maximum pulling force can be set by the operator; the speed range is 0 to 75 fpm and the force is adjustable from 200 to 800 lbs. The system includes puller mounting bracket, mounting chain and steel storage case. The actual pulling speed and force are continuously displayed electronically while the machine is running. The machine shuts down automatically when the maximum preset force is reached to prevent damage to the cable being pulled. A gradual restart after automatic shut down prevents quick snap stressing.

Extra ports are provided to link the machine to the optional strip recorder to permanently document pulling force or speed. The dual grooved capstans provide excellent rope control.

Fiber optic cable puller

Manhole Sheaves

Manhole sheaves are generally used when working in a manhole or a deep type pulling box where the cable is being pulled from below ground to above ground. The sheaves require an anchoring device of some type and also a hold-down device to firmly anchor the sheave when being used.

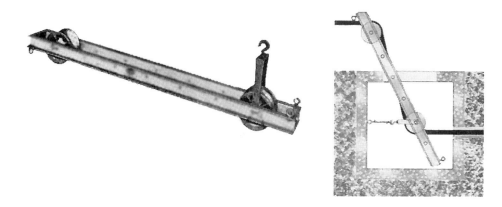

Manhole sheave

Right-Angle Twin-Yoke Sheave

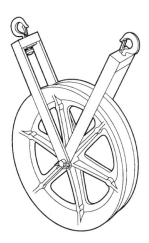

The right angle sheave has a minimum radius of $9\frac{3}{4}$ in. The sheave is constructed with strong welded steel yokes and has a cast steel closure-type hook with 1-in opening. The sheave is 5 in wide of aluminum alloy with self-lubricating bearings. The yokes rotate independently to fit various setups.

Right-angle twin yoke sheave

Tray-type Sheaves

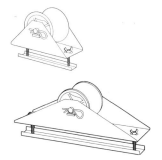

Tray-type sheave

The tray-type sheave has a steel frame with mounting that attaches to cable tray up to 2 in thick. The sheave helps in making roll and guide tray pulls easier to accomplish. The sheave is 5 in wide with self-lubricating bearings and is available in either 12 or 22 in lengths.

Feeding Sheave

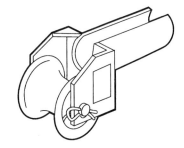

Feeding sheave

The yoke frame of the feeding sheave slides into the end of conduit and holds sheave allowing the cable being fed into the conduit to smoothly roll over the sheave. The sheave is 5 in wide aluminum alloy with self-lubricating bearings. The tube is split to permit easy removal after cable is installed. The sheave is built to withstand a pulling force up to 6,500 lbs. The feeding sheave is available to fit 2, 3 and 4 in conduit.

Greenlee Reel Jacks

The reel jacks are used to support the reel of cable or wire by use of a spindle placed through the center of the reel. The jacks are lowered enough to place under the spindle and then the reel is raised by use of the screw jack.

Another type of reel jack is the tip-up. With this type the spindle is placed through the reel, the jack is slipped on the spindle at an angle and then simply tipped to a standing position thus raising the reel to the desired height.

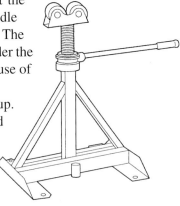

Reel jack

Fish Tapes and Wire Pulling Systems 33

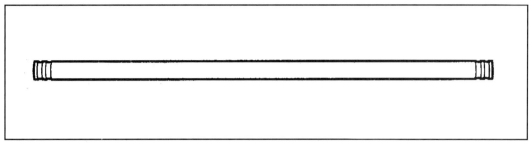

Heavy-duty spindle

Wire Dispenser

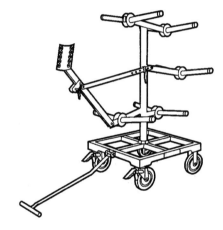

The dispenser is used to transport and pay out Romex, Greenfield and standard wire and cable. This model, pictured, handles reels in diamters from 24 to 40 in, and maximum width of 24 in.

Wire dispenser

Wire Cart

The wire cart is used to transport and dispense standard size 10, 12 and 14 wire from spools and will carry ten spools with built-in guides to separate the wires as they are pulled.

Wire cart

Greenlee No. 677 Wire Grip Package

The wire grips are used to secure the wire or cable to the pulling cable which greatly decreases the time required to set up and pull the cable or wire. The wire grips are sized by conduit size and available in the No. 674 for 2 and $2\frac{1}{2}$ in conduit, the No. 675 for 3 and $3\frac{1}{2}$ in conduit and the No. 676 for 4 and 5 in conduit. Each grip has four holes in the bottom with two set screws penetrating each of the four holes. To use, the cable is stripped and enough strands or conductors are cut so the holes in the grip can be filled 50% to 75% with wire conductor. The cable is secured in the grip by use of the set screws.

The wire grip is secured to the pulling rope by use of a clevis. The clevis may also be used to connect two ropes together as long as the first rope has an eye at each end.

Wire Pulling Sock

The terms "pulling sock" or "wire pulling basket" refers to many types of wiring pulling grips. In most cases the sock is made of woven steel wire, is round, has one open end and one eyelet end.

When the sock is compressed, cable may be inserted into it and when the sock is pulled by the eye, it tightens on the cable. (Much like a Chinese finger puzzle).

Another type cable grip is called the open basket grip or split grip. These are used when the end of the cable is not available. They are the same basic design as the sock grip with the exception that the basket has a steel rod

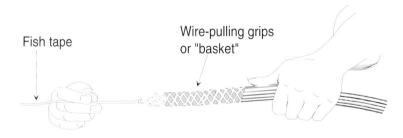

Wire pulling sock, sometimes referred to as "basket"

running the length of the basket. When the rod is removed, the grip opens up and can lie flat. It is then placed under the cable, wrapped around the cable and the rod reinserted gripping the cable in the middle.

Greenlee No. 435 Conduit Measuring Tape

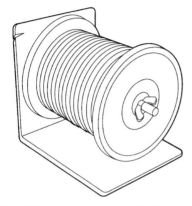

The conduit measuring tape is connected to the wire pulling device and fished through the conduit. Markings on the tape indicate the length of the pull so that enough cable can be supplied to fulfill the requirements of the pull.

Conduit measuring tape and holder

Chapter 4

Drills, Drill Bits and Screw Guns

The Occupational Safety and Health Act, commonly referred to as OSHA, has played an important part in the construction industry over the past several years. This act, which has become the law of the land, will continue to be an important consideration in the construction industry for years to come. Due to this fact, it seems pertinent that some attention be paid to the provisions of the act which apply to the use of the tools described in this manual. This manual will not include the full text of the act nor complete coverage of all the safety and health standards and other requirements of the law. Nor is this manual intended or represented to be a complete safety manual.

Under Section 1910.3012, entitled Hand Tools, the act states that contractors shall not issue or permit the use of unsafe hand tools. Impact tools shall be kept free of mushroomed heads. This includes drift pins, chisels and wedges. Wooden handles of tools shall be kept free of splinters and be tight in the tools. Under the title *Electric-powered hand tools*, Section 1926.300 and .302, the Safety Act states that electrically-powered and operated tools shall either be of the approved double-insulated type or grounded. Positive ON-OFF control switches are required for hand-held scroll saws, jig saws, drills, etc., with blade shanks $\frac{1}{4}$ in wide or less. Momentary contact On-Off control switches shall be on all hand-held power drills, saber saws, etc., and may have a lock-on control, provided that turn-off can be accomplished by a single motion of the same finger or fingers that turned it on. Constant-pressure switches are required on other hand-held tools such as circular saws.

Under Section 1926.401-F and 1926.402, under the heading, Extension Cords, the Safety Act states that extension cords used with portable electric tools and appliances shall be of the three-wire type. Extension cords shall be protected against such accidental damage as may be caused by traffic, sharp corners or projections, and pinching in doors or elsewhere. Extension cords shall not be fastened with staples, hung from nails or suspended by wires.

Anyone using drilling equipment should be aware of the correct procedures for safe operation of the equipment to protect both the user and others in the vicinity.

Drill Motors

Typical $\frac{1}{4}$-in drill motor

Prior to using a drill motor, the motor itself should be checked to ascertain that there is not an excessive arc at the brushes, that the cord is in good safe condition and that the switch releases promptly. If the drill is equipped with a switch lock, it should be checked for proper operation. When the switch is depressed and the lock engaged, the drill should remain running. The slightest pressure on the switch must release the lock and cause the drill to shut off when trigger pressure is released.

Some drill motors are marked as double-insulated drill motors and have a plastic, non-conductive case. If this type drill carries the proper UL label, a grounded cord is not required.

Cordless Drills

Cordless drill motor with recharger

The cordless drill can be used anywhere at anytime, because the power source is built in. There are no cords to store; nothing to plug in. The tool is perfect for on-site work where availability of electric outlets is a problem or where electricity may be a hazard. They are generally used for light drilling only with the $\frac{3}{8}$ in or smaller bit sizes.

This type of drill is powered by a self-contained power pack. A charger is also included as standard equipment. Some of the chargers and power packs will recharge in as little as two hours and others require up to sixteen hours to fully recharge. Ni-Cad batteries should be fully discharged before being recharged because they tend to develop a memory over time and if recharged when only partially discharged, they will only recharge to that level. This will also shorten the life of the battery because of the necessity for frequent charging.

Larger Drill Motors and Variable Speed Motors

The standard drill motors and the variable speed motors, both with and without reversing switches, are becoming very prevalent in the construction industry and also with do-it-yourselfers. They are time-saving tools and can be used to drive and remove screws and small taps in addition to the regular drilling operation.

Milwaukee $3/8$-in drill motor

These motors are very similar in appearance to the $1/4$ in drill motors, with the exception that they are generally slightly larger.

If the drill is rated as a variable-speed drill, it should be plugged into the power supply and the trigger slowly depressed to make sure the variable-speed feature is working. If the drill is also of the reversing type, a check should be made to determine that the reversing switch is working properly.

D-Handle Drill

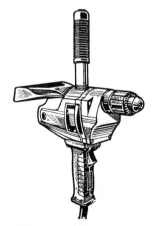

One type of D-handle drill

The $1/2$ in D-handle drill motor is much more powerful than either the $1/4$ in or $3/8$ in, and has a D-shaped handle plus an additional rod handle, because this type of drill requires both hands to hold it when it is being used. This drill runs slower than either the $1/4$ in or $3/8$ in drills, but is more powerful and will easily drill through heavy steel.

Hammer Drills

The hammer drill is used to drill in concrete or masonry and requires special bits with carbide tips. It also uses designated masonry bits and hammer-drill bits. However, masonry bits will quickly dull in a hammer drill, due to the hammer action. The hammer drill bits are hardened to a greater degree than the standard masonry bit and will remain sharp when subjected to the hammer and rotation of the hammer drill.

The smaller hammer drills are normally supplied with an adjustable guide rod to indicate drilling depth. The drilling depth can be predetermined by setting the rod that parallels the bit to the desired length. The rod will stop the bit from drilling deeper when it reaches the concrete surface, thereby setting the drilling depth.

Some hammer drills have a lever or slip collar which disengages the hammer action. This type of hammer drill can be used as a regular drill motor when in the disengaged mode.

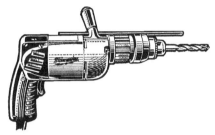

Hammer drill

Right-Angle Drill Motors

The right-angle drill is used extensively when installing electrical systems, particularly when drilling wooden studs, joists, or other wooden members in close quarters. The right-angle head allows this type of drill to be used in much closer quarters than a regular straight drill.

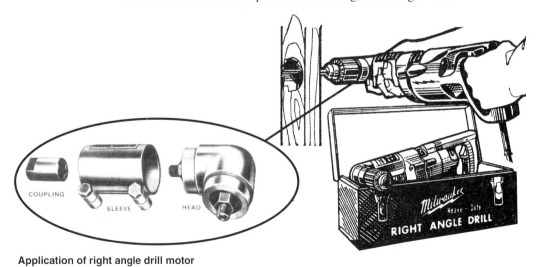

Application of right angle drill motor

Drills, Drill Bits and Screw Guns 41

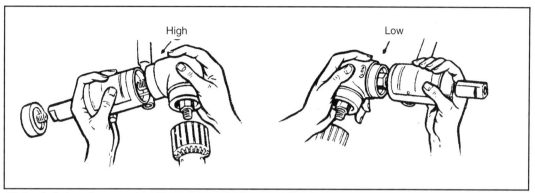

Method of changing speed and torque on a right angle drill

Two different drilling speeds are possible with the Milwaukee right-angle drill; a reverse switch is also provided. It is possible to change the torque rating of chuck transfer from low to high or vice versa. To change the speed and torque, it is necessary to remove the chuck, invert the right-angle fitting, and then replace the chuck.

A 33° angle long extension is available to assist residential electricians in drilling overhead joists for running wires. With this extension, the electrician can snap a chalk line on the bottom of the joists and do all drilling from floor level. The 33° angle gives the operator the proper angle to push or pull the bit.

Long extension used on the right angle drill to bore through overhead ceiling joists from floor level

Roto Hammer

The name roto-hammer applies to a drill that not only rotates a drill bit, but hammers at the same time. In addition to drilling for, say, setting bolt anchors, the tool may be used with chisels and points for chipping or breaking concrete. Flat and tapered bits are also available. The flat bit produces a flat-bottomed hole which is more suitable for anchors.

The tapered bits are generally used for drilling in cinder block or concrete masonry. When using the chisels and points, an adapter is used which prevents the bit from turning. The adapter is an internal hex bushing made of steel, with a rubber shield that clamps to the front of the hammer.

Proper oiling of the roto-hammer action is required — some of the tools have an oil reservoir and require that it be kept full of oil. Operation of these tools without sufficient oil will quickly cause the hammer action to fail.

Roto hammer in use

A bit-retaining spring on the front of the hammer will hold the bit firmly in the hammer if it is triggered without the bit being against a solid surface.

Carbide-tipped core bits are available for some roto hammers. The core bits are used for cutting larger-size holes, as the cutting action is only about $1/8$ in around the circumference of the bit. Drilling with core bits is more efficient than drilling with a standard solid bit and also causes less strain on the hammer.

Attachments are available for the large-type hammers such as clay spades for digging in hard clay or earth, and also ground rod driving adapters to drive ground rods.

Roto-Hammer Bits

Several different types of bits are available for use in roto hammers to perform various tasks. The names of the bits should be self-explanatory.

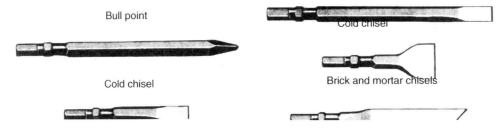

A few of the many roto hammer bits available to workers

Drill Press Stands for Hand Held Drill Motors

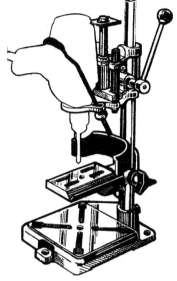

Drill press stand

Drill press stands are lighter than the standard drill presses which are used in the shop, and are principally used in the field. The drill motor is mounted to the drill press stand by a strap or roller chain clamp.

When using the drill press stand, it should be firmly mounted on a solid surface by use of bolts through the base of the stand. C-clamps may also be used to mount the stand.

Motor Driven Drill Press

The motor driven drill press is capable of performing heavier drilling jobs than the lighter weight drill press stand for hand held drill motors. When using the drill press, it is necessary to secure the work being drilled, as the extreme power of the press makes it impossible to hold the work by hand.

Most drill presses are capable of variable speeds, as it is important that the correct speed be used for the material being drilled. A rule of thumb is "the heavier the metal, the slower the speed." When using a drill press or any drill, the idea is to let the bit do the cutting, especially with the leverage available on a drill press. If the filings are turning blue, too much speed or pressure is being applied.

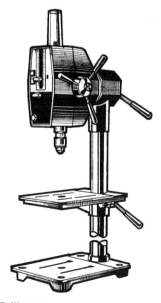

Drill press

Concrete Core Drills

The concrete core drill is used to drill into prestressed concrete, concrete floors, walls or ceilings. The machine is held in position on the surface to be drilled by large suction pads which are held in place by vacuum. There are rubber seals at the base of the suction pads that form the seal for the vacuum. The complete machine consists of the stand with wheels, jack screw, vacuum pump, vacuum reserve tank, vacuum hose, two vacuum pads, ammeter, bubble level, base leveling screws, cradle for mounting the drill motor, wheel feed control and water hose with fittings.

Concrete core drill

The vacuum reserve tank prevents water from entering the vacuum pump, maintains vacuum reserve in case of power failure, and stabilizes the amount of vacuum necessary for the vacuum pads to hold. The vacuum hose connects the vacuum pump to the reserve tank and the tank to the pads. The jack screw at the top of the drill stand is used to secure the unit to the surface to be drilled when the vacuum suction pads cannot be used. A piece of pipe is placed on the jack and the jack is raised to hold the pipe against an opposing surface such as a wall or the ceiling.

The ammeter indicates the power being applied to the diamond bits through the wheel feed control. The ammeter has a green and red indicator on the dial, with the green area indicating safe operating pressure. The bubble level and base leveling screws enable the operator to set the drill level and straight in relation to the surface being drilled.

The diamond bits require a constant flow of water for cooling during the drilling operation. The water hose and fittings are used for this purpose. If a source of water is not available at the location, a water storage tank is available to supply the required amount of water. A water collector pump and ring are accessories which provide a means of collecting the water used in situations where the water cannot be allowed to run off naturally. The rings are figure eight-shaped devices which allow water to be pumped into one of the rings and then pumped out of the other ring to a disposal area.

Core Bits

Core bits are available in two basic types:

- General purpose
- Open back

Core bit

For general core drilling in reinforced concrete, tile, masonry and marble, diamonds are surface-set. Functionally-designed waterways allow water to flush out cuttings, permitting longer bit life. The open back bits are available in sizes $1\frac{1}{2}$ in and larger, and require a three-piece friction drive for easy positioning of the bit and sleeve; this also permits easy bit removal. The bits are female threaded and require an adapter to permit them to be screwed onto the drill motor.

Bit extensions extend the drilling depth of the core drill. An example would be when the drill is placed at the top of a stairwell; an extension would be required to drill through a stair step.

Wood Boring Bits

Wood boring bit

Large diameter wood boring jobs are generally performed using a drill motor and a bit constructed in one piece, with the cutter blades welded to the shank. This type of bit is self-feeding and has a replaceable pilot screw that fits all sizes of the bit, eliminating the need to match parts to various sizes of bits.

The cutter blades may be sharpened on the job with a file (there are only two cutters to each bit) and will not separate when backing the bit out of the drilled hole. The bits are fast cutting, easy to use and rust resistant.

Self-Feeding Wood Boring Bits

Self-feeding wood boring bits are faster than an auger bit, and require less effort on the part of the operator. Furthermore, self feeding bits usually have a longer useful life. This type of bit cuts wet wood and hardwoods, producing a very clean hole in both. The bits are precision-ground, but can be sharpened with a file on the job site. They have replacement lead screws and indentations on the shank for extension bar set screws. The bit will not disengage from the extension during use. Extension shanks are available in $5\frac{1}{2}$-, 12-, and 18-in lengths.

Bellhanger bit

Bellhanger Bit

This type of bit is used to drill small holes for telephone, doorbell, or other low-voltage wiring. The bit is designed for drilling wood or plaster, but will not be severely damaged if soft metal, such as copper or aluminum, is encountered in the process. The bit is available in standard sizes of 3/16 in through $\frac{1}{2}$ in, and in lengths from 18 to 24 in. Extensions for the shank of the bit are available for use in either electric drill motors or in a hand brace.

Power Ship Auger Bits

The ship auger bit is designed for deep, fast, heavy-duty power boring. The design includes:

- Feed-screw pilot for easy, accurate starting.
- Properly-angled helical pitch to clear chips without clogging.
- Precision-angled chisel surfaces for a smooth, straight true bore and easy removal.

The bit should be heat-treated and tempered along its entire length to enable it to remain sharp, resist wear longer, and go right through nails without damage to the bit or distortion of the hole.

Ship auger bits are available in 6-in to 18-in lengths to permit drilling straight holes in heavy timber. This type of bit is used for heavy construction and repair in mines, on docks and bridges, and for electrical work.

The deep, open throat design effectively clears chips from the hole during the drilling process. The shanks are $\frac{3}{16}$ in hex shaped to fit $\frac{1}{2}$ in and larger three-jaw chucks.

Power ship auger bit

Screw Guns

The screw gun is also referred to as a screw shooter or drywall gun. Basically, screw guns are $1/4$ in variable speed drill motors with a special chuck that accepts a $1/4$ in hex shank bit. There is a clutch on the tool that prevents the screw from being overtightened, stripped, or broken.

Screw guns are used to fasten lather's channel, lighting fixtures, or any other application using screw-type devices for anchoring. Some brands of screw guns have adjustable chucks so that torque can be changed from one setting to another to accommodate various materials into which the screws are driven. In general, the torque setting is adjusted so the designated screw will firmly seat against the object being fastened, and to a point of torque to not break or strip screwheads.

Various attachments are available for use with screw guns. Phillips head, hex head, sheet metal or straight slot bits can be used, and most of the guns have a magnetic bit holder available.

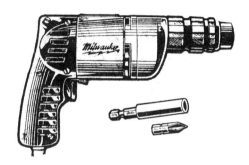

Screw gun with adjustable torque clutch

Whitney Punches

Small hand punches, often referred to as Whitney punches, are used to punch holes in light metal from $3/32$ in to $3/8$ in. The punches are supplied either individually or in sets, and with a straight handle or a ratchet handle.

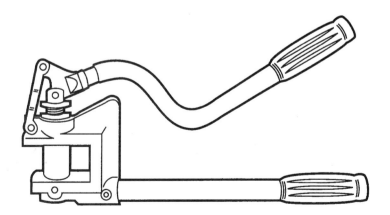

Typical hand punch

Knockout Punches

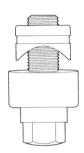

Typical knockout punch

In the electrical industry, it is often necessary to provide holes in metal panels much larger than those possible with the standard drill motor and bit. For these jobs, electrical workers use knockout punches, often referred to as KO punches.

The punches range in size from $\frac{1}{2}$ in to 5 in, with the stated size indicating the size conduit that can be inserted in the punched hole; that is, a punch designed for $\frac{1}{2}$ in conduit will actually cut a hole that is $\frac{7}{8}$ in diameter.

In use, all punches require a pilot hole, the size of which is determined by the knockout punch size. Pilot holes are drilled with a regular drill motor and bit. If a larger size pilot hole is needed after this, smaller sizes of knockout punches may be used to enlarge the hole for larger knockout punches.

Knockout punches $\frac{1}{2}$ in through $1\frac{1}{4}$ in may be driven by hand, using a standard wrench to turn the bolt that forces the punch through the metal. The hand driven punches have a thrust bearing that eases the pressure required to turn the drive bolt.

Ratchet knockout punch drivers provide enough leverage for the operator to ease punches from $\frac{1}{2}$ in to $1\frac{1}{4}$ in through the metal. Standard knockout punches are used with the ratchet unit, but the bolts are locked onto the handle, which in turn drives the bolt.

Hydraulic knockout punch set

Hydraulic knockout sets are used to punch the larger holes up to 5 in. These sets require a hydraulic hand pump which operates at very high pressure.

Greenlee 1731 KO Punch Driver

This punch driver has cutters for $\frac{1}{2}$-, $\frac{1}{4}$-, and 1-in holes. The tool has the appearance of a large C-clamp and has an 8-in throat. The die and punch adapter set is a time-saving device not requiring a pilot hole. A hydraulic pump, either hand operated or electric, is required to drive the punch.

Greenlee 1731 KO punch driver

Chapter 5

Saws and Blades

Portable Band Saw

Portable band saws are manufactured by all of the major tool companies such as Rockwell, Milwaukee, Ridgid, and Greenlee. All of their portable band saws are basically the same type tool.

The portable band saw is used for cutting conduit, plastic pipe, or Unistrut. Consequently, most material cuts are made with this tool. The blade tooth requirement for material cutting is the same as with other hacksaws. An 18-tooth blade will do almost all cutting, while a 24-tooth blade is recommended for cutting electrical metallic tubing (EMT), and a 14-tooth blade for cutting the heavier rigid conduit (GRC).

Portable band saw

The blade drive wheels on the band saw have a rubber tire on which the blades ride, and this should be kept clean of foreign material, bits of abrasion, metal filings, etc. The wheel bearings should be free-spinning and the mounting screws should be tight.

Some of the saws are two speed, high and low. Normal material cutting is done at high speed. The low-speed setting is used for problem cuts.

When cutting with a portable band saw, the saw stop should be against the work before turning the saw on; otherwise the blade will carry the saw forward and the operator with it, until it reaches the stop position. Whenever possible, be certain that the cut is started on the surface where the greatest number of teeth will contact the material being cut. For best results, the blade should not contact the work until the motor has started. The saw is designed so its weight alone furnishes proper pressure for cutting. It is not necessary to bear down on the saw during the cutting operation.

The two handles on the saw are provided so the saw can be accurately guided through the cut. On completion of the work, the machine should be held firmly so it will not fall against the work.

Band Saws

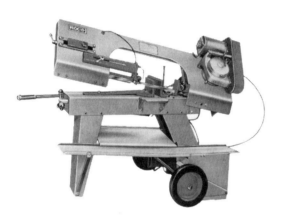

Table-mounted band saw

The band saw is a table mounted saw that also can be used in a vertical position. The saw is mainly used for cutting larger conduit or for production cutting of metal parts. This saw makes a straight 90° cut and will handle material up to 8 in wide with a capacity for rectangular objects 7 in × 11 in and round objects up to 7 in in diameter.

The saw has three operating speeds: 76 RPM, 141 RPM, and 268 RPM. A work table can be mounted on the saw when it is being used in the vertical position. The capacity of the saw in the vertical position is 6 in deep and 9 in high. The saw is mounted on wheels for mobility, with a pull-out handle to pull it by. When being used, the four feet of the machine should be on the ground or mounting surface. The wheels must be in the raised position to insure that the saw is stable and cannot move during operation.

Chain Saws

There are many brands of gasoline-operated chain saws available, but all of them require extreme caution when operating. Care must be taken when starting a chain saw. It should be secured so it cannot move during the starting phase. Saw chains should be kept sharp and the tension properly adjusted. Tension on the chain can be determined by taking the blade on the lower guide arm and pulling down on it. It should pull free of the guide

Chain saw

arm with a reasonable amount of effort and pressure, but should also, when released, return to the guide-arm. The oil reservoir on the chain saw is used strictly to maintain an oil level which allows oil to be expelled onto the chain while cutting is taking place.

Usually, the oil is expelled by thumb action, pressing the plunger while the saw is running to keep oil on the chain blade. Oiling the chain results in cooler cutting action and longer chain life.

The saw should be kept free of build-ups of chips and cuttings. When using oil, which is used at all times when cutting, the oil will cause these chips to stick and accumulate in the clutch and drive areas of the saw blade. The clutch and drive areas should be kept clean, washed out with solvent and blown out with air. If air is not available, debris in these areas should be brushed out.

Bench Saws or Table Saws

The table saw is used extensively for ripping and cut-off work. While they are very efficient, they are also an extremely dangerous piece of equipment. The operator's eyes must be protected with goggles when using this saw, and the saw blade guard must, without exception, be in place.

Many types of blades are available for use with the table saw, including plywood blades, cross cut and rip blades, combination blades and dado blades. Whichever blade is being used, it should be kept sharp. A dull blade is more dangerous than a sharp one.

If the blade binds in the work, the saw should be shut off before trying to free it. A wooden pusher should always be used to push or feed the stock being cut. Fingers should never be used to feed the stock into the saw.

Table saw

Sawzall

Sawzall is the brand name of a heavy duty reciprocating saber saw used to cut conduit close to the floor. It is not used as a general pipe cutting saw — just on special applications where a larger saw cannot be used. It can be used to cut out junction boxes or any type of sheet metal, and in hard-to-reach places. It is also useful for cutting drywall for installation of outlet boxes, and basic wood cutting such as 2×4 lumber.

Sawzall

The Sawzall is primarily useful for accomplishing rough work — a blade having 6 to 10 teeth per inch would be appropriate for this type work. For cutting most metals, a blade with either 18 or 24 teeth per inch should be used.

The saw standard equipment includes a blade assortment, two hex keys, hex key holder, blade clamp screw, steel carrying case, and an 8-foot heavy-duty cord. The case is designed to fit the Sawzall with up to a 6-inch blade installed and is constructed of heavy-duty 20 gauge steel.

Concrete Cutting Saw

Concrete cutting saws are powered by gasoline engines and some are equipped with battery driven starters instead of a rope pull starting system. The saws are mounted on wheels and are self propelled.

Two basic types of blades are available for use with these saws, carborundum or abrasive, and diamond. The carborundum blades are much less expensive to purchase but do not have near the life of the diamond blades.

Concrete cutting saw

The blades are supplied in different grades to facilitate cutting different material harnesses, such as old, hard, aged concrete as opposed to green concrete or asphalt. Asphalt cutting requires a different segment on the blade. The concrete cutting saw is not necessarily designed to cut a 4-in concrete slab to the depth of 4 in or to cut through to the bottom of the concrete. Usually the cut is made to a depth of perhaps $1\frac{1}{2}$ in and then the concrete is broken with a paving breaker. Using the more shallow cut greatly increases the life of the expensive blade, as the cutting life of the blade is not determined in only lineal feet cut but also by the depth cut.

All of the saws use water to spray on the surface being cut to keep the blades as cool as possible, so a constant water supply is required when using the saw.

Hole Saws

Hole saw

There are several manufacturers of hole saws with the basic difference being the construction of the mandrels which hold the cutting saw. Black and Decker is unique in that they have three sizes of mandrels fitting different sizes of hole saws, ranging up to 4 in or 5 in. The Victor company produces two different mandrels, one fits $\frac{1}{4}$ in drill and the other the $\frac{1}{2}$ in drill. Larger size hole saws from $1\frac{1}{4}$ in through 6 in require a mandrel to fit the $\frac{1}{2}$ in drill only because of the power required to drive the saw. All of the hole saws require the use of a pilot bit to stabilize the starting of the hole saw in the material being cut.

A straight high speed drill bit will not always work in a hole saw for a pilot bit. Most pilot bits in hole saws have a flat spot on the shank where a lock screw secures the pilot bit to the mandrel.

The hole saws are available in sizes ranging from $\frac{1}{2}$ in to 6 in cutting diameter.

Jig Saws

Jig saw

The small jig saws are used to cut in junction boxes and other outlet boxes or covers in various types of wood or metal, sheetrock, plaster, and many applications in plywood and light lumber. The saw is versatile in application, but the most important thing is to use the correct blade for the job. Metal cutting, wood cutting and scroll blades are available. The metal cutting blades are 14 tooth per inch for cutting up to $\frac{1}{8}$ in steel and a 12-tooth blade for thicker steel up to $\frac{1}{4}$ in. Eighteen- and 24-tooth blades are used for lighter types of metals. When cutting designs in wood, a scrolling blade is used. This is a thinner blade, not as wide as the heavy wood-cutting blade. The scrolling blade enables the cutting of smaller radius corners than can be turned with the wider blades.

General wood-cutting blades are 7 teeth per inch through approximately 10 teeth per inch. The more teeth per inch, the finer the blade and smoother the cut. The coarse (fewer teeth per inch) blades are used to cut heavier wood material.

Circular Saw

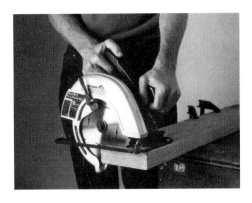

Circular saw

The circular saw is commonly referred to as a skilsaw. This saw is probably used more than any other type of saw because of its adaptability to so many sawing jobs. The light weight of the saw, compared to other power saws, allows it to be hand held while making vertical cuts as well as horizontal cuts.

While the saw is one of the most versatile saws, it is also a very dangerous piece of equipment. The high cutting speed of the blade seems to encourage the operator to force the saw into the cut, increasing the danger of the blade binding and causing the saw to kick back towards the operator. This can result in severe injury from the spinning blade coming in contact with the person using the saw.

A slight deviation from a straight line cut may also result in the blade binding, so it is imperative that the operator exercise care when using one of these saws.

Many types of blades are available for use with the circular saw. The wood-cutting blades are the most common, with the combination blade being used most often to make rip cuts or for cross-cutting. There is a rip blade used only for ripping with the grain of the wood, and a cross cut blade used only for cutting across the grain. Also available is a very fine-toothed blade for making smooth cuts in plywood, hardboard, masonite or thin plastics. This blade prevents chipping of the material. Metal-cutting blades and masonry-cutting blades are also on the market.

Coping Saw

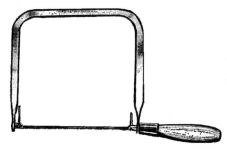

Coping saw

The coping saw has a walnut-stained hardwood handle and a high carbon-steel blade — hardened and tempered for durability.

Hand Hacksaw

Hacksaw

The hand hacksaw illustrated has a heavy-duty aluminum handle and rectangular steel crossbar. Spare blades can be stored inside its tubular back. The saw can be converted to a jab or compass saw. Tension of the blade is set by a lever to 28,000 psi (about twice that of ordinary hacksaw frames) with a steel tensioning nut. The blade holding pins are permanent hardened/serrated type.

PVC/ABS Plastic-Pipe Handsaw

This saw is of one-piece carbon-steel construction. The 2½ in wide blade is spring tempered, cuts on push and pull stroke, and resists flexing. The blade cuts plastic, wood, drywall, and nail-embedded wood.

Plastic-pipe hacksaw

Chapter 6

Cable Tools and Equipment

Cable benders, cable cutters and cable strippers are used to handle the heavier cables used in the electrical industry. These tools enable the electrician to be more efficient in installing large size cable. Some of these tools are hydraulic and require the same care as any high pressure hydraulic equipment. Dirt and water are enemies of any hydraulic system, so these tools must be kept clean.

The installation of high voltage cable is not a "guess at" job. These tools are manufactured with precision tolerances to perform a specific task.

Cable Benders

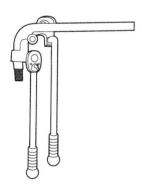

Hand operated cable bender

The hand-operated cable bender is sometimes referred to as a "cable hickey." It is used to bend larger sizes of wire and cable to make the 90° turns entering a service panel, or to bend the cable or wire to fit under a lug. Using a bender to shape the cable makes a neat-looking installation and sometimes is the only way the cable can be installed.

The hydraulic cable bender is equipped with several sizes of bending shoes to accommodate different-sized cable. The unit uses a foot-operated pump to supply the hydraulic pressure. The hydraulic unit eliminates the use of hand power, so it can be operated by one person. Cable benders are designed to be used on wire and cable, not for bending conduit or pipe.

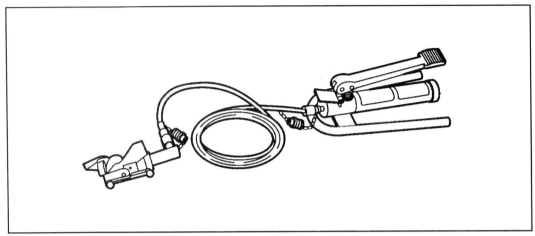

Hydraulic cable bender

Cable Strippers

Terminal cable stripper

Cable strippers are of two basic types:

- Terminal stripper
- Inline stripper

The Greenlee Model 1820 is for making a terminal strip only. The wire is inserted into the end of the tool. When the stripper is turned, the wire is stripped to the desired depth. This tool is equipped with a set of bushings that range in size to fit cable from 1/0 through 500 kcmil.

Cable Penciling Tool

Cable penciling tools are used to strip high voltage cable, cutting a tapered strip similar to the taper of a sharpened pencil, so connections can be made. They utilize a bushing to adapt to various sizes of cable. The cutters are marked to indicate the proper size cable to be used on. These

Cable Tools and Equipment

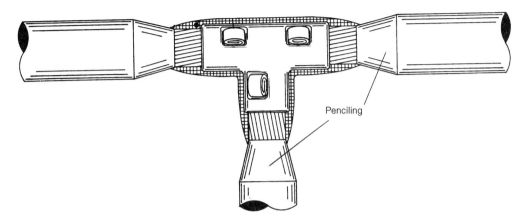

Penciling conductor insulation is a necessity for many high-voltage splices and terminations

tools make a precision tapered strip suitable for the installation of high voltage cable.

Cable Cutters

The cable cutters resemble standard bolt cutters. They are designed to cut aluminum or copper cable and are not to be used to cut steel of any type. The cable cutters are supplied as both a hand cutter and a hydraulic cutter.

Hand cable cutters

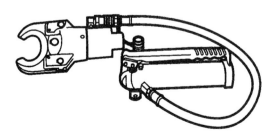

Hydraulic cable cutter

Compression Tools

These tools are used to compress lugs, sleeves, and for connecting wires or placing lugs on wire. They are also used for attaching one wire to the middle of a second wire, which would be a T-tap.

 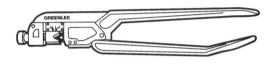

Hand-operated compression tools

The Thomas and Betts company (T&B) manufactures a series of compression tools with identifying codes such as TBM-2, which indicates a 2 ton compression tool. The 5 and 8 ton tools would be TBM-5 and TBM-8.

The lugs to be used with these tools are color-coded to match the dies supplied with the tool. The TBM-2 has four dies built in, which are simple to change by turning the handle. There are four dies per die head on the 5 and 8 ton tools, with six other die heads, each containing four dies per head, available for each.

The TBM-2 will handle copper lugs and connectors from #8 wire to #2 wire, taps from #10 though #6, and aluminum lugs from #12 wire to #6. The TBM-5 will handle #8 wire through 250 kcmil. The TBM-6 is suitable for copper wire from #8 to #400 kcmil, and splices and taps from #12 through #300 kcmil. The TBM-8 tool will handle copper lugs and connectors from #8 wire to #500 kcmil, and aluminum from #12 to #350 kcmil.

The TB-12 and TB-15 have dies which are color-coded and marked by number to indicate size. These larger size tools require a hydraulic pump of either the hand or electric type.

Some of the hydraulic tools have an option which insures positive com-

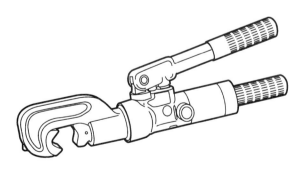

Hydraulic compression tool

pression. This will not allow the tool to be opened until it has been fully closed when compressing a lug. This eliminates the chance of a loose compression or loosely tightened lug to the wire.

The 12 ton hydraulic tool has the capability to compress #8 through #750 kcmil in copper and from #8 through #750 kcmil in aluminum.

The 15 ton hydraulic tool can handle all copper and aluminum connectors up to #1000 kcmil.

The TBM-2 through TBM-8 tools are supplied with complete die sets. The 12 and 15 ton tools have optional dies which must be ordered separately.

Cadweld or Thermoweld

Cadweld tools

The terms cadweld and thermoweld refer to an exothermic type welding process which requires no other source of power or heat. The reaction of the process is reduction of copper oxide by aluminum producing molten super-heated copper. The welding heat is obtained from the super heat in the molten copper. Because the reaction occurs quickly and with good thermal efficiency, it is ideally suited for joining copper conductors. The flow of molten copper into conductors is "cadwelded," causing them to be melted and fused into a solid, homogeneous mass.

The cadweld connection is completed before the heat is dissipated by the copper conductors. The principle of this process is to apply high temperatures for a short time. Total heat input is generally much less than that involved in brazing or soldering the same conductors. This is important, as the conductor insulation must be protected.

Other products that can be used in a cadweld mold or welded by this mold are common steel, stainless steel, copper-clad steel such as ground rods, galvanized steel, wrought iron, brass, bronze, and nickel chrome, to name a few.

The cadweld method cannot be used to connect aluminum cable. This method is used to join two pieces of copper conductor to a third piece, or one piece to a solid strand to make a T-type connection. Two pieces welded together make a butt splice or splice-type connection.

The components and processes required to make a weld are:

- Handles
- Mold of the correct size
- Flint gun
- Cadweld cartridges
- Disc
- Clean-out tool

The clean-out tool is used to scrape the inside of the mold after discharge. If a number of shots are being made, this tool is of great help. The tool is a tapered flat piece of steel, attached to a 90° handle, used to scrape the mold internally clean after each use.

Various accessory tools are available, such as the MT2-65 clamp, which will clamp three conductors in making a tap-and-run connection, thus holding the conductors in place while the weld is being prepared and made. The molds have an identification plate to indicate the size cartridge required, and the cartridges are packaged with a number, such as #45, #90, #115, #150, or #200 to indicate their size. Thermoweld is considered the same as cadweld.

Chapter 7

Welders and Torches

Welding equipment and its use and care occupy an important part of the Occupational Safety and Health Act. Concerning the storage of welding gas cylinders, the Act states that the cylinder truck, chain or other steadying device shall be used while the cylinders are in service. Cylinders shall be stored in an upright position to be serviced. Valve protection caps shall be in place and cylinder valves closed.

Before the regulator is connected, the cylinder valves shall be opened slightly, not more than $1\frac{1}{2}$ turns, and closed immediately, while the employee stands to one side. When a wrench is required, it shall be left in position.

Nothing shall be placed on top of fuel gas cylinders while in service. Fuel gas shall not be used from cylinders through torches or other devices which are equipped with shut-off valves without reducing the pressure through a suitable regulator attached to the cylinder valve.

When a regulator is removed, the cylinder valve shall be closed and the gas released from the regulator. Leaky cylinders shall be tagged and removed from the work area.

Welding hoses shall be inspected for defects at the beginning of each shift. Oxygen and fuel gas hoses shall not be interchanged and shall be readily distinguishable. Not more than 4 in in 12 shall be covered with tape. Hoses shall be kept clear of passageways, ladders and stairs.

Torches shall be inspected at the beginning of each shift for leaky shut-off valves, hose couplings, tip connections and clogged torch tips. Torches shall be lighted by a friction lighter or other approved device.

Electrode holders shall be specifically designed for this purpose, and welded, with grips and outer surfaces of the jaws insulated. When electrode holders are left unattended, the electrodes shall be placed so they cannot make electrical contact with an employee or a conducting object.

Welding cables shall be insulated, flexible and in good condition, with no splices within 10 in of the holder.

When practical, arc welding and cutting operations shall be shielded by non-combustible or flame-proof screens to protect employees working in the vicinity. Appropriate eye and face protection shall be provided and used by welders.

Gasoline-Powered Arc Welders

Most gasoline-driven welders are called generator welders, indicating they have both generating and welding capabilities. With electrical power, the unit, under certain circumstances, can be used as a generator. In most instances, the output will be direct current (dc), which can operate certain lights and hand-powered tools.

The larger gasoline-powered units are either trailer-mounted or built on their own frame, with wheels. Smaller units, very similar to portable generators, are mounted on wheels or sled-type frames with handles for pushing or carrying.

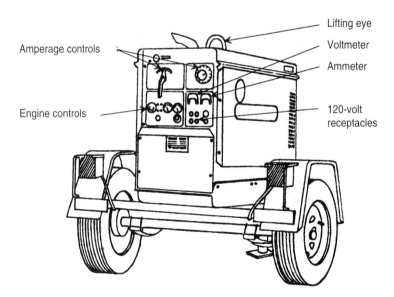

Gasoline-powered arc welder

In addition to the correct welding rod for the project, equipment, including a welding hood for face and eye protection, gloves, a chipping hammer, a wire brush, and required lengths of welding lead, is required to complete the system.

Smaller and more compact than the gasoline-driven welders, the electric welders are connected to a source of electrical power. Those that produce in the range of 225 A require a 220-V source. Small units producing 100 A may be connected to a 110-V source.

Electric-Powered Arc Welders

The electric welders require the same welding rod for any given project, plus the same equipment needed for use with the gasoline-driven units — a welding hood, gloves, chipping hammer, wire brush and lengths of welding lead.

225-A electric welder

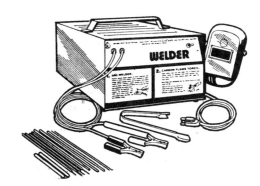

100-A electric welder

Spot Welder

Spot welder

For making small spot welds on lightweight metal, the spot welder is ideal. The unit requires a 220-V electrical source and produces 2.5 kV. The spot welder is hand held and easily moved.

Cutting Torches

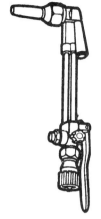

Cutting torch

Cutting torches are used for cutting steel to the length required, or cutting holes in steel plate or steel flooring. A complete torch set consists of the torch, one twin hose, two regulators, each with two gauges, one oxygen tank, one acetylene tank, a striker, and generally a tank cart.

Accessories required when using a cutting torch are goggles, gloves, tip-cleaning rods, and a wrench to fit the valve on the acetylene bottle. The goggles are a dark sunglass type, but are readily seen through. Two types of goggles are available, standard $1\frac{3}{4}$ in diameter round, and the coverall style, which fits over regular eye glasses.

The tip-cleaning rods are graduated in size, and used to clean chips and debris out of the cutting tip of the torch. The oxygen bottle has a hand-operated valve which does not require a wrench.

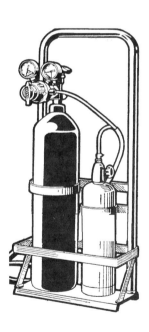

Industrial welding/cutting tank set with cart

The cylinders should be secured upright, in a cylinder cart designed for the purpose, and should also be transported in an upright position. If oxygen cylinders are removed from the cart, protective caps supplied by the cylinder dealer must be installed. It is against safety rules, and state and federal regulations, to transport the oxygen cylinders without the protective caps on, when not in the cutting outfit, or secured to cutting-torch stands. Each cylinder requires a regulator with two gauges on it. One gauge indicates the amount of gas in the bottle. The second gauge indicates operating pressure. Turning the adjusting screw adjusts the operating pressure on the smaller-numbered gauge. The acetylene gauge is scaled from 0 to 30, though this may vary with different manufacturers. The oxygen-operating pressure gauge reads from 0 to 100. The higher-scaled gauge on each regulator indicates the contents of the cylinder.

Bernzomatic Torches

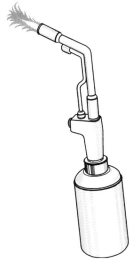

Small propane torch

Small propane torches are used for soldering, heating PVC conduit, and sometimes for removing paint. Various tips and torch heads are available and include:

- Pencil tip
- Flare head
- Round head
- Other accessories

Chapter 8

Generators, Power Distribution, Lighting

Portable Generators

Small residential portable generator

Generators are rated by wattage output in sizes 1500 watts (W), 2500 W, 3500 W and 4000 W, etc. The machines supply temporary power and will not supply more power than their watt rating. A 110-V unit rated at 1500 W will supply about 15 A, and a 3500 W unit puts out about 35 A. The 110/220-V combination generators' amperage output is reduced by one half or more. A 2500-W combination will produce about 10 A.

The value of the portable generator is to produce power to operate drill motors, power saws and other types of power equipment when no other source of electrical power is available on the job site. Most of the portable generators are started with a pull rope, but some of them have battery-driven starting motors.

The portable generators that are designated heavy-duty come with twist-lock outlet plugs, continuous-duty capability, circuit

Handbook of Electrical Construction Tools and Materials

Most portable generators designated as heavy-duty usually have twist-lock receptacles for both 120 and 240 V

breaker protection and up to a 50-A full capacity outlet.

More advanced units will have electronic voltage regulation, an electric starting full-pressure oil system, dual mainline and panel-mounted circuit breaker protection, low oil-pressure shutdown, and high-temperature shutdown.

The portables designated medium-duty are most often designed to operate intermittently.

Typical portable generators are available in sizes from 1500 to 4000 W

Trailer Mounted Generators

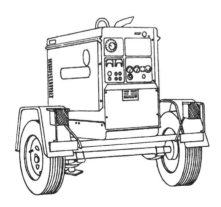

Trailer mounted generator

Large generators may be mounted on a trailer, with higher amperage output, an automobile-type gasoline engine, and large wheels to allow easy mobility. The large generator may be rated at 30 kW or higher.

Generators, Power Distribution, Lighting

Commercial and Residential Standby Generators

The stationary standby generator will sense a power outage and automatically start up to produce electric power needed to keep a business running, preventing a loss of sales, refrigerated inventory or production time. Residential installations produce enough power to supply the entire home.

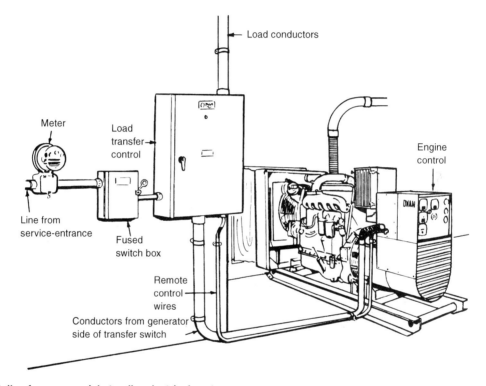

Details of a commercial standby electrical system

Both the commercial and residential models are available that operate on either diesel fuel, gasoline, natural gs or LP gas, have a transfer switch, and usually have a 120/240-V single-phase output. However, 120/208-V, three-phase units are also available for commercial electrical systems. The units are designed with automatic low-oil shutdown features, overcrank protection after starting, automatic voltage regulator with over-voltage protection, and 2-A timed trickle battery charger. The main circuit breaker has manual reset, the field circuit breaker automatic reset.

Temporary Electric Services

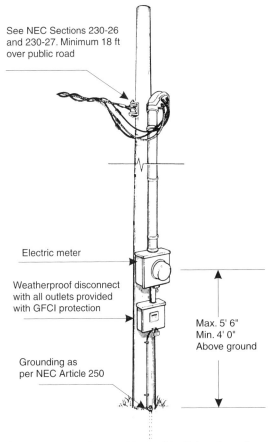

Typical temporary service used for residential and small commercial construction

OSHA regulations have put an end to the disorganized way that electrical power was supplied on the job site a few years ago. The regulations require that all job site power be supplied using ground fault interrupters and specified types of outlets fitting only approved cord cap types.

Most electrical contractors have developed their own style of temporary power boards incorporating the requirements of the regulations. The boards must supply 120 V, 220-240 V and in some instances 440-V power all the time in compliance with OSHA regulations.

Sizes of temporary services can vary from a 120/240-V, single-phase, 100 A service for residential and small commercial projects to 1200 A or more for a large commercial project, especially where electrically-operated construction cranes are being used.

Extension Cords

Extension cord

The electrical contractor will be required to distribute electrical power on the job site using many, approved, extension cords. The cords, depending on length and power demand, will be of sufficient wire size to provide the required power. The insulation must be durable enough to withstand the heavy traffic and usage common to construction sites. Lengths of 50 or 100 ft are common.

Temporary Lighting

Temporary lighting

Almost every construction project has a need for temporary lighting, which will be supplied by the electrical contractor. The lights are supplied in 100-ft lengths of heavy-duty cord, with lamp sockets spaced at 10-ft intervals. The lamp sockets must be provided with wire lamp guards for most construction projects.

Heavy-Duty Trouble Lamps

The heavy-duty trouble lamp is used to produce light in a closed area where the ordinary room lighting is insufficient to allow perfect vision, such as when working on conductors or overcurrent protective devices inside of large panelboards and motor control centers.

The lamps are available with cords of 25 and 50 ft lengths, constructed of 18 or 16 gauge wire, and suitable insulation. The lamps to be used on a job site must meet OSHA requirements.

Trouble lamp

Twist Lock Cord Adapters

Twist lock cord adapters have the twist lock cord cap on one end and a standard grounded plug on the other. This allows a cord with standard grounded cap to be used.

Portable Flood Lights

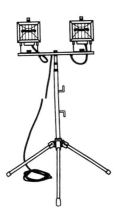

The portable flood light is used to brightly light a large area on the job site. One or two of them will often provide sufficient light for an entire room. They use a 500 or 1000 W quartz lamp for efficiency.

Portable flood lights

Chapter 9

Digging and Compacting Equipment

Two-Man Gas-Powered Post Hole Digger

Two-man gas-powered post hole digger

This machine is commonly referred to as "ground hog," or "hole hog." Two persons are required to operate the post hole digger because of the power developed by the gearing on the auger when the machine is digging. The machine has a safety clutch, which will slip if the auger binds while digging. This safety feature prevents injury to the operators.

Automatic clutches enable the unit to idle without the auger rotating. The centrifugal clutch locks in when the motor is accelerated. Several safety checks should be made before operating this dangerous machine. The auger must be firmly attached to the gear box drive shaft to prevent accidental disengagement and possible injury to the operators. The cutting blade attached to the auger should be sharp and in good condition for cutting. The handles should be straight, smooth, and firmly attached to the unit.

Manual Digging Tools

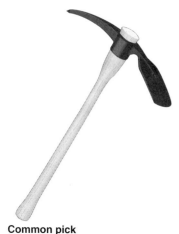

Common pick

The common pick and mattock are among the most used of the manual digging tools. The pick, with its sharp points, and the mattocks, with the sharp vertical and horizontal blades, can be driven into extremely hard earth to break it up into chunks small enough to be shoveled away.

Before using either of these tools, the wooden handle should be examined for cracks, straightness, and splinters. The handle should lock tightly in the head of the tool to prevent the possible sliding of the head down the handle when the tool is raised above the head of the operator.

Digging Bars

The long steel digging bar has a tapered end, useful in helping dig post holes. It also may be used as a pry bar to move heavy objects.

Digging bar

Shovels

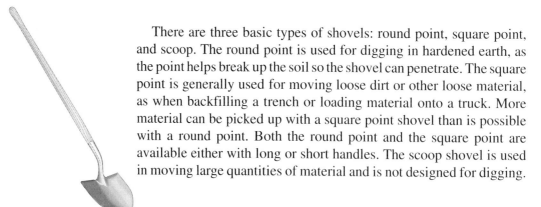

Round point shovel

There are three basic types of shovels: round point, square point, and scoop. The round point is used for digging in hardened earth, as the point helps break up the soil so the shovel can penetrate. The square point is generally used for moving loose dirt or other loose material, as when backfilling a trench or loading material onto a truck. More material can be picked up with a square point shovel than is possible with a round point. Both the round point and the square point are available either with long or short handles. The scoop shovel is used in moving large quantities of material and is not designed for digging.

Digging and Compacting Equipment 77

Square point shovel

Post Hole Spoons

The post hole spoons are either of scoop or straight spade design. The straight is used for cutting into soil and the spoon type is used for scooping out loose dirt. The digging head should be kept sharp and the handles should not be cracked or splintered.

Hand Post Hole Digger

Post hole spoon

The hand post hole digger is designed for one person use and consists of two hinged blades attached to two handles. The tool is raised and then driven down into the ground by hand. The handles are then pulled apart, which closes the blades and allows earth to be removed from the hole.

Hand post hole digger

Sharpshooter

The sharpshooter is a sharpened straight spade tool used to drive into and break up hard soil, principally when digging post holes.

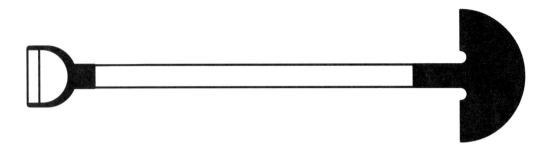

Sharpshooter

Long Handle Scraper

The long handle scraper has a straight sharp hoe blade about 5 in wide, used to scrape material sticking to pavement and floors.

Long-handle scraper

Mortar Hoe

The mortar hoe has two large holes in the blade and is used to manually mix mortar in a wheelbarrow or mortar box. The long handle enables the operator to reach to the end of the mortar box and pull the hoe through the mix.

Mortar hoe

Compactors, Whackers, Ground Pounders

These are power machines used to tamp and/or compact earth, such as in backfilling a trench and compacting asphalt patches. Since these machines come in various sizes, with varying capabilities, it is necessary to determine which machine will do a specific job.

Whacker

The term "Whacker" is a brand name, which also has become trade terminology, covering much compacting equipment. The Whacker-brand machine has a two-cycle engine. The clutch to start the tamping action is controlled by acceleration. When tamping, the machine develops considerable vibration. It should never be started on a solid surface such as concrete or asphalt pavement. Due to the automatic nature of the clutch, the machine should never be started in other than idle speed.

Ground Pounder

Ground pounder

The ground pounder has a vibrating plate either 20 or 24 in wide used to compact areas 20 in or wider in width. When in operation, extreme vibration is created, so the machines should never be operated on a solid surface. If the machine has a water holding tank, it is recommended for patching blacktop. The water is slowly trickled onto the blacktop and prevents the newly-laid blacktop from adhering to the tamping plate.

Front End Loader

The front end loader is mounted on a gas or diesel powered tractor. The front-loader and the rear mounted scraper are hydraulic powered and very useful in moving large quantities of various materials, loading trucks, large cleanup jobs, and to some extent, scraping jobs. Due to the size of the

Front end loader

machine and the power involved, the tractor and equipment should only be operated by an experienced operator.

Chapter 10

Material Handling Equipment

There have been tremendous developments in material handling machinery, but the human body is still probably the greatest material mover ever developed. As with the machine, improper use or abuse can damage the body. Generally, this damage is expressed in one of two forms, either a hernia or back strain. Both injuries are usually the result of over-stretching certain muscles, and can generally be avoided by using the proper lifting techniques.

Hernias are not generally caused by one great effort expended when lifting a heavy object. The average person, for example, could try to lift the biggest building in town once in his lifetime, and probably never suffer any noticeable effects of hernia.

However, continued extreme exertion, especially when done in a manner antagonistic to the mechanical structure of the body, is liable to eventually result in a hernia.

The other type of lifting injury that is far more frequent and just as uncomfortable as a hernia is back strain. Both injuries can be avoided, and in many instances, can be overcome.

The average abdominal hernia, sometimes called a rupture, is merely a portion of the abdominal contents pushed through the muscle wall of the abdomen. Ruptures are known to physicians as *hernias* when in the groin. Other types of hernias are also possible. Most hernias occur along the big blood vessels of the thigh. Umbilical hernias happen at the navel area.

The bulging generally associated with hernias, caused by the protruding loop of abdominal contents, may remain small for a long period, but it will more often tend to become larger.

The fundamental difference between a strain and hernia is that a strain is simply the stretching of a muscle, weakening it and causing pain. This does not necessarily result in a hernia. However, a rupture may result if stretching has occurred so frequently that the muscles are weakened and become unable to hold back the abdominal contents.

The primary cause of hernia goes back to infancy. In every male baby there are weak spots in the abdominal wall at birth. Within a short time, as the baby grows, these weak spots are usually strengthened. In some cases, however, this strengthening does not occur to the degree that it should, and this is where hernias may happen.

In addition, when abdominal muscles become weak and flabby, as generally happens with increasing age and lack of physical activity, the muscles are easily strained, especially around the site of the potential weak spot. Any sudden strain, even a violent cough or sneeze, may cause muscle stretching in these areas. In other words, hernias are not necessarily the result of heavy work. When the potential hernia site has been sufficiently weakened, the noticeable bulging of the abdominal wall takes place.

It is not true that hernias affect only workers in heavy industry. Of course, they are more likely to occur in workers doing heavy physical work than in those whose work, for the most part, is light, but even clerical workers are susceptible to hernias when their physical condition and unsafe lifting practices permit over-stretching of the abdominal muscles.

The basic preventive measure is in knowing how to perform safely those activities that ordinarily might cause a straining or stretching of the abdominal muscles. Of course, no one has developed a procedure for coughing or sneezing without effort. It is possible, however, to describe how a person can lift heavy objects with comparative safety.

The Right Way To Lift

First, get a good footing. Second, place your feet about shoulder-width apart. Third, bend at the knees to grasp the weight. Fourth, keep your back straight. Fifth, get a firm hold. Sixth, keeping the back as upright as possible, lift the weight gradually, by straightening the legs. When the weight is too bulky or too heavy for you to lift comfortably, get help.

In general, muscle strains due to lifting occur when an object is raised incorrectly. The object may even be a relatively light one, for it is not extraordinary weight alone that causes strain. It is when an object is lifted improperly — when the load placed on the body is poorly distributed, that excessive strains are apt to result. Because of the way the body is constructed, when the back muscles are called on to do work, the abdominal muscles must work, too. Therefore, an improper overloading of the back muscles can also strain the abdominal muscles.

Using your head can save your back. Strains of the abdominal or back muscles are frequently due to improper lifting. To suffer from these injuries is particularly unfortunate when the safe way to lift is at the same time the easy way to lift. Why? Take a look at the mechanics of the body. Your body is a mechanical system composed of a series of levers and hinges, actuated by cables, just like many machines. The levers are long bones, the hinges are the joints, and the muscles are the cables. Nature intended each cable, lever and hinge for specific jobs. As in every mechanical system, the part functions most satisfactorily when it is doing the work intended. Overload it suddenly or frequently and it will fail. The abdominal and back muscles each have a very important job. They must act as supports for the trunk, just as you would place guy wires on either side of a tall derrick to support it. So do these two sets of muscles support you. When you bend in one direction, it is the guy wire muscle on the opposite side that supports your body in that position, and when it contracts, it straightens you up. Now, if you were to place an excessive load on the guy wire on one side of the derrick and there was a weak spot somewhere along its length, it would snap.

The muscles on either side of your spinal column are composed of many fibers, each acting like a guy wire. When the guy wires get taut in the back or abdominal wall, there is an increase in the load or strain that they must assume. If the strain is too great, something must give. In an improper position, the weak spots give.

When you lift a weight incorrectly, keeping the legs apart and straight and bending at the waist to pick up the object, you place an undue strain on certain important muscles, the muscles in your back. When it is in this bent-over position, functioning in a manner similar to the cable that operates the boom of a crane, the more the boom is lowered, or the nearer to the horizontal that the back is bent, the greater the strain on the cable or muscles. Moreover, because nature in this case designed the body rather poorly, the abdominal muscles must also do part of the work. Therefore, if there is a weak spot in either one of these two sets of muscles, when they are subjected to undue stress, a strain can occur at the weakened place. Each strain weakens the muscle just a little bit more.

Rather than lift in this manner, the smart way is to use your powerful leg muscles and try to keep the back as upright as possible, the way nature intended. The leg muscles are stronger because they are bigger and given more use and exercise.

The reason most people bend at the waist, of course, is to reach down to grasp an object. They do it unthinkingly, and then have to literally lift with the back to raise the object. Instead, if they would bend the knees when reaching for the object, and then lift it by straightening the legs, keeping the back in as nearly a vertical position as possible, there would be less strain on their backs and abdominal muscles. The safe method of lifting heavy objects from floor level to a higher level is to try to keep a good balance while you lift, and above all, keep the object close to you. This applies whether you are lifting something off the ground or merely picking up a heavy object off a work bench or table. Even trying to lift 10 or 15 lbs off a bench at arms' length requires unnecessary effort. Get close to what you are lifting or hold it close to you so your center of gravity is close to the object. If you find the object is too heavy or bulky, ask someone to help you.

When lifting an object with another person, make certain you both lift at the same time and let the load down together. Don't drop it suddenly without warning the other person and cautioning him to do the same.

Any conception that it is only the physically weak who are prone to injury from improper lifting posture is a wrong conception. The strongest of men as well as the weakest suffer from back strain and ruptures. Each improperly-performed lift may cause a slight muscle strain that continually widens until a pronounced injury finally occurs. Be smart. Save your energy and save yourself from injury. Lift the safe and easy way.

Pallet Jack

The pallet jack is designed to move wooden pallets in situations where forklifts are not practical and also where price is a consideration. They are easily maneuvered for moving material and can be used to help set electrical gear and transformers, etc.

The foot-operated lever control valve located by the jack tower of the unit controls the raising and lowering of the forks.

Pallet jack

Steel Slings, Nylon Slings, Shackles

U-shaped shackle

Cable and nylon sling

Slings and shackles are used in all phases of lifting operations. Slings are manufactured in a variety of sizes and weights, such as 3/8-in cable sling six feet long or eight feet long, to accommodate any size or weight load.

The U-shaped shackle is used to attach the loop ends of slings by utilizing a pin or bolt to close the end of the U. The size of the shackle is determined by the diameter of the pin. A 3/8-in shackle would have a pin size of 3/8 in.

The shackle is rated at low capacity, and different-quality shackles have different standards, but basically, a 3/8-in shackle will handle a one ton safe working load. Most lifting equipment is designed with a 5 to 1 safely ratio, but the rated capacity should be observed in use.

The nylon slings are rated by size and weight capacity. Most are identified by a leather tag attached to the sling. There are various types of so-called endless or basket nylon slings. These consist of one piece of webbing sewn together in a continuous band, used as a circle or placed underneath a load and interlocked to form a basket. More commonly-used is the eye-to-eye sling, a straight sling with an eye at each end. The eye-to-eye nylon sling is referred to as a choker. The straight sling is threaded through one eye, forming a slip knot which

4-hook shackle

tightens around the load being lifted. The nylon slings are rated according to their designed use, either choker, vertical or basket. For example a 1-in-wide sling is used as a choker, rated for 1100 lbs. Used as a vertical, it is rated for 1400 lbs. As a basket, it is rated for 2300 lbs.

Rollers and Equipment Rollers

Rollers are sometimes referred to as cat rollers or roller skids, and are used to move heavy objects such as large switch gear or transformers. The two types available are straight, where the platform on top of the rollers is stationary, and swivel, where the platform turns to allow guiding.

Generally the rollers are used in sets of three, two straight and one swivel. A handle is attached to the front of the swivel roller to allow steering. A lifting device such as a house jack is used to lift the load to be placed on the rollers.

Four Wheel Dollies

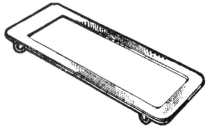

Four wheel dolly

Four wheel dollies consist of a platform mounted on two swivel casters and two stationary wheels. Efficient moving of electrical panel cabinets or other medium-weight objects is possible on the dollies. Most dollies are equipped with solid rubber-tired wheels to prevent damage to the floor surface.

Small Hand Trucks

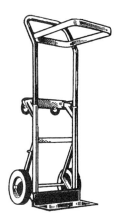

Hand truck

Hand trucks come in a multitude of sizes rated by load capacity. Overloading is the primary enemy of hand trucks, and will cause welds to break, lifting plates to bend and distort, and wheels to disintegrate. Most hand trucks are equipped with either solid or pneumatic rubber tires to permit easy mobility. The handles generally have either rubber or plastic hand grips.

Johnson Bars

Johnson bar

A Johnson bar is a long wooden-handle device with a steel load-bearing plate at the bottom and two heavy wheels, used as lifting aid to move heavy equipment. The steel load-bearing plate is inserted under the object to be lifted and the load raised by lowering the handle. The long handle gives enough leverage to lift very heavy loads which may then be placed on a set of equipment rollers. The tools are generally used in sets of at least two.

Material Handling Equipment

Wheelbarrows

Wheelbarrow

The contractor wheelbarrow is designed to haul about five to seven cubic feet of material, should not be overloaded and should be pushed, not pulled. When the operator is pushing, the barrow can be steered properly, is much easier to maneuver and less prone to accidental dumping. When using a wheelbarrow, the handles should be checked for cracks and the inflation of tires should be as recommended by the manufacturer. Underinflated tires make it more difficult to move the barrow with a load.

Appliance Carts

Appliance carts are similar to the hand truck, but have more sophisticated features such as protective guards on the face of the truck to protect the load from being scratched, and high/low wheel positions to aid in climbing stairs or resting the load. A nylon strap is included as part of the equipment to secure the load to the cart. The strap is tightened around the load by a ratchet roller bar on the rear of the cart.

Industrial Wagons

Industrial wagon

Industrial wagons have the general appearance of a heavy-duty child's coaster wagon. The industrial wagon is constructed of medium-weight steel, has heavy-duty wheels and tongue along with substantial tires.

The wagons are useful for moving materials, tools, or any small- to medium-sized items, saving time and trips for the worker.

Gang Boxes

Gang boxes constructed of metal are the most durable, which is advantageous because of the hard usage they are subjected to on the job site. The boxes are used for storage of tools and materials in isolated locations. They

88   Handbook of Electrical Construction Tools and Materials

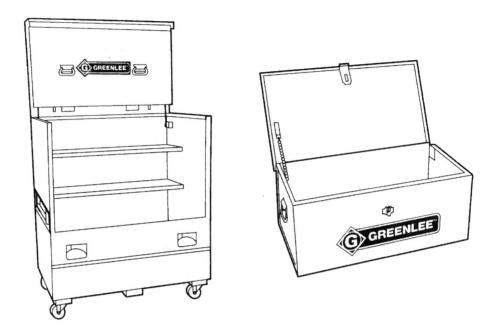

Two of the many types of gang boxes

have locking capability to prevent loss by theft. Several configurations are available, some with work benches attached on the front or ends of the boxes. Others have sets of metal shelves built into the interior to increase storage space and improve organization of the contents.

Tug-It Hoists

The tug-it hoist is generally used to lift and aid in setting heavy switch gear or transformers. The load limits marked on each unit are important — a hoist rated at $1\frac{1}{2}$ tons will not safely lift 2 or $2\frac{1}{2}$ tons. Tug-It is a brand name which is applied to many other hoists. The vertical lift capability of the hoist is determined by the length of the chain. Most of the hoists have a finger-tip control handle marked Up and Down, which determines the direction the chain will move. Some of the hoists are capable of allowing the chain to free-wheel,

Tug-it-hoist

which permits the chain to be pulled through the unit without using the crank handle.

Chain Hoists

The chain hoist performs the same function as the so-called Tug-it hoist, but is operated by a continuous loop of chain moving over a pulley. Pulling the chain in one direction moves the lifting chain up, and the other direction moves it down. There is a brake on the lift chain which prevents load free fall.

Cable Hoists

Ridgid cable hoist

Cable hoists are constructed similar to the Tug-it hoist, discussed previously. They are designed for numerous applications requiring lifting, lowering, pulling, stretching, and tightening. A rewind handle for fast cable rewind and tightening is standard on many models. Most types of cable hoists also offer double line capability for added capacity.

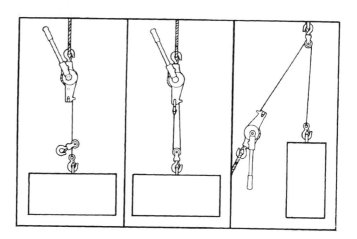

The cable hoist may be rigged in three different ways for many applications

Safety Precautions

Some safety precautions should be observed when using hoists of any type. These are:

- Do not load the hoist beyond its rated capacity.
- Do not use the hoist to lift personnel.
- Conduct periodic inspections for damage and wear, paying particular attention to the cable, chain and hook.
- Do not use hoist chain or cable for slings.
- Never leave an unattended load suspended in the air.
- Stand clear of all loads.
- Never travel a load over the heads of personnel.
- Give ample warning of your intention and direction of travel.
- Be sure the load is correctly seated in the hook.
- Be sure there are no kinks or twists in the chain.

Chapter 11

Scaffolding

The two most prevalent types of scaffolding available for use in the electrical industry are aluminum and steel. The aluminum is much lighter in weight than the steel, plus it requires somewhat less maintenance. Proper maintenance is important because people's lives may depend on the scaffolding being properly maintained and erected.

In addition to proper maintenance, safe use of scaffolding requires adherence to twelve basic safety rules:

1. Do not permit anyone to use scaffolding unless they are familiar with safety rules.

2. If in doubt about the assembly of scaffolding, seek help from a knowledgeable person. Do not take chances.

3. Never use any scaffold that has been damaged or is not properly erected. Never force parts. Parts should fit easily.

4. Before climbing a scaffold, be sure the wheels or caster brakes are locked correctly.

5. Never move a scaffold when anything or anyone is on it.

6. Be sure scaffolding is level at all times. When a leg is adjusted, be sure the locking device is completely engaged. Never make leg adjustments when anyone or anything is on the scaffolding.

7. Do not try to stretch platform heights with adjustable legs. If additional height is needed, add scaffolding sections. Save the leg adjustments for leveling.

8. Never lean a ladder against a scaffold, nor place ladders on the platform of scaffolding. Never push or lean against the wall or ceiling when standing or sitting on a scaffold, unless the scaffold is securely tied to the building.

9. Make sure all locking hooks are firmly in place, and that spring-loaded locking pins are functioning properly. These hooks appear at each end of separate horizontal and diagonal braces at the lower end of all stairways on walk-up scaffolding.

10. Before scaffolding is used with folding braces, be sure latches and all locking hinges are locked and in proper working condition.

11. Always install safety railing when the platform will be used at heights of 6 ft or more.

12. When the height of the scaffold platform exceeds three times the minimum base dimension, the base must be enlarged by the use of outriggers, or the scaffolding must be tied to the building.

Span Scaffolding

The term "span scaffolding" applies to most scaffolds erected with frames and detachable braces. There are four types of span scaffolding:

- Standard scaffolding, containing two types: narrow width and double width
- Walk-up span scaffolding
- High-clearance span scaffolding

The standard narrow-width span scaffolding is 29 in wide and is commonly called 2 ft wide scaffold. The standard double-width span scaffolding is approximately 50 in wide and is referred to as 4 ft wide scaffold. The narrow-width scaffold will require a base of two frames 2 ft wide × 6 ft high with casters. To erect the two frames, three braces are needed — two diagonal, one horizontal and one scaffold plank. These components, when properly assembled, will make up one base section.

The scaffolding planks come in standard sizes, 6 ft, 8 ft or 10 ft and determine the length of bracing required. If the plank is 10 ft long, three horizontal braces would be the same length as the plank. The diagonal braces would be approximately 4 in longer than the horizontal braces from center to center of the end cups on each end of the brace.

A section of standard narrow width (2 ft wide) scaffold using a 10 ft plank and two base sections with the three applicable braces and casters would be referred to as a 2×10 base section. The working platform would be 2 ft wide, 10 ft long and 6 ft high. To increase the height of the working platform, extensions in either 4 ft or 6 ft lengths would be added to the basic frame. The extensions require the same bracing arrangement (two diagonal braces and one horizontal) as the base section.

Scaffolding erected over 6 ft high requires the installation of guard rails and toe boards. The guard rail sections must have two horizontal braces, thus building a cage or a complete rail around the man on the scaffold. The guard rail must be 40 in or higher, making the guard rail approximately waist high on a man standing on the planks. The toe boards are 4 in high metal or wooden plates, depending on the manufacturer of the scaffold. The toe boards are placed around the plank area to prevent items lying on scaffolding planks from being kicked or falling off, injuring someone below.

Outriggers are required if the scaffold height exceeds three times the width of the base. The scaffold extensions require the use of interlock clips to lock the extension to the base, and extension to extension.

Standard double-width 4 ft wide scaffolding requires twice as many braces as the narrow 2 ft wide scaffold — that is, four diagonal and two horizontal braces. Two planks for the platform are required opposed to the single plank on the narrow type. Extension sections used to increase the height on the double-width scaffolding require twice the bracing necessary on the narrow scaffold.

The rule for toe boards is the same in the double width as in the narrow span scaffolding. Height three times the width of the base requires the use of outriggers. Flooring planks used in both types of scaffolding have two coupling fittings on each end of the plank for firm attachment to the scaffolding base.

94 Handbook of Electrical Construction Tools and Materials

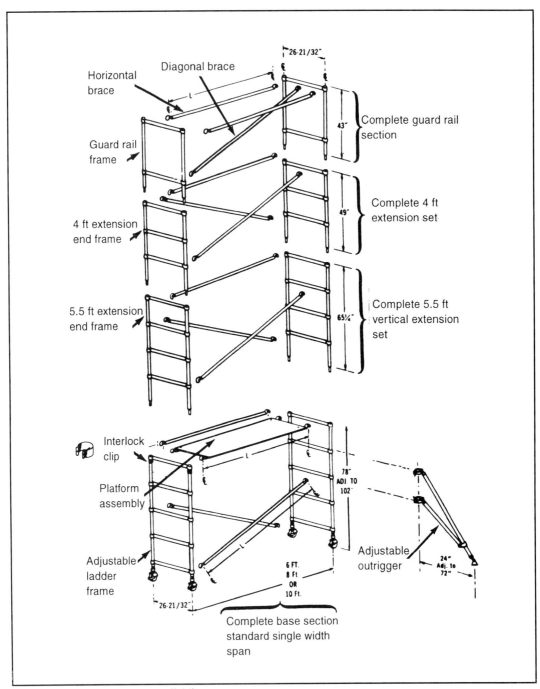

Standard single-width span scaffolding

Scaffolding

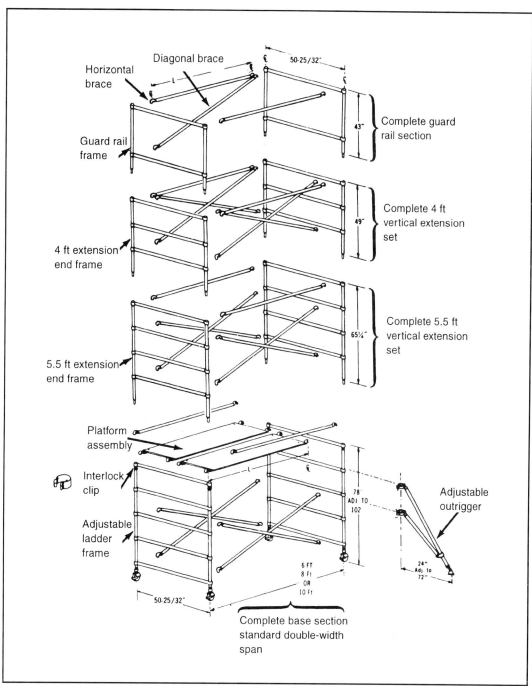

Standard double-width span scaffolding

Span Scaffolding Safety

Some special safety rules apply to span scaffolding and should be followed at all times:

1. The platform is designed to carry a maximum distributed load of 500 lbs with a safety factor of 4. Do not exceed this load.
2. The horizontal brace of the span scaffolding should never be installed at the same level as the intersection of the diagonal braces. Either install it higher or lower than the intersection or cross point.
3. Never climb a span scaffold which does not have at least two diagonal braces and one horizontal brace properly installed in the bottom of the section.
4. Double-width span scaffolding requires double bracing.

Walk-Up Scaffolding

Walk-up scaffolding

Walk-up scaffolding is similar to standard-span scaffolding with the exception that it has a stairway for climbing. The walk-up scaffolding requires the same bracing, toe boards, guard rails and outriggers as those required in standard-span scaffolding.

The stairway on walk-up scaffolding runs on a diagonal, from the bottom end of one section to the top end on the opposite side of the scaffolding. One scaffolding plank is required for every section to enable a man to reach the bottom of the next stairway going up. The stairways for each section start at the same point as the stairway in the lower section. If the bottom of the first section is on the west end of the section, the bottom of the stairway for the second section would be on the west end, etc.

The planks on walk-up scaffolding are made of plywood $5/8$ in with four holes drilled in them to lock onto vertical locking pins built into the scaffolding base sections.

Special safety rules apply to walk-up scaffolding:

1. Never use the stairway to work from, as it is only designed for personnel to walk up and down between platforms. Stairways take a weight of 200 lbs with a safety factor of 4. However, they are not designed to take heavier loads or any abuse.

2. Never climb up outside a stairway scaffolding. Always use the stairs for access.

3. The platform of the stairway scaffold must be located on four locating pins of the floor braces. Whenever used outdoors, exposed to wind or up-draft, the platforms must be fastened with wind lock clips or tied down, and the scaffolding secured to the building.

4. The platform of the stairway scaffold is designed to carry a maximum distributed load of 750 lbs with a safety factor of 4. Do not exceed the load rating.

5. When bridging between scaffold with planks or ladder stages, place the ends of planks or stages on scaffold platform across both floor braces to distribute the load. Other braces of scaffolding are not designed for heavy loads. Floor braces are of heavier material and have vertical pins for locating on plywood platforms.

6. When erecting or dismantling upper sections of the scaffold, stand in the center of the platform below it and keep a firm hold on the section.

VX Folding Scaffold

The cross braces on the base section of folding scaffolding are built in and can be folded at the base section brace, making the base section self-contained and complete, with all the necessary braces built into it. The unit contains a ladder which is attached at the top of the section. The ladder can be released from the bottom of the unit to facilitate folding, but it is permanently affixed at the top of the scaffolding. The braces run opposite in direction to the ladder and form an X, intersecting at the middle of the square or rectangle formed by the scaffolding sections.

An intermediate platform frame is available which permits the use of double planking. This platform interlocks to the frame of the unit, permit-

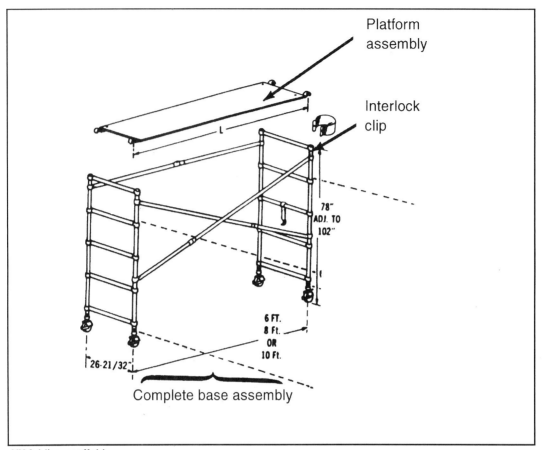

VX folding scaffold

ting full planking of the top section. Any extensions on the VX require bracing the same as the narrow or wide widths of span scaffolding.

High-Clearance Scaffolding

High-clearance scaffolding differs from regular span scaffolding by the design of the plank assembly. Braces are built into the plank on the high-clearance scaffold, eliminating the need for additional bracing on the base section. The adjustable bracing underneath the plank gives higher ground clearance, thus the term "high-clearance scaffolding."

The high-clearance scaffold is satisfactory for use in spanning aisles in grocery stores, department stores and factories where it is necessary to keep

the aisles open. Height extensions used with high-clearance scaffolding require the same bracing procedures as the standard-span scaffolding.

Caster wheels are available for all types of scaffolding, generally in the 5 in and 8 in diameter sizes. The wheels are equipped with either hard rubber or pneumatic tires. The rougher the terrain or base supporting the scaffold, the larger the wheels should be.

There is no room for error or misjudgment when scaffolding is used because of the extreme danger of collapse and injury. The scaffolding manufacturer's specifications for erection and load limits should always be observed.

High-clearance scaffold

Aluminum Scaffolding Planks

Aluminum scaffolding planks 20 in wide and 24 ft long with secure locks to fit aluminum scaffold are available in addition to the standard wooden scaffolding planks. The aluminum planks offer a very stable walking/working surface.

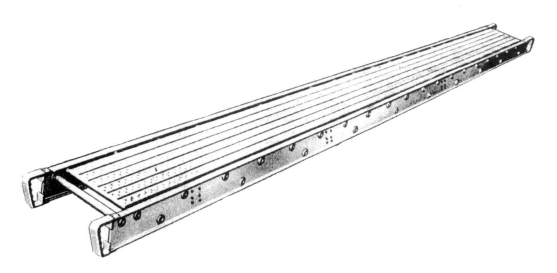

Aluminum scaffold plank

Steel Scaffolding

Steel scaffolding is available in heights of 2 ft, 3 ft, 4 ft, 5 ft and 6 ft. Generally, the width of steel scaffolding is 5 ft, with the total length being determined by the bracing used. The horizontal brace is referred to as the "spacer." The cross braces consist of two steel braces pinned in the center, unfolding like large scissors to form cross bracing. One brace is actually a pair, and two pairs are required for each section regardless of section height. The length of the braces varies with the height of the section, requiring reference to charts furnished by the manufacturer to determine "correct bracing."

The braces may be attached to the section frames with special U-shaped hooks with a locking device, or, depending on the brand of scaffolding, a nut and bolt going through to the frame.

The wheels for steel scaffolding may be inserted and used in any of the steel frame sections, as opposed to the wheel arrangement on aluminum scaffold, where the wheel-coupling devices are permanently attached to the base section. Aluminum scaffolding base sections cannot be used as extensions, but steel sections may be used as either base or extensions.

The erection of a base section of steel scaffolding requires two sections of scaffolding, four casters with lock pins for the wheels, two braces of the folding scissors type, and, if available, spacer bars or horizontal braces to stiffen the scaffolding. The addition of an extension frame to increase the height of the scaffolding requires a coupling pin. Coupling pins come with holes and may be equipped with locking devices. Some require a locking pin and some have spring clips built into the coupling pins. These are needed to couple extensions to base sections. There must be one for each corner of the scaffold, or four for each extension added to the base.

The requirement for guard rails and toe boards is the same for both steel and aluminum scaffolding.

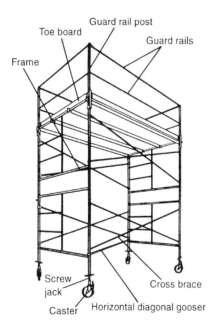

Steel scaffolding

Motorized Electric Scaffolds

Large amounts of overhead work such as hanging lighting fixtures, or installing cable tray and hanging conduit in warehouses or other commercial buildings, demand more efficient man-lifting devices than is afforded by standard scaffolding. Self-propelled scaffolding, with the capability of being steered from the working platform, has fulfilled part of this requirement.

The motorized electric scaffold base section has an overall width of 7 ft 7 in, its length is 12 ft 2 in and the base section height is 5 ft 11 in. Standard steel scaffolding extensions, bracing and planks are used to extend the height and provide a working platform. Wheels are equipped with pneumatic tires. The tongue attached to the front allows the scaffold to be pulled and manually steered into position.

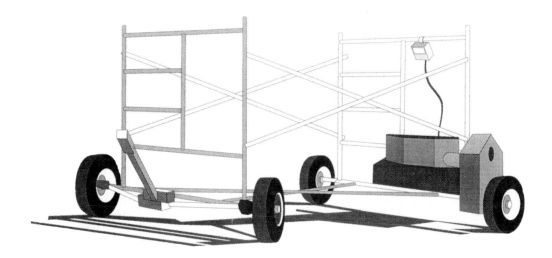

Motorized electric scaffolding

Hy-Jacker Lift

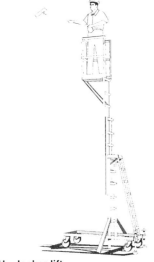

Hy-Jacker lift

The Hy-Jacker platform will raise and support a maximum load of 300 lbs at a height of 15 ft maximum. Equipped with four 6 in rubber-tired wheels, the unit is easy to move around the job site on relatively flat surfaces. Two hinged outriggers for stability are welded to the frame. The working platform is reached by a series of built-in ladders.

Selma Lift

Selma lift

The Selma lift provides a 25 ft maximum working platform height with a large working surface. The unit is powered by either a gasoline or propane engine, which supplies hydraulic power to the lift mechanism, steering system, and drive wheels.

The unit is controlled by the person on the working platform. Due to the size of the unit, it is used in larger areas which require the height available with this lift.

The wheels are automotive-type, equipped with standard automotive tires. Wheel covers are available which reduce the possibility of damage to the floors where the unit is being used.

Self-Propelled Battery-Operated Lifts

Self-propelled battery-operated lift

The battery-operated lift offers a large working platform with a maximum height of 27 ft. The unit is mounted on pneumatic tires, guided by the operator on the platform, and powered entirely by batteries. It is suitable for use in enclosed areas.

This lift is particularly useful for work in very high, large areas such as warehouses, airports, shopping malls, etc.

Tel-Hi-Scoper

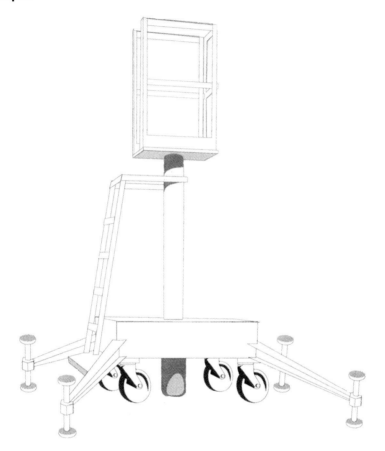

Tel-Hi-Scoper

The Tel-Hi-Scoper uses a battery hydraulic system to power the lift. The machine is 6 ft 9 in high in the lowered position × 35½ in wide. An attached ladder enables the operator to reach the lowered working platform and then ride the lift to its maximum extended height of 40 ft. The Scoper has four built-in outriggers with a screw stabilizer arrangement to firmly adjust the outriggers to the supporting base.

Tiger lift

Tiger Lifts

The Tiger lift is a much smaller version of the self-propelled battery-powered lift and has many of the same features. It too is battery-powered, self-propelled and steered from the working platform. The advantage of the Tiger Lift is the fact that it is narrow enough to allow transport through a standard door opening. With a maximum working platform height of either 14 or 18 ft, the unit is better designed for lower overhead work than the larger lifts. In addition to the standard working platform, both the 14 and 18 ft Tiger lift models are available with a cantilever 6 ft platform extending from the front of the machine, which allows the worker to reach areas beyond the capability of the regular work platform.

Ladders

Basically, there are three types of ladders: standard step ladder, extension ladder and extension trestle ladder. For all practical purposes, ladder size is determined by the number of steps, counting the top step. A 5 ft step ladder has five steps, including the top step. Most ladders used in the

Step ladder

Extension ladder

Trestle extension ladders

Types of ladders in common use

Scaffolding

electrical industry are constructed of fiberglass, due to its insulating qualities and the durability inherent in the fiberglass construction.

Adherence to the basic safety rules is imperative when using ladders. The rules are:

- Set the ladder feet on a firm level surface.
- Avoid dangerous over-reaching.
- Do not let belt buckles catch on steps or other parts of the ladder.
- Do not sit on the ladder top or back section.
- Do not straddle a ladder.
- Avoid overloads.
- Never allow more than one person on a ladder at a time. They are not built for two.
- Do not apply a side load.
- Never push or pull anything while on a ladder.
- Never drop or apply an impact load to a ladder.
- Do not attempt to "walk" a ladder. This means to not stand on a ladder and twist it to make it move beneath you. Get off and move it by hand.
- Always face the ladder when ascending or descending.
- Stand on the center of the rung or step and distribute your weight evenly.
- Keep a firm grip with both hands while climbing hand over hand, grasping rungs or steps and not side rails.
- Do not climb on a ladder from the side or from one ladder to another.
- Never climb above the third rung at the top of an extension ladder, nor above the upper support point.
- Never set up an already-extended ladder. Raise it hand over hand to the work location and then raise the extension sections. Set the ladder up at 75° by placing the base of the ladder a distance equal to one quarter the total length of the ladder away from the base of the vertical support.

- Erect an extension ladder with a minimum of 3 ft extending above the roof and tie the top at the support point.
- Extend only from the ladder bottom, never from the top.
- Never bounce or try to pull a ladder into a more-extended position. The minimum overlap of sections should be 3 ft on a 36 ft ladder. Overlap should be 4 ft on ladders over 36 ft in length. This is the overlap where the extension meets the base section.
- All ladder legs should be secure and on level ground or floor. Do not shim one side of a ladder for temporary support, to increase the height, or to adjust for uneven surfaces.
- Do not place ladders in front of doorways.
- Never paint rungs or steps.
- Check your shoes before using a ladder and make sure they are free from excessive oil, mud, grease, etc.
- Check the extension ladder safety shoes. Be sure the bolts that connect them to the ladder are correctly adjusted. This means that the shoes can still move, but are not excessively loose.

Chapter 12

Test Equipment

The use of test equipment is an important part of every technician's job. Selecting the proper instrument to be used in a specific application will help the worker to fully perform his or her task. The electrician will use the information provided by these instruments to help evaluate the work being accomplished.

Digital Meters

The accuracy of digital meters, in many cases, far exceeds that of analog or dial-type meters. Better than 0.5 percent accuracy is fairly standard for most digital meters. This improved accuracy is due to the carefully calibrated electronic circuitry contained in digital meters. Because of the readout on a digital meter, the precision of the reading accuracy by the operator is exact — not a close estimate.

Digital meters also offer an advantage in poor lighting conditions when the face of an analog meter might be hard to see. LEDs (light-emitting diodes) displays can easily be seen in low-light areas. They do however, consume more power and, therefore, the batteries may not last as long.

Digital meters are also smaller and are often less expensive to manufacture than the bulky analog meters of the past. Consequently, digital meters are quickly replacing the dial-type meters and you will probably see more and more digital type meters appear on job sites than any other types.

Pocket-Size Circuit Tester

Pocket-size circuit tester

The pocket-size circuit tester is a handy pocket-clip tester for checking line circuits. It can be used on 100-500 V circuits, ac or dc. A neon glow lamp indicates the hot side of the line. The screwdriver-shaped probe easily fits into receptacles. The probe length is $5/8$ in with an overall length of $4\frac{1}{2}$ in.

E-Z Check Plus GFI Circuit Tester

E-Z check plus GFI circuit tester

The tester checks for the following conditions: correct wiring, open ground, reverse polarity, open hot open neutral, hot on neutral, hot and ground reversed with open hot and ground fault interruption. By depressing the black button, the actual GFI breaker is intentionally overloaded to test its mechanical operation in addition to checking for proper wiring sequence.

Vol-Con Voltage/Continuity Tester

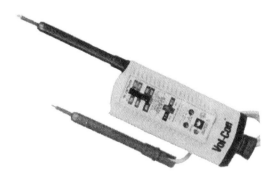

Vol-Con voltage/continuity tester

The Vol-Con is a unique, multi-function, solenoid-operated device which is three testers in one. The unit is a voltage tester, low-voltage tester, low-voltage indicator and non-switching continuity tester. It is a heavy-duty voltage tester up to 600 V ac or dc, and low-voltage indicator as low as 5 V ac or dc. It tests for circuit continuity, identification of ac or dc, 24 to 60 Hz frequency, dc polarity, blown fuses, grounding, and current leakage. The non-switching continuity tester is automatically protected against damage from accidental contact with live voltage.

Vol-Con Lite Voltage/Continuity Tester

This lighter model combines all the features of the previously mentioned standard model with a new smaller size and patented automatic switching function between voltage and continuity testing. This tester may be the smallest, toughest solenoid-type voltage/continuity tester in the industry. Dual function eliminates the need to carry two tools.

Vol-Con 249 Digital Voltage/Continuity Tester

The digital tester is microprocessor-controlled with a digital display, testing voltage/continuity with audible signals. The high-input resistance makes this universal tester safe even at high voltages up to 999 V, ac or dc. The LCD display has 3.5 digits, a measuring cycle of five per second, with an accuracy of 1.5% of reading + or - digit.

Power-Glo Pocket Tester

The power-glo tester checks for voltage from 12 to 440 ac or dc without contact on electrical lines. The sensor locates breaks along live, insulated wires and power cords, checks for energized ac circuits, distinguishes polarity and detects leaks in ungrounded devices.

Power-Glo pocket tester

Current Master Testers

The Fluke 33 Current Master offers the combination of true-rms measurements and rugged, reliable performance needed to troubleshoot problems associated with both traditional and non-linear loads.

Today's highly-computerized environment creates special demands on electrical professionals. Adjustable-speed drives, computers and some electronic ballasts draw current in a non-linear fashion, making standard industry tools obsolete. True-rms measurement capability is now needed to accurately measure the current drawn from these devices. The Fluke 33 will indicate reliable true-rms measurement.

Frequency (Hz): Helps to detect the presence of harmonic currents in the neutral of a 3-phase system. This function can be used to determine the output frequency of an adjustable-speed motor drive or to set the governor of an engine generator.

Display Hold: Captures and holds any reading for later viewing when used in difficult-to-reach areas. Its rugged angular jaw can safely pry into tight spaces. The jaw opening will handle two parallel 500MCM cables. A safety-designed hand guard helps prevent accidental contact with conductors. Its heavy-duty construction withstands job site abuse.

Some distinctive features of this tool include:

- Crest: Measures the instantaneous peak value of the current waveform with a capture time of two ms. This measurement is used to determine crest factor, which can indicate the presence of harmonics.

- Smooth: Displays a running three-second average, allowing accurate measurements of machinery that draws variable current.

- Min./Max. Record: Used to measure the starting current of motors. Records over 24 hours, monitors major fluctuations in current or frequency, and determines the average value over time.

80 Series Analog/Digital Multimeters

Analog/digital multimeter

These are high-performance meters with 11 functions and 40 ranges for a wide variety of electronic and industrial applications. They have Min./Max. average recording mode with Min./Max. alert, and frequency, duty-cycle and capacitance measurements. They have 1000 V rms input protection, an Input Alert detects wrong input jack connections, and the $3\frac{3}{4}$-digit, 4000 count digital display updates four times per second. The analog display updates 40 times per second in $4\frac{1}{2}$-digit mode, and it has a 1 ms peak hold backlit display and true-rms measurements (87 only). It includes Touch Hold and Relative modes, comes in a splash-proof and dust-proof case, and is EMI shielded. The Fluke 87 is a true-rms ac meter, and offers $4\frac{1}{2}$-digit, 19,999-count high-resolution mode, plus 1 ms peak Min./Max. hold. The backlit display assures readability in poorly lit settings. The light shuts off after 68 seconds, prolonging battery life.

Infrared Comparative Thermometer

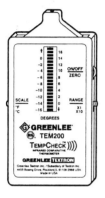

Infrared thermometer

The infrared thermometer is used to measure heat generated in electrical panels to determine if the internal temperature is within prescribed specifications. These instruments are used mainly to detect hot circuit breakers, switches and transformers. It may further be used for testing motor bearing failure, steam leaks, defective electronic components, heating and cooling leaks, adequate wall and ceiling insulation, abnormal temperatures and more.

After zeroing the unit to a reference temperature, infrared sensing instantly measures the chosen surface temperature without contact.

Cable Locator/Fault Finder

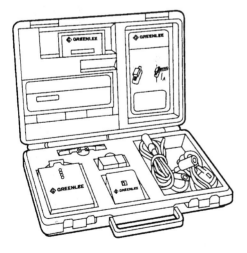

Cable locator/fault finder

The cable locator/fault finder is used to locate cable enclosed or hidden and determine the location of any possible ground faults. Features usually include:

- Indicates whether GFCI is acceptable at 3mA, 5mA, and 7 mA
- Tests and indicates conditions of GFCIs and circuits
- Indicator lights for wiring errors
- Four-second interval mA test impulses.

Megohmmeters

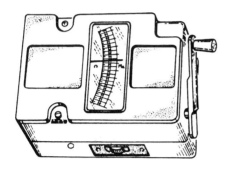

Megger

A typical megohmmeter (megger) is composed of a hand-driven or motor-powered ac generator and/or a transformer with voltage rectified to 100, 250, 500, and 1000 V dc, a cross-coil movement with 0 to 20,000 ohms (Ω) and 0 to 1000 megohms (MΩ) scales, a carrying case, and test leads. The megger is used to measure the resistance in megohms to the flow of current through and/or over the surface of electrical equipment insulation. The test results are used to detect the presence of dirt, moisture, and insulation deterioration. The instrument also typically measures resistances up to 20,000 Ω.

The test set and the sample to which it is connected are sources of high-voltage electrical energy, and all persons making or assisting in the tests must use all practical safety precautions to prevent contact with energized parts of the test equipment and associated circuits. Persons actually engaged in the test must stand clear of all parts of the complete high-voltage circuit unless the set is deenergized and all parts of the test circuit are grounded. If the set is properly operated and all grounds are correctly made, no rubber gloves are necessary. As a routine safety procedure, however, some applications require the use of rubber gloves in making connections to the high-voltage terminals and in manipulating the controls.

The instruction manuals accompanying the megger contain detailed instructions about preparing for tests and connecting the megger to various types of equipment.

Insulation Resistance Testing

To prepare for an insulation resistance test, first take the equipment or circuits to be tested out of service. Check between the equipment terminals and ground using the megger voltage ranges to be sure there is no voltage present. If possible, disconnect all leads to the unit being tested. When a motor or circuit is not completely isolated, make sure you are aware of all the components that will be tested when the megger is connected. Should an interconnected circuit be overlooked, the megger reading may be lower than expected.

The testing of wiring can be performed on all types of systems if two rules are kept in mind:

- Be sure all wiring is deenergized.
- Know what wiring is included in the test and make a record card of it.

When a distribution is present, check the entire system to ground by attaching one megger lead to the dead post of the open main power switch and the other lead to a grounded conduit or grounded metal housing.

Individual circuits are tested to ground by opening distribution panel switches, fuses or circuit breakers and testing each circuit in turn.

Multiconductor cables may be tested in several ways. For instance, measurement of insulation resistance can be made between the wire and lead sheath. Various other measurements can be made, such as wire to ground, wire to wire, wire to braid, and wire to sheath.

Keep in mind that when testing wiring that is connected to any panelboard or equipment there may be appreciable leakage between terminals, which will show in tests as lowered insulation resistance. If previous test and record cards were made with the panels connected, continue any future test in the same manner.

Frequency Meter

Frequency is the number of cycles completed each second by a given ac voltage; usually expressed in hertz; one hertz = 1 cycle per second.

The frequency meter is used in ac power-producing devices like generators to ensure that the correct frequency is being produced. Failure to produce the correct frequency will result in heat and component damage.

There are two common types of frequency meters. One operates with a set of reeds having natural vibration frequencies that respond in the range being tested. The reed with a natural frequency closest to that of the current being tested will vibrate most strongly when the meter operates. The frequency is read from a calibrated scale.

A moving-disk frequency meter works with two coils, one of which is a magnetizing coil whose current varies inversely with the frequency. A disk with a pointer mounted between the coils turns in the direction determined by the stronger coil. Solid-state frequency meters are also available.

Power-Factor Meter

The principle of operation of the power-factor meter is the principle of ratio meters which is also applied to synchroscopes, frequency meters, and some electrical thermometers.

Power factor is the ratio of the true power to the apparent power, and it depends on the phase difference between the current and the voltage. The power factor of a circuit can be obtained by taking simultaneous readings with an ammeter, a voltmeter, and a wattmeter and then dividing the wattmeter reading by the product of the voltage and current (volt-amperes); that is,

$$\text{PF (single-phase circuit)} = \frac{\textit{True Power (wattmeter reading)}}{\textit{Apparent Power (volt–amperes)}}$$

Much time, however, can be saved by using a single instrument — the power-factor meter. This instrument directly indicates the value of the power factor. For example, the single-phase power-factor meter is so constructed that the rotating field is produced by the line voltage, and at a power factor of 100 percent, or unity, there is no torque on the moving coil and the pointer rests at the center of the scale, or 100. Depending on whether the current is leading or lagging when the power factor is less than unity, the pointer swings to the left or to the right of the scale, indicating directly the value of the power factor.

Three-phase power-factor meters are also available and are quite common on industrial switchboard installations. Since most industrial establishments are charged a penalty if the power factor falls below 90 percent, industries try to maintain a high power factor at all times. A power-factor meter will continuously indicate the power factor so that a correction may be made at once, if necessary.

Power companies supplying power to industrial customers usually install two meters — one to measure active kilowatt hours and the other to measure reactive kilovar-hours; both are provided with 15-minute attachments. These demand attachments register the demand in kilowatts and kilovolt-amperes reactive for the two respective meters

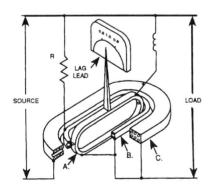

Interior components of a power-factor meter

if sustained for a period of more than 15 minutes at any time. If the power-factor meter on the industrial switchboard shows that the power factor is below the 90 percent mark, an attempt can be made to correct it at once to avoid the penalty. Two other reasons for trying to maintain a power factor near unity are:

- The reduction of reactive current provides more capacity for useful current on the mains, feeds, and sub-feeders.
- A high power factor provides better voltage regulation and stability.

Synchroscopes

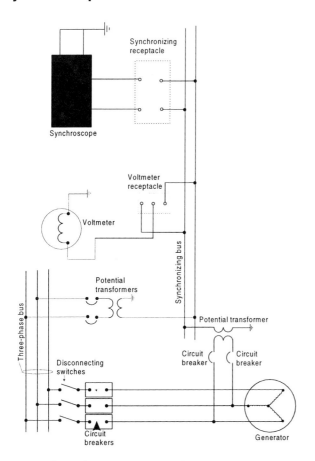

Synchronizing high-voltage generating system

When two alternators are about to be connected in parallel, the voltages of the two must be approximately the same; their voltages must be exactly in phase; and their frequencies must be approximately the same. If these differences are too great, the alternators are likely to pull entirely out of phase, thus causing a complete shutdown. Therefore, some means of showing the difference in phase and frequency between the voltages of the two alternators is necessary.

The synchroscope normally provides the following functions:

- It indicates whether the generator is running too slow or too fast.
- It indicates the amount by which the generator is slow or fast (the difference in frequency) at any instant.

It indicates the exact time of coincidence in phase relationship between all generators connected to the system.

When the synchroscope shows that both alternators are synchronized, the switches may be closed to allow both alternators to work together in parallel. The ideal indication, on most synchroscopes, is when the pointer either stops or moves very slowly at mid-scale.

Tachometers

A tachometer is a device used to indicate or record the speed of a machine in revolutions per minute (rpm). Several designs are available which include the following:

- Centrifugal
- Eddy current
- Surface speed
- Vibrating reed
- High-intensity stroboscope or photo tachometer

Some of the tachometers mentioned are connected to a motor, generator, or other machine by means of belts or gears while others are hand held. All have various scales from 0 rpm to over 50,000 rpm. Most of the better hand-held models have two buttons; a stop button to hold readings and a second button to releases the pointer to 0. The ball-bearing spindle on these devices is placed against a rotating object on the machine such as the motor-drive shaft, and the speed of the shaft is read directly on the dial of the tachometer.

Totally enclosed rotating equipment may be checked for speed by using a vibrating-reed tachometer. These tachometers operate on the well-known and time-tested principle of resonance. The instrument is simply held against the motor, turbine, pump, compressor, or other rotating equipment and the speed is shown by the vibration of a steel reed which is tuned to a certain standard speed.

A photo tachometer uses a light that is aimed at the rotating shaft on which there is a contrasting color such as a mark, a chalk line, or a light-reflective strip or tape. The rotational speed in rpm is conveniently read directly from the indicating scale of the instrument. This tachometer design is especially useful on relatively inaccessible rotational equipment such as motors, fans, grinding wheel, and other similar machines. This is a definite advantage in many tests where it is difficult, if not impossible, to make contact with the rotational unit. Most photo tachometers can be recalibrated in the field, if necessary, and are powered by long-life mercury batteries.

An electrical tachometer consists of a small generator that is belted or geared to a unit whose speed is to be measured. The voltage produced in the generator varies directly with the speed of the rotating part of the generator. Since this speed is directly proportional to the speed of the machine under test, the amount of the generated voltage is a measure of the speed. The generator is electrically connected to an indicating or recording instrument which is calibrated to indicate units of speed such as rpm, fps, fpm, etc.

Footcandle Meter

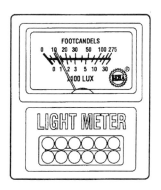

Footcandle meter

There are a number of large and elaborate devices used in laboratories for making exact tests and measurements of lamps and lighting fixtures. But for practical use in the field, a portable light or footcandle meter is quite satisfactory.

A typical footcandle meter consists of light-barrier layer cells and a meter enclosed in a suitable covering that is capable of reading light intensities from 1 to 500 footcandles or more.

To use the footcandle meter described, first remove the cover. Hold the meter in a position so the cell is facing toward the light source and at the level of the work plane where the illumination is required. The shadow of your body should not be allowed to fall on the cell during tests. A number of such tests at various points in a room or area will give the average illumination level in footcandles. Readings are taken directly from the meter scale.

Phase-Sequence Indicator

A common phase-sequence indicator is designed for use in conjunction with any multimeter that can measure ac voltage. Most can be used on circuits with line voltages up to 550 V ac, provided the instrument used with the indicator has a rating this high.

To use the phase-sequence indicator, set the multimeter to the proper voltage range. This can be determined (if it is not known) by measuring the line voltage before connecting the phase-sequence indicator. Next, connect the two black leads of the indicator to the voltage test leads of the meter. Connect the red, yellow and black adapter leads to the circuit in any order and check the meter for voltage reading. If the meter reading is higher than the original circuit voltage measured, then the phase sequence is black-yellow-red. If the meter reading is lower than the original circuit voltage measured, then the phase sequence is red-yellow-black. If the reading is the same as the first reading, then one phase is open.

Chapter 13

Miscellaneous Tools

Air Compressors

Air compressor

Small and medium-sized electrically or gasoline motor-driven air compressors are used as the air supply for spray paint units, to blow out lines or conduit, to inflate pneumatic tires or any other purpose requiring high pressure air.

Most machines have an air storage tank that maintains the constant required air pressure. An automatic pressure regulator shuts the unit down when the proper air pressure is reached.

When an air compressor is used as an air supply for paint spraying, a regulator should be utilized to insure proper control of air pressure.

Air Handling Equipment

Construction, during the summer months in almost any locale, involves men working in trenches or enclosed buildings with little air circulation. These warm areas require some form of air supply fans or blower-type air handlers to provide some degree of comfort for the work force. The air handlers

Floor-standing fan

supplied should be easily transportable and capable of being focused on the work area.

Large (24 in) heavy-duty, stand-type revolving fans are most often used. Safety regulations require that any fan or blower must have a cage-type cover over the moving blades or blower fins to prevent the possibility of injury to the workers.

Paint Spray Guns and Equipment

Paint spray outfit

Many types of paint spray guns are available but the most common type used in general construction requires the use of regulated pressure compressed air. Spray equipment should only be used by experienced operators and then only for the intended purpose of spraying paint and for cleaning purposes.

The most common cause of failure of paint spray guns and equipment is improper cleaning after each use. Spray equipment should be cleaned with the proper thinner for the type of paint used. Oil-based paint requires the use of turpentine or mineral spirits, lacquer requires lacquer thinner and water-based paints require water and sometimes soap. The paint supply container should also be thoroughly cleaned using the same material as for the spray gun.

Paint to be sprayed should always be strained to prevent small particles blocking the nozzle of the gun and causing a distorted spray pattern.

Wrenches Used In The Electrical Industry

Spud wrench

Spud Wrench

The spud wrench has a tapered handle and one open-end wrench. They are commonly used in the electrical industry to tighten the bolts securing large lighting support poles to their base. The wrenches are available in graduated sizes up to 3 in. The larger the wrench, the longer the handle will be to afford enough leverage to secure the nut.

Open-End Wrench

The open-end wrenches, so called because of the open configuration of the end of the wrench, start at approximately 7/16 in and proceed to 1/2 in, 9/16 in, 3/4 in, 13/16 in, 7/8 in, 15/16 in, and the 1 in. The wrenches are available individually or in complete sets.

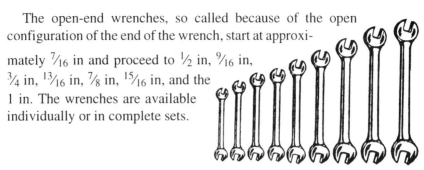

Open-end wrenches

Box-End Wrench

The closed-end or box wrench is available in the same sizes as the open-end wrench. The closed end helps to prevent the wrench slipping off the nut when pressure is applied and in some cases is easier to slip over a nut than the open-end wrench. They are available individually or in sets.

Box-end wrench

Combination Wrenches

The combination wrench (as the name implies) combines the advantages of the open-end wrench and the box wrench. Common sizes are 3/8 in to 1 in with both ends of the wrench being the same size.

Combination wrenches

Socket Wrenches

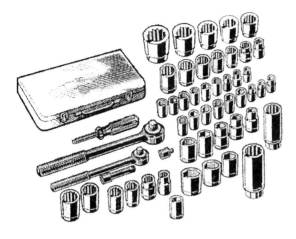

Socket wrench set

Socket wrenches are handy, efficient tools often used in the electrical industry. They are available in sets of various sizes that usually contain a ratchet and a 6 and 10 in extension. The socket sizes range from $\frac{1}{4}$ in to $1\frac{1}{8}$ in depending on the drive size. The drive size range is $\frac{1}{4}$ in, $\frac{3}{8}$ in and $\frac{1}{2}$ in.

The $\frac{1}{4}$ in drive sockets generally start at $\frac{1}{4}$ in upward to $\frac{1}{2}$ in while the $\frac{3}{8}$ in drive set will contain sockets from $\frac{3}{4}$ in to $\frac{7}{16}$ in. The $\frac{1}{2}$ in drive set contains sockets that start at $\frac{7}{16}$ in and go upward to $1\frac{1}{8}$ in. Deep-well sockets are available in larger sizes when the situation warrants such a tool.

Sockets and Accessories

The sockets available in sets are also available individually. Both are available in the standard USA in sizes, Metric sizes, all drive sizes, and in the standard length and deep length.

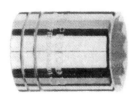

Several depths of sockets are available: standard length (left) and deep length (right) are common

Miscellaneous Tools 125

Ratchet drives are available in several lengths

The ratchet drives commonly found in socket sets are available individually in all drive sizes plus various lengths ranging from 7 in to 20 in. To allow maximum ultimate torque, some ratchets provide up to 15 degrees of ratcheting angle combined with a thin head design to permit access into tight work areas.

Hinge handles are available in several lengths

The hinge-handle socket drives permit using the sockets in very tight work areas where it is necessary to use the handle at an angle. The handle has a hole at the top to allow use of a T-bar to increase leverage. The drivers are available in both short ($5\frac{1}{4}$ in) and long ($18\frac{5}{8}$ and 20 in) lengths.

Universal joint drives come in all drive sizes and when coupled with drive extensions permit even sharper angle working positions. Universal joint length depends upon the drive size and varies in length from $1\frac{5}{16}$ to $3\frac{5}{8}$ in.

Universal joint drive

Speed handle drive

Speed handle drives have a bearing-mounted knob on the top and a crank bend in the middle of the shank to permit an easy quick driving motion. Most of them are approximately 18 in long.

The bar on the sliding T-handle drive slides back and forth to fit about any available space and allows the use of both hands for increased leverage on the socket. The bar lengths are 8 in and 20 in.

T-handle drives are also available in many sizes

Drive extensions in all drive sizes and lengths from 1¾ in to 36 in are readily available. The extensions combined with any of the drives allow the driving of a socket in about any situation.

Drive extensions are available in many different lengths to fit almost any conceivable application

Allen wrench

Allen Wrenches

Allen wrenches are hex-shaped hardened steel rods with a turned over portion to make a handle. Sizes ranging from $\frac{1}{32}$ in to $\frac{5}{8}$ in are commonly used with the wrench size corresponding to standard bolt sizes. Sizes proceed by $\frac{1}{16}$ in with the $\frac{1}{4}$, $\frac{5}{16}$, $\frac{3}{8}$, $\frac{7}{16}$, $\frac{1}{2}$, $\frac{9}{16}$ and $\frac{5}{8}$ in fitting most lug bolts, mechanical lugs and wire-tightening devices used in the electrical industry. Depending upon the manufacturer, the size is sometimes stamped on the wrench, otherwise the wrench size may be determined by measuring across the body between two of the flat sides.

Allen Wrench Socket Set

Allen wrench sockets are available with the Allen permanently mounted in a standard socket allowing the use of a standard socket wrench extension and rachet to turn the wrench. This set-up allows considerably more power to be applied to the Allen wrench plus being a much faster method than turning the wrench by the short handle part of the wrench. These socket Allen sets are readily available in sizes $\frac{1}{8}$ through $\frac{5}{8}$ in.

Allen wrench socket set

Nut Drivers

Nut driver

The nut driver has a screw driver handle and a regular screw driver shank with a socket-like wrench on the end of the shank. Some of the wrenches are welded to the shank and other types are separate sockets which fit a square $\frac{1}{4}$ in drive universal shank. Socket sizes available are $\frac{1}{4}$, $\frac{5}{16}$, $\frac{11}{32}$, $\frac{3}{8}$, $\frac{7}{16}$ and $\frac{1}{2}$ in. The nut drivers are easy to carry and very handy for light tightening jobs.

Electric or Air-Driven Impact Wrench

Impact wrench

Electric or air-driven impact wrenches are useful in production-type situations where large numbers of bolts are to be tightened or loosened. One common use of this tool is in tire changing to either remove or tighten lug bolts. The impact wrench requires special sockets made of hardened steel.

The impact sockets generally do not use the small spring-loaded ball common to regular socket sets to hold the socket to the tool. A small pin that sticks out of the drive shank must be depressed and then allowed to fit in the hole on the socket thus securing the socket to the drive shank of the wrench. To remove the socket, it is necessary to depress the shank pin.

Torque Wrench

Torque wrench

The torque wrench has either a scale or a gauge on the back of the tool marked in the pounds of torque from 0 to 150 lbs. The scale differential reading may vary according to the manufacturer or the wrench model number. The wrench is used to tighten bolts to the specified torque on various types of equipment such as automotive engine heads, wheel lugs or any other equipment where correct torque is critical. The torque wrench is a precision, calibrated tool and should be treated as such. The accuracy of the calibration should be checked periodically by an authorized service center.

Wrecking Bars and Crow Bars

Wrecking bar

Wrecking bars and crow bars are almost used interchangeably with the wrecking bar being larger and is used for heavier work. Both are useful in opening crates and for many other pry-type requirements. The crow bar has one end flattened with a center slot suitable for pulling nails. The other end is slightly curved, flat and fairly sharp to allow driving it under an item to be pried. The wrecking bar is larger, longer, and without the nail pulling capability, it is more adaptable to heavier prying or lifting.

Wonder Bar Pry Bar

Wonder Bar

The "Wonder Bar" pry bar is constructed of forged high-carbon steel, heat-treated for extra toughness with polished and beveled cutting edges. The contoured bar is ideal for pulling nails, prying, lifting and scraping. It has a beveled nail slot on both ends.

Builder's Level

The builder's level is used to determine grade levels in any area and often used in the electrical industry to define the grade level of floor duct or concrete boxes installed prior to pouring concrete.

The level is mounted on a tripod and requires the use of a grade stick. The self-contained protractor and bubble level assure that the proper angles and altitude are attained.

Builder's level

Brass Bound Level

The solid mahogany straight-grained wood, brass bound, 4 ft level has open hand holds and a hole for hanging while in storage. The $\frac{3}{8}$ in acrylic vials are bent for increased accuracy, easy to read, and are doubled to facilitate use of both sides of the level. The vials are designed to be replaced on the job without loss of accuracy. The brass binding protects the wood frame while the hard rubber end caps protect the ends of the level from damage. The vial stations are equipped with white plastic bezel to improve reading ease and the stations are protected with replaceable plate glass.

Brass bound level

Lighted Torpedo Level

Torpedo level

This level has three 360° vials that light up making it easy to read even in the dark. Light is controlled by an on/off button with an automatic shut-off after about two minutes. The frame is heavy-duty aluminum and I-beam extrusion with water-resistant casing for durability. The top read level vial and the plumb vial areas are highlighted in white for easy reading. Power for the light is provided by two AAA batteries.

Combination Square

The combination square has a 12 in movable grooved blade graduated in 8ths, 16ths, and 32nds. The handle has a scratch awl and level vial built in along with the adjusting knob which locks the blade into position. The blade is coated to resist rust and the adjusting knob and scribe knob are solid brass.

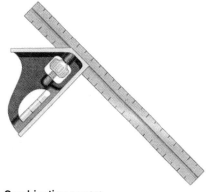

Combination square

Quick Square Layout Tool

The moving locking arm insures fast, accurate and repeatable measurements plus a large frame to span wide boards. This tool may be used as a power saw guide, protractor, rafter square and combination square.

Quick square layout tool

Hardwood Try Square

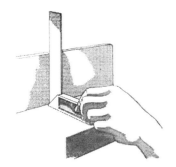
Try square

The 8 in try square has a hardwood handle with brass faceplates. The blade is marked in $\frac{1}{8}$ in graduations and the markings are filled either in white or yellow fluorescent color for easy reference.

Carpenter's Square

This square has a 24 in × 2 in body with a 16 in × $1\frac{1}{2}$ in tongue. Tables and scales are: Lumber Scale—board-feet equivalent, decimal equivalent table—fractions to decimals, metric conversion table—inches to meters, formula for squaring a foundation, 45°, 60° and 30° angle markings, volume and area formulas, wood screw gauge table—drill sizes, depth scale, $\frac{1}{8}$ in graduations on face and back. The high-tempered aluminum carpenter's square has useful tables deeply embossed on both sides to assure long-term visibility. Body is

Framing square

black with yellow-filled graduations and tables for superior readability. This type of square is often referred to as a "framing square."

Folding Wood Rules

Folding rule

Folding wood rules are hardwood sticks with regular marking, graduations on all edges in inches and $\frac{1}{16}$ of an in. Stud markings are indicated every 16 in. Steel locking joints are brass-plated. Protector plates and tip protectors are brass. The rule is 6 ft long when extended.

Tape Rules

Tape rules are available in blade widths of $\frac{1}{2}$ in, $\frac{3}{4}$ in and 1 in. Rule lengths of 6, 12, 16, 20, 25 and 30 ft are available depending on blade width. These rules are suitable for a multitude of measuring jobs. Tapes are spring loaded to automatically retract when the self-contained blade locks are released. Most of the rules have a belt clip attached for easy carrying and availability.

Tape rule

Long Tape Measure

Long tape measure

Long tape measurers may have either a steel or fiberglass tape in 50 – 100 ft lengths. They are equipped with a spurred hook on the end which enables one man to take many measurements without assistance. These tapes are retracted by use of a hand crank on the side of the case.

Cold Chisels

Cold chisel

The cold chisel is the most widely used type of metal-cutting chisel. It is designed to cut any cold metal softer than the chisel itself such as wrought iron, cast iron, steel, brass, bronze, copper, aluminum, bolts, nuts and rivets. The chisels are available either in sets or individually in sizes ranging in width from $\frac{1}{4}$ in to $1\frac{3}{16}$ in.

Drive Pin Punches

Drive-pin punches are designed to drive out straight and tapered pins. They are also referred to as rivet punches because of their use in driving out rivets. The punches are issued in sets or as individual punches.

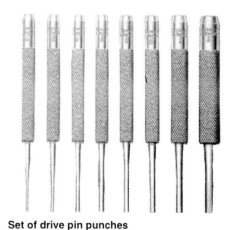

Set of drive pin punches

Center Punch

The center punch is used for marking intersecting center lines, hole locations, pilot drill starting points and in metal working layout. Center punches look very similar to drive-pin punches except the center punch has a hardened sharp point on the end for marking.

Drift Punch

The drift punch is the best tool for installing and removing pins and shafts. Drift punches look similar to drive-pin punches except the shafts on drift punches usually show more of a taper.

Wood Chisels

Set of wood chisels

Wood chisels should be of the highest quality available, otherwise the cutting edge is easily dulled or chipped rendering the chisel useless. A top-quality chisel will have a tough handle designed for balance and grip with a cap capable of withstanding hammer blows. The blade should be one-piece, of crown shaped forged cutlery-grade steel with a high polish to help prevent rust and deterioration. Chisels are available individually or in sets of 4, 6, or 9.

Flooring/Electrician's Chisel

The electrician's chisel is used to cut holes for installation of boxes. It is a solid steel/polished bit with black enamel finish.

Flooring/electrician's chisel

Putty Knives and Scrapers

Putty knife

Putty knives are primarily used to shape putty or caulking but can also be used as scrapers. A good putty knife will have a hardened, tempered and polished full tang blade permanently fastened to either a high-quality wood or shatter-resistant nylon handle. The blade is very stiff.

The joint knife is a much wider version of the putty knife and has a flexible high-carbon steel blade, hardened, tempered, taper-ground and polished. The knife is usually used to smooth finish dry wall joints or light scraping projects.

The razor blade scraper uses single-edge razor blades which are changeable and retractable. The tool is handy for light scraping jobs.

Utility Knives

Utility knife

The swivel-lock utility knife body swivels open for quick and easy blade change — no screwdriver required. The retractable blade may be extended in three positions—full, half, and one-quarter out. The body of the knife is constructed of die-cast aluminum with blade storage in the handle.

The fixed-blade utility knife is constructed of heavy-duty die-cast aluminum, has blade storage in the handle and is recommended for heavy-duty cutting operations. The blade does not retract but is locked in cutting position.

The retractile blade utility knife has a three-position retractable blade — full, half, and one-quarter cut. Die-cast aluminum handle is lightweight and serves as storage for the blade.

Pocket Knives

The various pocket knives are used by electricians for skinning insulation on wire and cable. The slitting blade pocket knife has an extra-large sheepfoot slitting blade $2\frac{1}{2}$ in long that is excellent for skinning in heavy-duty linework. The blade locks in the open position.

The lineman's plastic-handle skinning knife is extra strong for line work. The flat, tempered steel blade with hawksbill tip makes it easy to scrape copper wire. The large molded-on plastic handle has built-in finger guard and a handle ring.

The cable-splicer's knife has a special purpose short blade and heavy-duty handle. The coping-type blade is made of finest cutlery steel and is very durable. The handle is dipped in a non-slip plastic to improve stability.

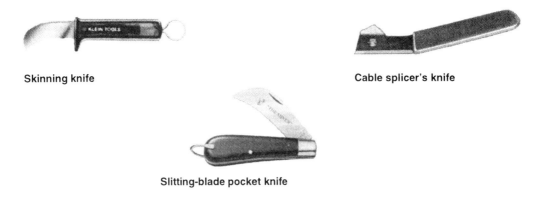

Skinning knife

Cable splicer's knife

Slitting-blade pocket knife

Types of pocket knives used by electrical workers

Tool Pouches

Typical tool pouch

General purpose pouches have an extra-large utility pocket for assorted tools and other items, a smaller utility pocket, three small pockets, and a tape thong. Loops fit belts up to $2\frac{1}{4}$ in wide. They are constructed of heavy canvas with a heavy, Ultra Hyde™ back for comfort and stability.

Multi-purpose pouches provide space for a number of tools. There are four large, gusseted utility pockets, three pliers pockets, two screw driver pockets, hammer loop, knife snap pocket, special partitioned pocket for 9-in torpedo level, special pockets for wire stripper, awl, steel tape, and two pockets for additional tools.

Tool Bags

Canvas tool bags are made of one-piece No. 8 white canvas. The bottom and 3 in of the lower sides are covered with one-piece Naugahyde, cemented and lock-stitched. The bottom is protected with steel studs. The mouth has a steel frame and the handles are harness leather. There are two retaining straps and buckles. Bags are 6 in wide and available in lengths from 12 to 24 in.

Canvas tool bag

Hilti Gun

Many types of fasteners and anchors are required in the electrical industry to mount panels to walls or floors, mount boxes, and secure various other types of equipment in a multitude of situations. The Hilti gun drives pins or studs into concrete in one operation without the use of a drill motor. Driving power is supplied by an explosive shell much like a rifle shell without the bullet. The pin or stud is inserted into the gun, placed against the supporting structure or surface and driven securely into place when the trigger of the gun is pulled.

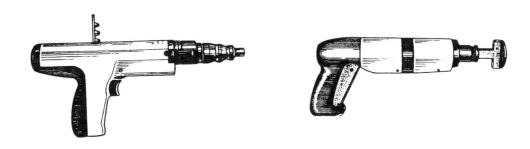

Two types of Hilti guns. The DX 350 (left) is for pins only; the DX 400 (right) is for both pins and studs

The pins used with the power-actuated guns have a $5/16$ in head and are available in shank diameter sizes $9/64$ in, $1/8$ in and $5/32$ in with lengths of $1/2$, $3/4$, 1, $1\,1/4$, $1\,1/2$, $1\,3/4$, 2, $2\,1/2$, and 3 in.

The piston studs come with a $5/32$ in diameter shank, lengths ranging from $1/2$ in to $1\,1/4$ in and threaded length of $3/4$ in to $1\,1/4$ in.

Pop Riveting Tool

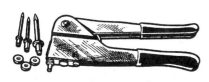

Pop riveting tool

The pop riveting tool permits the use of rivets to fasten objects together without using a hammer and back-up to set the rivet. The rivet shank is inserted through a pre-drilled hole in the object being fastened and then into the jaw of the tool. When the handle is compressed, the jaw locks onto the rivet shank pulling the rivet tight against the surfaces of the object being fastened. The shank is then snapped off by the tool and the rivet is set.

Staple Gun and Staples

The staple gun is spring-actuated when the handle is depressed. It is used to secure coverings such as carpet to wood, hang signs on the job site or for any fastening job requiring the use of staples.

Staple gun

Bolt Cutter

Bolt cutter

The bolt cutter may replace the use of a hack saw when it is necessary to cut bolts to free a bolted object. The $\frac{1}{4}$ in capacity cutter is 14 in long, the $\frac{5}{16}$ in capacity is 18 in, and the $\frac{3}{8}$ in capacity is 24 in.

Engraver

This tool is used extensively in the electrical industry. It will engrave plastic, metal tags, signs, panel markers and the like. It comes with a large assortment of brass letters and numerals that are inserted and locked in the order required on the pattern plate.

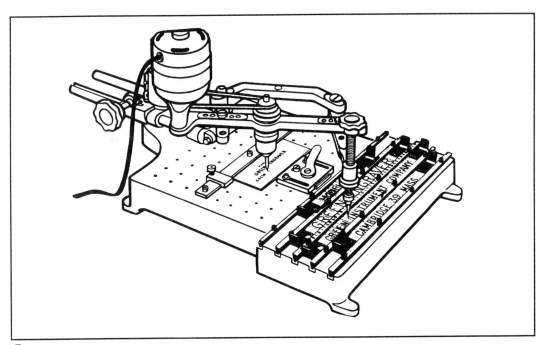

Engraver

The font point size is controlled by adjustments to the arm holding the engraving point.

Every large service panel should be identified and the various parts marked with the engraved markers for safety and convenience.

Metal Shear

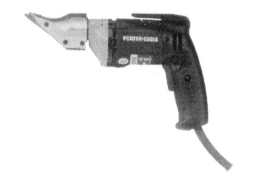

The metal shear is electrically driven for heavy-duty use. The shear has a minimum cutting radius of $\frac{3}{4}$ in and will trim cut 14 gauge mild steel, 16 gauge stainless and 12 gauge aluminum. For an inside cut the maximum gauges are mild steel 16, stainless 18, and aluminum 14.

Metal shear

Chapter 14

Electrical Materials Introduction

The electrical construction industry utilizes over 200,000 items of material to make electricity available for the use of mankind. Of course, all of the items are not used in any given situation but a great number ranging from conduit to lighting fixtures are used each time an electrical installation is completed.

As the complexity of electrical projects has increased the necessity for improved methods of installation, improved material designs, along with improved productivity has become apparent. The cost of skilled labor overshadows the cost of the material being used during this period of growth in the use of electrical power.

Here is a breakdown of how total manhours (actual studies done by industrial engineers) are expended on a typical electrical project.

- Direct productivity — 32 percent
- Indirect operations — 15 percent
- Material handling — 20 percent
- Ineffective activities — 26 percent
- Miscellaneous activities — 7 percent

Direct productivity is defined as actual direct installation of electrical material, on-site prefabrication operations, and testing operations. Indirect operations consist of job set-up operation, talk about the job, issuing and receiving instructions, planning the job, and getting out, setting up and putting away tools. Material handling involves unloading trucks, hoisting and moving materials and tools to installation area. Ineffective activities consist of workers starting late and leaving early, taking early lunch and returning late, coffee breaks and other rest periods. The miscellaneous seven percent was spent getting coffee for the crew and other go-for activities.

Material handling on the job site can be improved by careful planning and implementation of more efficient methods of providing material to the actual site of installation. Actual improvement requires project-by-project and operation-by-operation examination and planning. Some of the suggested improvements will involve the general contractor's superintendent, major subcontractors, project owner and project engineer. The responsibilities of each of the players should be specified before the project starts as these will affect the coordination/scheduling of the total project.

The effective improvement of the material handling operation on the job site requires that access by delivery trucks to and from the site be convenient and not require excessive waiting or maneuvering by the drivers. An unloading dock should be provided if at all possible.

Plans should include the delivery of supplies and materials in compact cartons or pallets for easy unloading and storage. It would also be helpful if scheduled deliveries of materials by the supplier could be made to coincide with the time it is actually required.

Multistory projects should have a man-and-material hoist and possibly a tower crane to aid in lifting heavy items of material. The available times for use of the hoist and/or crane should be established and agreed to by the general contractor to aid in planning for the movement of material.

The facilities for movement of men and material within the building or on site should be carefully studied. Providing ramps instead of stairs would allow the use of small trucks to move materials to various building levels. Pallets could be moved by forklifts. The foreman should determine the materials for the job the day before they are needed and make sure delivery is made from the warehouse or storage area. This eliminates lost time.

Housekeeping and garbage removal from the work site is critical to the movement and safety of workers and material. Clutter and debris cut down on efficiency of the workers and inhibit the delivery of material. Studies show that a clean work area can increase direct productivity by 8 percent.

The lost time attributable to ineffective activities can be reduced 3 percent by providing a lunch area on the job site for the workers. This would eliminate the need for the workers to go off the job to eat. Set times should also be established for coffee and rest breaks.

With management defining the limits and providing comfort facilities on the job site, the 3 percent improvement in ineffective activities and 4 percent improvement in time lost attributable to miscellaneous activities, the direct productivity will be increased another 7 percent thus bringing the percentage of direct productivity to 47 percent.

The 15 percent lost time due to indirect operations as described earlier could be reduced by 5 percent due to good planning. This 5 percent added to the 47 percent brings the direct productivity up to 52 percent of the total hours.

Accomplishment of all of the above requires cooperation and a commitment of each worker to work towards a continuous improvement in productivity. This does not mean to work harder but means to work smarter.

Chapter 15

Conduit

When discussing electrical installation or the electrical contracting industry, the phrase "Listed by Underwriters" and references to the "Code" will be used. It is important to understand both of these terms. Uncontrolled electricity can lead to great harm and damage. It is under control when the right kinds of materials are properly installed. Electricity is a powerful force. When under control, it safely performs a great variety of work for us. It is uncontrolled when the wrong kinds of materials are used or when the right kinds of materials are used or installed improperly.

Uncontrolled electrical power can cause fires, kill people and lead to a multitude of other costly results. When these things happen as a result of the wrong use of electrical power, individuals as well as businesses and insurance companies suffer great losses. Insurance companies have led the way in the setting up of minimum standards of quality in the electrical industry and electrical materials. They have also been instrumental in setting up minimum standards of installations.

Experiments and experience, over the years, have led to maximum standards of usefulness with the least amount of danger. The Underwriters' Laboratories, Inc., (UL) is a non-profit organization set-up to establish, maintain and operate laboratories for the investigation of material, devices, products, equipment, construction and methods and systems with regard to hazards affecting life and property.

The manufacturer who wishes to do so may submit samples of his products to one of these laboratories for testing in accordance with established standards. If the product meets the minimum safety requirements, the manufacturer's name and identification of the product are listed in the UL official published list. In addition, the product may then use the UL listing mark.

The tests applied by UL and, in some instances, by the manufacturers are much more severe than any conditions likely to be encountered by the merchandise during actual use.

If approved electrical devices or materials of high quality are used, but installed in a haphazard fashion without regard to relation of one device to another or the total electrical load they may be called upon to carry, the complete installation may be dangerous. Therefore, it is necessary to establish standardized methods which have been found in practice to be safe.

Through experience and experiments, these standardized methods that are shown to be correct are set down in a form which has come to be known as the National Electrical Code. The chief objective of the Code is to promote safety in electrical installations. Many cities, counties and states have adopted the provisions of the National Electrical Code into law which must be observed in the area involved.

In most localities, it is necessary to get a permit from city, county or state authorities before an electrical job can be started. These fees for permits are generally used to pay the expenses of the electrical inspectors whose work leads to safe, properly installed electrical jobs. The utility supplying power to the installation will generally not supply the power until the inspection certificate has been issued.

In addition, many cities, counties or states have adopted laws which require that any person engaging in the business of electrical wiring or electrical contracting cannot do so without a license. To obtain this license, in most instances, it is necessary to pass a written examination covering your knowledge of electrical wiring and installation. All these regulations or laws referring to the Code have been adopted for the safety and protection of the public.

There are many different wiring systems requiring a wide variety of material and tools to accomplish the installation. Five types of conduit are currently regularly employed:

- Rigid nonmetallic (PVC)
- Rigid steel conduit or galvanized rigid conduit
- Electrical metallic tubing (EMT)
- Flexible metallic tubing
- Aluminum rigid conduit

PVC Conduit

PVC conduit

PVC is the abbreviation for polyvinyl chloride and is an extruded plastic composition conduit. Sizes range from $\frac{1}{2}$ in to 6 in inside diameter furnished in standard 10 ft lengths with other custom lengths available on special order. The conduit wall thickness must meet UL standards and be of a specific weight for electrical construction use. PVC is available in two standard wall thicknesses referred to as Schedule 40 heavy wall and Schedule 80 extra heavy wall.

This type of conduit can be used in almost all applications that rigid or EMT conduit is called for and is particularly adaptable to direct or concrete embedded burial because it is impervious to various soil conditions. Conduit is furnished with a coupling either formed on or glued on one end. PVC conduit can be readily cut with a hacksaw, tube cutter or special PVC cutters. This conduit can be stored outside but care must be taken to protect it from extreme heat which may cause warping.

Fittings are affixed to the conduit with either a clear or gray PVC cement. The clear sets up almost immediately while the gray type is slower acting allowing for some adjustment of the fittings after application.

PVC Fittings and Accessories

PVC coupling

PVC Coupling: This is a smooth unthreaded sleeve coupling with a slight ridge in the center, and is of the same inside diameter as the outside diameter of the conduit to be spliced or terminated. It is used to couple an additional length of the same size conduit or PVC ells. The cement is applied to the interior of both ends of the coupling, the coupling slipped onto the ends of the conduit or ell to be joined, given a half turn and allowed to set for 10 minutes before applying pressure. Prior to affixing the coupling or other fitting to the conduit, a coating of PVC primer should be applied to the conduit to assure a permanent bond.

PVC Female Adapter: This fitting is similar to a coupling in appearance. However, one half is threaded in its interior which identifies it as a female adapter. It is designed to couple PVC conduit to EMT, GRC or rigid aluminum threaded fittings or conduit. Generally, this adapter is used to stub up from underground or slab to a rigid 90° ell or any application requiring a connection between PVC and another type of conduit.

PVC female adapter

PVC male adapter

PVC Male Adapter: The PVC male adapter is one-half standard coupling, and the other half is threaded on its exterior. This fitting is used to terminate the PVC conduit run by connecting to a junction box or other terminating application. The smooth end of the adapter is affixed to the PVC conduit and the other end is usually fastened with a lock nut. The threaded end may also be used to connect with a threaded GRC coupling.

PVC Elbow: The 90° elbow is used to make 90° turns in a regular PVC conduit run. The elbow is affixed to the conduit with a coupling cemented as described previously under couplings. The 45° elbow is used to make 45° turns and the 30° used to make even less sharp turns in the conduit run.

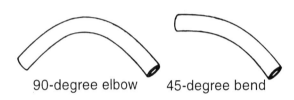

90-degree elbow   45-degree bend

Typical PVC elbows

PVC ells are made for all PVC conduit sizes above degree bends and in 26 in to 36 in sweeps. Where bends of different degrees are required, a PVC bender that heats the conduit to a pliable texture can be used. PVC ells are not used to stub up from underground runs because of strength requirements.

Rigid Aluminum Conduit

Rigid aluminum conduit is supplied with both ends threaded and in standard ten foot lengths. One end usually has an aluminum coupling, threaded, and the other end is protected with a plastic cap. It is available in inside diameter sizes of $\frac{1}{2}$, $\frac{3}{4}$, 1, $1\frac{1}{4}$, $1\frac{1}{2}$, 2, $2\frac{1}{2}$, 3, $3\frac{1}{2}$, and 4 in.

Rigid aluminum conduit may be used in almost all applications. When odd lengths of rigid aluminum conduit are needed, it is necessary to cut the conduit with a pipe cutter and a new threaded end must be cut with a pipe threader. When the conduit has to be bent, special rigid bending tools are used. Aluminum conduit is not suitable for direct burial or concrete encasement due to the electrolytic reaction on the conduit. Rigid aluminum conduit sizes of $\frac{1}{2}$, $\frac{3}{4}$, or 1 in can be bent with a manual bender referred to as a rigid "hickey". Larger sizes require the use of special benders that use "shoes" and are either operated by a hand driven ratchet or electric hydraulic pump.

Rigid aluminum conduit is specified in some plants, factories, and hospitals because of its resistance to corrosive action from the atmosphere and chemicals. It is considerably lighter and easier to install than rigid steel conduit but requires the use of a compound called "Noalux" which prevents "seizing" when the conduit is being coupled and provides better continuity and bond.

Rigid Aluminum Conduit Fittings

The use of aluminum conduit requires the use of aluminum couplings and ells because of the electrolytic action developed when dissimilar metals are connected together.

Rigid Aluminum Coupling: Rigid aluminum couplings are threaded all the way through internally. They are used to join one length of conduit to another with the coupling being screwed onto the end of the conduit. Couplings are supplied in all conduit sizes from $\frac{1}{2}$ to 4 in.

Rigid aluminum coupling

Rigid Aluminum Ells: Aluminum ells are available with 90° and 45° bends in all conduit sizes. The radial span, extension length, and overall bend length are determined by the conduit size. Radial span identified as "A" ranges from 4 in for the $\frac{1}{2}$ in size to 16 in for the 4 in size. Extension length "C" is from $2\frac{1}{2}$ in to 7 in. Overall bend length "B" runs from $6\frac{1}{2}$ in to 23 in.

Galvanized Rigid Conduit

Galvanized rigid conduit (GRC) is a steel product with both ends threaded and it is available in 10 ft lengths. One end comes with a rigid coupling, threaded, and the other end is usually protected with a plastic cap. It comes in inside diameter sizes from $\frac{1}{2}$ in to 6 in. Sizes up to 2 in are available in $\frac{1}{4}$ in increments and sizes from 2 in to 6 in in $\frac{1}{2}$ in increments.

GRC may be used in almost all applications. However, if it is used in direct burial situations or encased in concrete, it must be wrapped in plastic tape or ordered with a pre-applied polyvinyl chloride jacket to prevent corrosion.

Rigid conduit use is compulsory in explosionproof installations. When used in such locations, special explosionproof fittings are required by the National Electrical Code.

When odd lengths of rigid conduit are needed, it is necessary to cut the pipe with a pipe cutter and a new thread must be cut with a pipe threader. Special bending tools are used to bend rigid conduit. Sizes up to 1 in can be bent with a manual bender referred to as a rigid "hickey". Bending of

GRC Fittings

Standard GRC couplings

GRC Coupling — Standard: The standard GRC coupling is a sleeve, threaded all the way through internally. It is used to join one length of GRC conduit to another.

GRC Coupling — No Thread: There are two types of no thread GRC couplings — (a) compression and (b) set screw. When it is not possible to use a standard rigid coupling, either a compression type or set screw type can be used. The compression coupling is used where a concrete tight connection must be made. Otherwise, the set screw coupling is acceptable.

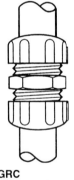

GRC compression coupling

The compression coupling is a steel sleeve-like device which incorporates a tapered washer slipped over and then tightened down onto the ends of the conduit being joined, thus creating a sealed bond between conduit runs. The compression coupling is only available up to the 3 in size.

The set screw coupling is a steel sleeve, screw holes drilled and threaded at both ends, and when the screws are tightened onto the conduit, a firm connection results.

Types of set screw couplings

GRC elbow

GRC Elbows: Preformed GRC elbows, both 90° and 45°, are available in all conduit sizes up to 4 in. The minimum workable radius is determined by the conduit size of the elbow varying from 4 in on the $\frac{1}{2}$ in size to 16 in on the 4 in size. The length of the straight leg of the elbow (each end) varies from $2\frac{7}{8}$ in on the $\frac{1}{2}$ in conduit size to $6\frac{15}{16}$ in on the 4 in size.

EMT Conduit (Electrical Metallic Tubing)

Due to its comparative light weight and ease in bending, EMT conduit is widely used in the electrical construction field. The conduit is a thin-walled steel tube supplied in 10 ft lengths in sizes from $\frac{1}{2}$ in to 4 in. EMT is not threaded and is easily cut with a hack saw or portable saw.

EMT is used extensively in branch circuit applications, can be bent readily with manual EMT benders up to the $1\frac{1}{4}$ in size and is considered relatively simple to install. Bending larger sizes require the use of shoe-type benders.

EMT conduit is not generally used in concrete encased or direct burial installations but is available with a corrosion resisting polyvinyl chloride jacket to be used where corrosion is a problem.

Conditions that cause corrosion are acids, alkalies, moisture, saline conditions, soil acids, etc., and are encountered in oil refineries, chemical plants, foundries, dairies and many more locations.

EMT Conduit Fittings

EMT Couplings: There are three distinct types of couplings for use with EMT conduit.

- Compression
- Set screw
- Indent

The EMT compression coupling is also a metallic sleeve but is made in five sections. The middle section is threaded at each end and a compression washer slipped over each end of the conduit to be spliced. A capped lock washer is then tightened to make the compression washer tighten down onto the conduit. The compression coupling is available in sizes to 2 in.

Couplings over 2 in must be the set screw type.

The set screw EMT coupling is popular and easy to use. The sleeve-like coupling is furnished with a screw on each end. It is slipped over the end of the conduit and the screws tightened to secure the coupling to the conduit. This coupling is not concrete-tight and is normally used in indoor locations.

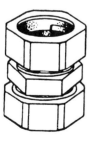
EMT compression coupling

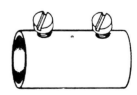
EMT set screw coupling

150   Handbook of Electrical Construction Tools and Materials

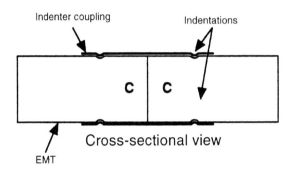

EMT indent couplings and indenter tool

The indent coupling is a divided (marked in the middle to indicate proper conduit placement) sleeve-type coupling. It requires a special indenting tool of the same size as the conduit being used that allows fastening to the conduit. The sleeve is slipped onto one end of two lengths of conduit, and then the tool is placed around the sleeve and compressed until an indentation is caused in both sleeve and conduit. The indent coupling is available in ½, ¾ and 1 in sizes.

Flexible Conduit

Flexible conduit, commonly referred to as flex conduit, is manufactured in four types.

- Regular steel flex
- Seal-tight flex, liquid-tight
- Regular aluminum flex
- Nonmetallic, liquid-tight

Flexible metal conduit

Regular steel flex is a spiralled steel tubing, capable of being bent with ease, which is used in various applications such as motor control connections and other locations where vibration is a factor. It is also used in locations where the installation of rigid type conduit is impractical. Steel flex is manufactured in sizes ranging from $^{15}/_{16}$ in to 4 in inside diameter.

Seal-tight liquid-tight flex is identical to the regular steel flex with the addition of a plastic covering. It is designed to be used in waterproof installations such as connections to roof air conditioners, outdoor motor controls, etc. Sizes available range from $^{3}/_{8}$ to 4 in. Various additional features are provided in this flex, such as extra flexibility, extreme temperature resistance and extra resistance to oil. Sizes $^{3}/_{8}$ in through $1^{1}/_{4}$ in have a continuous copper ground wire built into the core. Sizes $1^{1}/_{2}$ in through 4 in must be installed using a separate ground wire to conform to Article 351 of the National Electrical Code under "Liquidtight Flexible Metal Conduit."

Liquidtight flexible conduit

Regular aluminum flex is identical to regular steel flex in application and size. However, a lightweight aluminum flex is from 7 to 157 lbs per 100 ft lighter than corresponding sizes of conduit made with galvanized steel. This conduit is generally used where weight and/or corrosive environments are a problem. Many times it is installed in residential apartment construction where it can be protected inside of walls.

The smooth inner core, nylon reinforcing and rugged outer jacket are bonded together on the nonmetallic liquidtight flex. This provides protection against abrasion, mechanical damage and oils when extreme vibration is present. The conduit conforms to Article 351 of the National Electrical Code under "Nonmetallic Liquidtight Flexible Conduit." It is listed with Underwriters' Laboratories File No. 113116. It is approved for use in continuous temperatures from 176° F dry, and 144° F wet.

Chapter 16

Conduit Fasteners

The National Electrical Code requires that when conduit is installed, it must be supported at certain intervals, not to exceed the following:

- ½ in and ¾ in conduit every 10 ft
- 1 in conduit every 12 ft
- 1¼ in and 1½ every 14 ft
- 2 and 2½ in conduit every 16 ft
- 3 in and larger conduit every 20 ft

In addition, conduit must be supported within 3 ft of every box, fitting or cabinet. To support conduit, there are many different kinds of straps, clips and conduit hangers to apply to any installation situation.

EMT Clip, One Hole, Medium

EMT clip, one hole, medium

This line of clips is manufactured by the Minerallac Company. It is a one-hole type strap for use on thin wall or rigid conduit or pipe where the load is too heavy for the standard Minerallac clip. These clips have an inverted rib providing more strength at the bend of the clip. The inverted rib also provides a snap-on feature, which holds the clip on the conduit to allow both hands to be free to align and fasten clips to the structure, either with screws, bolts or welding.

The medium- and heavy-duty clip can be distinguished from the standard clip by the inverted rib and by the size and type of conduit being noted on it, in addition to the Minerallac identifying number on the clip. The clip is available up to 2 in for both EMT and GRC conduit. The Minerallac identifying method for their straps is to use the initials "TW" for thin wall or EMT, and the letter "P" for pipe or GRC. The standard packaging for this medium-type clip from $\frac{1}{2}$ to $1\frac{1}{4}$ in both in thin wall and GRC is 100 each. The packaging standard for the larger sizes is 50 each.

EMT and GRC Malleable Clip

EMT and GRC malleable clip

This is a one-hole EMT or GRC strap with the fastener made of malleable iron and hot dipped galvanized to minimize corrosion. Because of its sturdier construction, it is used in applications where additional fastening strength is required. It is available in sizes of $\frac{1}{2}$ in through 6 in rigid. 3, $3\frac{1}{2}$ and 4 in rigid malleable iron strap can also be used to fasten the same diameters of EMT conduit.

The strap is manufactured by various companies, such as Gedney and Thomas and Betts. They all have the same general appearance, in most cases, and the sizes of conduit and types are molded on the strap as part of the casting process. The standard packaging is 100 each up to the $\frac{3}{4}$ in size, 50 each in the 1 in size, and 25 each for the $1\frac{1}{4}$ in and $1\frac{1}{2}$ in sizes. The larger sizes are individually ordered.

One-Hole Standard EMT Jiffy Strap

The term "Jiffy" is a trade name used by the Minerallac Manufacturing Company to identify their particular line of conduit fastening straps and clips. The one-hole Jiffy strap is a curved piece of metal with the inside diameter corresponding with the outside diameter of the size of conduit to be secured. There is a single hole to provide a securing method to the structure.

The Minerallac line of one-hole Jiffy clips for EMT are identified for size such as 130 for $\frac{1}{2}$ in EMT, 145 for $\frac{3}{4}$ in, and larger sizes are marked with actual conduit size such as 1 in TW, etc. The standard packaging for the EMT line of clips is 200 each for $\frac{1}{2}$ in and 100 each for $\frac{3}{4}$ to $1\frac{1}{2}$ in.

The standard EMT Jiffy clip does not have the snap-on feature which allows it to be snapped onto the conduit. Minerallac does produce a clip called Min-E, which is a snap-on clip made of zinc-plated steel.

EMT and GRC Two-Hole Straps

EMT and GRC two-hole strap

Both the EMT and GRC two-hole straps are designed for use on loads too heavy to be supported by one-hole Jiffy clips or straps. One set of two-hole straps is manufactured in thin wall and ranges from $\frac{1}{2}$ to 2 in. After the 2 in size, the GRC or heavy-duty two-hole straps are used, both for EMT and GRC. The two-hole heavy-duty or GRC line of straps is manufactured up to 6 in diameter. The GRC strap is distinguished by a stamp with the letters "IP" after the conduit size marking. The IP stands for iron pipe or GRC.

EMT and GRC Naylon Strap

EMT and GRC Naylon strap

The Naylon strap is a fish-hook shaped device designed to fasten conduit to wooden structures utilizing its self-nailing feature. The inside curve of this self-nailing strap is of the same diameter as the outside diameter of the conduit it is designed to support. It is only available up to 1 in size for thin wall or rigid conduit. A flat protrusion on the end of the strap provides the self-nailing feature.

EMT and GRC Shoot Up Strap

In appearance, this strap resembles the standard EMT or GRC one-strap minus the hole. A Star Gun or other powder-actuated gun is required for shooting a pin to affix this strap to a structure. The shoot up strap is constructed of a gauge metal somewhat heavier than the standard Jiffy strap but not quite as heavy as the heavy-duty one-hole strap. The strap has a snap-on feature which permits it to be attached to the conduit prior to being shot onto the wall.

Minerallac Conduit Hanger

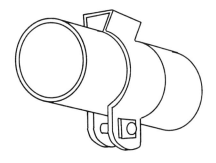

Minerallac conduit hanger

The Minerallac standard conduit hanger (also known as MIN-E hanger) is a one-piece device designed to clamp around the conduit and is held firmly in place with a nut and bolt. It has a hole at the top of the hanger to permit it to be fastened from the ceiling by a screw or piece of threaded rod and nut. The hanger is offered in steel, aluminum or stainless steel material.

The hanger size is marked with numbers from 0 to 10 referring to the size of conduit. The 0 size accommodates $\frac{3}{4}$ in and $\frac{1}{2}$ in rigid or $\frac{1}{2}$ in EMT. No. 4 size applies only to $1\frac{1}{2}$ in rigid while No. $2\frac{1}{2}$ applies to $1\frac{1}{4}$ in EMT. It is necessary to refer to the manufacturer's catalog to determine the correct hanger with the bolts and nuts to be used with a specific conduit.

Unistrut Hanger

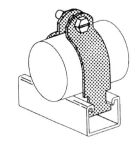

Unistrut hanger

This particular conduit fastening device is a two-piece unit designed to be used exclusively with metal framing channel systems. The fastener consists of two curved shaped pieces of metal with notches on either side to fit into the channel. Each piece has a hole at the other end to permit insertion of a machine screw and nut to fasten it around the conduit.

The pair of straps is tightened around the conduit with the machine screw and nut, the notches are engaged in the inside grooves of the support channel and a firm fastening is then assured.

The Unistrut hanger is available from $\frac{1}{2}$ to 2 in for EMT conduit and from $\frac{1}{2}$ to 6 in for GRC conduit. As sizes of the pairs of straps vary for larger conduits, the screw and nut associated with each pair will differ in length and diameter.

Unistrut Uni-Clip

The Uni-Clip for both EMT and GRC conduit is a one-piece stainless steel spring device notched at one end and curved to fit the particular size conduit for which it is designed. The fastener is then placed over the conduit. Pressure applied to the sides of the clip permits

Unistrut Uni-Clip

engaging the notched portion of the clip into the lips of the channel, thus supporting the conduit. The clip is furnished in sizes from $\frac{1}{2}$ to 2 in for EMT and GRC conduit.

GRC Clamp RC

GRC clamp RC

This clamp is designed for the installation of conduit on a 90° angle to the beam it is to be anchored. The clamp is marked "RC" with the conduit size following the letters. It is constructed of malleable iron, but is also available in a steel version with a hardened steel insert that bites into the structural member for maximum security. These inserts have a single-edged gearing surface to insure good seating on beams of varying taper and maximum penetration of the beam flange when tightened.

The clamp is supplied in sizes to fit from $\frac{3}{8}$ in to 4 in conduit in the malleable iron style and $\frac{1}{2}$ in to $1\frac{1}{4}$ in in the steel.

GRC Clamp PC

This clamp is designed for hanging conduit parallel to the structural beam. The PC indicates parallel hanging with numbers following the PC to indicate conduit size. The PC clamp is similar in construction to the RC style clamp, furnished in sizes $\frac{3}{4}$ to 4 in and available in malleable iron or steel.

Another GRC clamp very similar to the previous clamp is available with the nomenclature "EC" designed to mount conduit vertically across the beam edge. This clamp is furnished in sizes from $\frac{1}{2}$ to 3 in.

E-Z Grip Hanger

E-Z grip hanger

The E-Z grip hanger is a light-duty, one-piece steel band for use with all types of tube and conduit including EMT and GRC from $\frac{1}{2}$ through 2 in sizes. The load rating on this hanger is 150 lbs with a safety factor of 5 lbs. It is one-piece construction with a captive locking insert for fast, low-cost installation. It is basically designed to be used with $\frac{1}{4}$ in hanger rods.

Rigid Conduit Clamp

Rigid conduit clamp

This is a versatile clamp that holds pipe or conduit snugly against any type of beam, channel, angle or column with the pipe or conduit in a parallel or right-angle plane.

The steel cup pointed set screw bites into the structural member for maximum security. The clamp is offered in three sizes. The first will take $\frac{1}{2}$ and $\frac{3}{4}$ in conduit, the second takes $\frac{1}{2}$, $\frac{3}{4}$ and 1 in conduit, the third size will support $\frac{1}{2}$ through $1\frac{1}{2}$ in.

U-Bolt Pipe Hanger

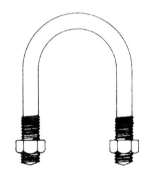

U-bolt pipe hanger

U-bolt pipe hangers are constructed of steel rod in various rod sizes such as $\frac{1}{4}$, $\frac{3}{8}$, $\frac{1}{2}$, $\frac{5}{8}$ and $\frac{3}{4}$ in. The $\frac{1}{4}$ rod U-bolt is designed to support pipe sizes from $\frac{1}{4}$ 2 in. The $\frac{3}{8}$ in size will support pipe $\frac{1}{2}$ to 4 in and the $\frac{1}{2}$ in rod pipe sizes from 2 through 6 in. The larger rod sizes are not generally used in the electrical industry.

The U-bolt may be ordered with or without a flat steel spacing bar to fit across the threaded ends of the bolt. Two pairs of hex-head nuts of the same thread size as the bolt are supplied.

EMT Clip No. BCC-25-100

EMT clip No. BCC-25-100

Three different manufacturers offer the same basic clip design. This particular clip is a combination of two separate devices. The first is a beam clip and attached to it is a conduit support clip. Both items are available independent of each other. The beam clip is a BBC 250 designed to be hammered onto flanges $\frac{3}{16}$ to $\frac{3}{8}$ in thick. The conduit pipe clip is a basic BCC 1000 constructed for the support of 1 in EMT.

The combination clip is variable in applications allowing installation of the conduit, either vertically or parallel to the support beam. The Caddy company supplies this clip style to hang $\frac{1}{2}$ to 2 in EMT.

EMT Clip, Caddy Type-K Series

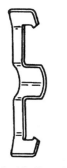

EMT clip Caddy Type-K series

This conduit support clip is a flat spring steel clip with two large squeeze easy tabs for secure mounting, greater rigidity and horizontal or vertical positions without use of tools. The tabs are squeezed to fit over the flanges of beams, columns, open web studding, rods or wire where the clip then supports itself.

The steel loop of the clip supports conduit, armored cable, bell wire, roll necks, copper water tubes, control tubes, telephone cable, flex metallic tubing, etc. The static load rating of the clip is 50 pounds.

Flange Mount Clip Series "P"

Flange mount clip Series P

This clip is available with conduit clip bottom- or side-mounted. Sizes are furnished to fit $1/2$ through 1 in EMT, aluminum, or rigid GRC conduit. The clip fits flanges $1/8$ thru $3/4$ in thick and requires only a hammer to install. The conduit clip will pivot thru 360°. Static load indicated is 25 lbs in the vertical position and 15 lbs in the horizontal position. Conduit is snapped into the clip and secured by steel prongs.

Flange Mount Clip Series "M"

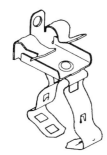

Clip Series M

This clip is similar to the "P" series in that it is driven onto the supporting flange and is furnished with the conduit clip either bottom- or side-mounted. The conduit is secured by a snap close feature which completely surrounds the conduit. The clip fits beam flanges $1/8$ thru $3/4$ in thick with sizes to fit conduit $3/8$ thru 2 in EMT, aluminum, or rigid GRC conduit. The conduit clip will pivot through 360°.

Conduit/Box Support

Conduit box support

This support is designed to support a junction box above acoustical T-bar and in combination with a standard conduit clip to support $\frac{3}{4}$ in conduit. The flat-bottom half of the clip is divided to allow it to be placed over the rib of the acoustical T-bar and secured with a furnished $\frac{1}{4}$-20 screw. The top of the support is threaded to fit a $\frac{1}{4}$-20 screw to permit direct attachment of the conduit clip or outlet box. When used in combination with a conduit clip and outlet box, it provides exact alignment — eliminating an offset bend in the conduit.

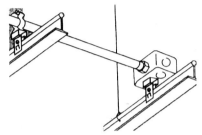

Application of conduit box support

EMT Conduit Clip

EMT conduit clip

The standard EMT conduit clip is offered in sizes for conduit $\frac{3}{8}$ through 2 in. The clip may be used with a universal clamp for both vertical and horizontal installation of conduit, mounted on a hanging rod or mounted directly to a stationary surface. The clip will be furnished with either a plain mounting hole for $\frac{1}{4}$ in bolt or a $\frac{1}{4}$ in-20 thread impression.

The clips have a load limit of 100 lbs in the vertical hanging position and a limit of 25 lbs in the horizontal position. The finger close-and-lock capability allows the clip to be attached to conduit without the use of tools.

Unistrut Flex Hanger

Unistrut flex hanger

The Unistrut hanger is a U-shaped device designed to be installed with one hand leaving the other hand free to position the flex conduit or cable. There is a flat flange under the set-screw type fastener which prevents damage to the flex conduit or cable. Each leg of the "U" has a notch on opposing sides to allow the hanger to be affixed to the Unistrut by inserting it in the channel and twisting to secure.

Snap Clip Flex Hanger

Snap clip flex hanger

The snap clip is used to attach flex to metal studs and does not require any tools to apply. The spring-like tongue in the curved hanger portion of the clip allows it to be snapped onto the flex conduit while the same type tongue in the flat side of the clip permits a firm attachment to the cutout on the stud.

Flex Conduit Stud Fastener

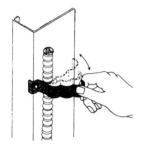

Flex conduit stud fastener

This fastener is constructed of spring steel allowing it to be removed from the metal stud and then snapped back in place as additional flexible metal conduit runs are added. The fastener keeps cables and flex snug against the stud and will hold 1, 2, or 3 runs of 14-2 through 12-3 MC or AC flex. The crimped end fits squarely on the side of the stud and the rounded corner end allows easy removal and reinstallation.

Flexible Cable Clip

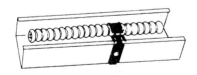

Flexible cable clip

This clip is designed to position flex conduit or cable in the open face of metal studs and installs with just a hammer. It accommodates 12-2 through 10-3 flexible conduit without causing a bulge in the drywall. The pound on extension has a locking prong which is pressure held against the side of the stud and the flex snaps in the fitted semi-circle holder.

"Z" Purlin Clips

The slotted attachment body eliminates the need to drill purlin to allow fastening with bolts. The angled slot is designed to compensate or purlin angles and locks firmly onto the purlin. The conduit fastener can be

Z purlin clip

snapped closed by hand and the positive self-locking feature holds the conduit in place. Removal of the conduit clip allows the "Z" clip to be used to mount a box in perfect alignment with the conduit.

Acoustical T-Bar Clips

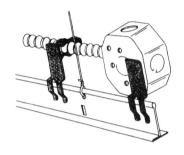

Acoustical T-bar clip

This is a specialized clip for supporting flex conduit or cable above suspended ceiling acoustical T-bar grids. The fastener snaps and locks over the bulb of the T-bar. The conduit clip may be closed and locked with finger pressure and no tools are required for installation. Removal of the conduit clip allows a box to be supported in line with the flex conduit.

Beam Hanger Clamps

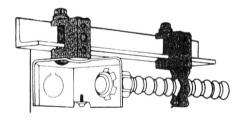

Beam hanger clamp

The flexible clamp adapts to bottom- or side-mounting situations. The hanger clamp will accommodate $1/8$ through $3/4$ in thick flange and pivots through 360° to speed up installation. The conduit clip is positive locking and may be closed over the conduit with only finger pressure without the use of tools.

Multifunction Clips

These are designed to support flex and/or cable from drop wire or plain and all thread rod. The clips are two-piece devices with a self-locking conduit slip and a spring steel finger slotted on each end which locks onto the drop wire or rod. No tools are required for installation as the conduit

Conduit Fasteners

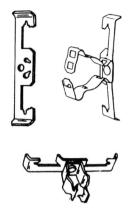

clip is designed to be closed with finger pressure and the steel attachment fingers are easily spring onto the wire or rod.

Multifunction clips

Beam Clamp

The U-shaped beam clamp is for beam flanges up to $\frac{1}{2}$ in thick and has gripping teeth which fit under the beam to prevent the flex conduit from sliding during installation. The set screw included is tightened onto the side of the beam flange for security. The flex conduit clip has the positive locking feature which is easily closed with finger pressure. Without the conduit clip, the clamp is adaptable to mounting a box on the beam flange. The clamp installs with a screwdriver.

Beam clamp

Trapeze Hangers

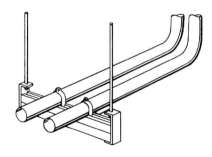

Trapeze hangers are constructed in several different ways. One method is to use a piece of scrap EMT conduit suspended from a beam flange and then using beam clamps and all thread rod. A self-locking fastener then snaps onto the EMT with finger pressure and remains stationary while flex conduit is being enclosed in the conduit clip. All parts of the trapeze hanger attach without the use of tools. Conduit may be supported above or below the trapeze which will support up to 100 lbs.

Trapeze hanger in use

Another method is to use Unistrut with appropriate attachments to form the bar of the trapeze. All thread rods and beam clamps are then used to hold and secure the Unistrut.

Standard MC/AC Cable Clip

This clip is ideal for both vertical and horizontal installations and fits $\frac{1}{2}$ in and $\frac{3}{4}$ in cable. It is available with $\frac{1}{4}$ in-20 thread impression for threaded rod installation. The clip can be attached to block, wood or concrete by use of a screw or power driven attachment pin or device. The clip is locked onto the conduit by closing the self-locking clip with finger pressure.

Standard MC/AC cable clip

Flexible Cable Support

The flexible support allows the bundling of up to 6 runs of flex conduit through one fastener. When used in conjunction with other fasteners, bundled cable can be supported from main and substructure. No tools are required for installation. The static load limit is 5 lbs.

Through-Stud Flex Conduit Support Clip

Designed to be installed with a screw gun and specifically for horizontal runs, the clip eliminates noisy rattle between drywall. The clip complies with the National Electrical Code. The self-locking feature is put in place with finger pressure and holds the conduit firmly in position away from the surface of the stud.

Through-Stud Conduit Support

The design of this support allows flex conduit to be pulled through the stud. The simple support holds the cable firmly in place and prevents metal to metal rattle. Easy installation requires only a self-tapping screw.

Through-stud conduit support

Chapter 17

Conduit Fittings

There are conduit fittings manufactured to fit practically every possible installation requirement encountered in the electrical industry. It is necessary that the proper fitting be used for the particular installation for which it was designed. The use of the wrong fitting could result in an extremely hazardous situation.

Conduit Fittings

Rigid split coupling

Rigid Split Coupling: The rigid split coupling is manufactured in all GRC and aluminum conduit sizes. It is used as a union for joining conduit on indoor installations where conduit cannot be turned or where space is limited. Sizes $\frac{1}{2}$ through $1\frac{1}{2}$ in require only one bolt to attach the coupling. Sizes 2 in through 6 in require two bolts for firm attachment.

Rigid Erickson coupling

Erickson Coupling: This three-piece coupling device has become known in the trade by the registered trademark of "Erickson". The trademark name belongs to the Thomas & Betts Manufacturing Company (T&B), which produces a line of these couplings in $\frac{3}{8}$ in through 6 in sizes. The GRC or aluminum IMC coupling is concrete-tight and is used to couple two lengths of conduit when neither conduit can be turned.

Rigid compression connector

Rigid Threadless Compression Connector: As the name implies, this is a connector for GRC or IMC conduit without threads, and it is equipped with a compression ring to lock it into place on the conduit. It is used when there is no threading equipment available or where there are damaged threads on conduit stubs. This no-thread connector, in conjunction with the no-thread set-screw coupling or rigid no-thread compression coupling, permits the use of odd lengths of conduit to connect junction boxes without the need for threaded nipples. This fitting, considered by the UL as concrete-tight, is available in diameters from $\frac{1}{2}$ to 4 in.

Rigid Threadless Set-Screw Connector: This connector is used similarly to the compression connector, with the exception that it is fastened, or attached to the conduit length by the use of a set screw. It is used on conduit where the threads are damaged and it requires no special tools other than a screwdriver, wrench or pliers. This connector is available from $\frac{1}{2}$ through 4 in diameter conduit size.

Rigid set-screw connector

Rigid Meyers Hub: This is another example where a name given to a particular device by its manufacturer has become synonymous with a type or class of fitting. The Meyers Hub is a connector with a sealing ring for water-tight applications. It is basically a 3-piece fitting consisting of a male and a female body, a sealing O-ring and a locknut. The combination of parts provides a water-tight threaded hub on enclosures.

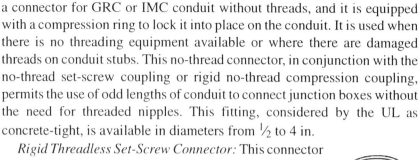

Rigid Meyers Hub

Chase Nipple: Chase nipples are used to connect conduit couplings to box knockouts or with a locknut to hold two boxes back-to-back or side-to-side, and with a locknut to connect fixture housings in continuous runs. They are available either insulated or noninsulated, in $\frac{1}{2}$ to 1 in sizes with an overall length of 1 in.

Chase nipple

Gutter Connector: The gutter connector derives its name from its customary application of mounting junction boxes or other types of switches or load centers from run of gutter. It is a three-piece device consisting of what appears to be a long chase nipple which, in turn, has a threadless sleeve inserted over it, and a locknut. The threadless sleeve-like portion of this acts as a spacing device between the gutter and whatever will be mounted on top of it. Gutter connectors are furnished in sizes from $\frac{1}{2}$ to 2 in.

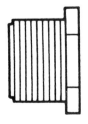

Gutter connector

Rigid close nipple

Rigid Close Nipple: A rigid close nipple appears to be a short length of all thread conduit with the exception that the threads do not continue the full length but stop in the middle of the nipple. This stopping of the threads assures that each half of the nipple is used for a balanced connection.

The nipple is a specific length of GRC or IMC conduit which is threaded at each end. They are supplied in standard factory lengths in conduit sizes $\frac{1}{2}$ through 3 in in lengths of $1\frac{1}{2}$ in and proceeding in $\frac{1}{2}$ in increments to 4 in. Nipples longer than 4 in are 5 in through 10 in in 1-in increments and finally 12 in length. Conduit nipple sizes 4 in to 6 in are available in 2 in increments up to 12 in from the factory.

Conduit nipples are generally used to terminate an odd length of conduit into a junction box or other terminating device, or used as spacers between various enclosures, gutters or junction boxes. When rigid nipples are incorporated in a conduit run, a total of four locknuts and two insulating bushings are required to complete the assembly.

Rigid Offset Connector (Nipple): An offset connector is an S-shaped device available with either a $\frac{3}{4}$ in or $1\frac{1}{2}$ in offset. It is generally a

Rigid offset connector

male-to-male fitting, although offset couplings are available similar to the connector but with female-to-female threads at each end. The offset connector for rigid conduit is supplied in $\frac{1}{2}$ in through 2 in conduit sizes in the male-to-male style and $\frac{1}{2}$ in to 1 in in the female-to-female coupling style. The Thomas and Betts Company offers an offset connector with a male thread on one end and a female thread on the other in $\frac{1}{2}$ in, $\frac{3}{4}$ in and 1 in sizes.

Locknut

Rigid Locknut: All manufacturers of locknuts for rigid conduit offer them from $\frac{3}{4}$ in through 6 in conduit size. The locknuts are notched on the outside rim to permit easy tightening with a screwdriver. The locknuts are threaded with special locknut taps to allow for maximum number of conduit variations. The notches are concave-shaped to assure perfect ground with the box. Locknuts are constructed of steel-malleable iron or aluminum. They are also available with a screw for grounding and bonding purposes in sizes $\frac{1}{2}$ in to 4 in diameter.

Most GRC, EMT, IMC and flex conduit fittings are supplied with applicable locknuts, the exception being the chase nipple, conduit nipple, gutter connectors and short 90° elbows.

Plastic bushing

Metallic insulated bushing

Metal bushing

Capped bushing

Bonding bushing

Plastic Bushings: The National Electrical Code requires that whenever a rigid conduit run, rigid nipple, or other rigid terminating device is used, a plastic bushing must be provided in addition to the locknut, for insulating purposes. The insulating bushing is used to provide protection and prevent damage to the wire being pulled into the junction box or other terminating location. The bushings are available in all conduit sizes from $\frac{1}{2}$ in through 6 in.

Rigid Metallic Insulated Bushing: A metallic insulated bushing is similar to the plastic insulated bushing with the exception that it is all metallic with an insulated throat. This bushing is specified for those installations where additional strength is required and the tough smooth insulating ring prevents strain or wear on wire or cable insulation. The metallic bushing is supplied in sizes $\frac{1}{2}$ in through 6 in.

Rigid Metal Bushing: The rigid metal bushing is similar in appearance to plastic and metallic insulated bushings, the difference being that there is no insulated throat and the construction is all metal. The bushing is generally used in conjunction with either plastic or metal pennies in capping conduit stubs to prevent the introduction of foreign matter, dirt, cement, etc. into the open conduit. It is supplied in all conduit sizes $\frac{1}{2}$ through 6 in.

Rigid Capped Bushing: The capped bushing has a snap-in fitted plastic cap which allows it to be used to seal off the open end of a conduit run. The bushing is supplied as a two-piece unit available from $\frac{1}{2}$ in through 2 in sizes. Large size conduit requires the use of a metal bushing and metal penny to close off the open end.

The capped bushing is appropriate for use in applications where there is no great danger of damage to the plastic bushing. Areas exposed to heavy construction movement require the use of the metal bushing.

Rigid Bonding Bushing: The bond bushing has an insulated throat, and a swivel type solderless connecter designed for use in grounding or bonding purposes. It is provided with a set screw to assure that it remains locked into position after being attached to the threaded end of the conduit or terminating fitting.

Bonding bushings are made in all rigid conduit sizes from $\frac{1}{2}$ in to 6 in. Use of the bonding bushing eliminates the requirement for use of the plastic insulating bushing.

Rigid Threadless Bushing: The no-thread bushing is used when it is impossible to create a thread on the conduit stub-up. The bushing is fastened into place with set screws with which it comes equipped. There are 3 set screws used to secure the bushing and a screw type connector for use with a grounding wire. The no-thread bushing is supplied in sizes $\frac{1}{2}$ in through 4 in.

Reducing bushing

Reducing Bushings: This bushing is threaded on the interior and exterior. Sizes are expressed in the outside diameter thread and the inside diameter thread such as 1 in to $\frac{3}{4}$ in. The purpose of the bushing is to reduce the conduit or hub size from the outside diameter to that of the inside diameter. There are at least 38 different combinations of reducing bushings available, beginning with $\frac{1}{2}$ in through $\frac{3}{4}$ in up to 4 in to $3\frac{1}{2}$ in. The most commonly used RE bushing sizes are $\frac{3}{4}$ in to $\frac{1}{2}$, 1 in to $\frac{3}{4}$ in and 1 in to $1\frac{1}{2}$ in.

Male enlarger

Male Enlarger: The application of a male enlarger is just the opposite to what the RE bushing does—that is, it will take a smaller hub opening and enlarge to the next size up. For example, if the junction box had only $\frac{3}{4}$ in hubs and it is desirable to continue with the 1 in rigid conduit run from that box, the use of the enlarger will provide attaching the 1 in conduit to the hub by the use of a $\frac{3}{4}$ in to 1 in male enlarger. The male enlarger is only available in sizes $\frac{1}{2}$ in to $\frac{3}{4}$ in, $\frac{3}{4}$ in up to 1 in, 1 in to $1\frac{1}{4}$ in, and from 1 in to $1\frac{1}{2}$.

GRC 90° Short Elbow: The short elbow has a female thread on one end and a male thread on the other. This style ell is available from $\frac{1}{2}$ in to 2 in conduit size. It has a very short radii, known as a short ell usable in very limited spaces.

90° short elbow

90° threadless ell

GRC 90° Threadless Ell: The no-thread ell has an insulated throat and a screw type no-thread locknut and compression ring. Its application is where available space makes it impossible to thread conduit to which it will be attached. The insulated throat permits it to be terminated into a junction box directly allowing wire to be pulled through it without damage to the wire.

GRC 90° Bushed Elbow: The bushed elbow is often referred to as the telephone elbow and has a long radii. It is also called a long rigid ell. The bushed ell is made only in the ½ in, ¾ in, 1 in, and 1¼ conduit sizes and is available with or without an insulated throat.

90° **bushed elbow**

90° **EMT short elbow**

EMT 90° Short Elbow: Available in conduit sizes ½ through 2 in with a short radius ranging from 4 in for the ½ in to 9 in for the 2 in size, the EMT 90° short elbow is used in applications where space is at a premium. There are no threads on the elbow necessitating the use of standard EMT compression or set screw fittings for installation.

GRC 90° Corner Adapter Ell: The adapter ell is offered in either malleable iron or aluminum. This is a 4-piece device known as a female to male, consisting of a cover, gasket, locknut and main body. The screw-down removable cover and gasket makes it accessible from either direction for wire installation and the gasket feature makes it a waterproof-type fitting for use in outdoor installations. This ell is also available for use with seal-tight flexible conduit. Sizes ½ in through 2 in are manufactured.

90° **corner adapter ell**

GRC 90° Corner Elbow: This type is a female to female — that is, the connecting ends are internally threaded. The 4-piece construction — main body, cover, gasket, and locknut allows access from either direction for wire installation and the gasket feature makes it waterproof for use in outdoor installations. The corner ell comes in sizes ½ in through 2½ in.

90° **pull-in ell**

GRC 90° Corner Pull-in Ell: This three-piece ell is for indoor installation where weather is not a problem. The screw-down cover permits wire installation from either direction but a gasket is not supplied. The connecting ends are internally threaded in the female type. The ell is furnished in sizes ½ in through 2 in.

Die-Cast EMT Fittings

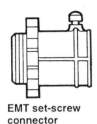
EMT set-screw connector

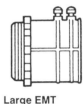
Large EMT set-screw connector

EMT Set-Screw Connector: The set-screw connector is slightly larger than the corresponding conduit size which allows it to be slipped over the end of the conduit. The set screw is tightened onto the conduit thus securing the connector to the conduit. The other end of the connector is threaded with a locknut installed against the body of the connector. The threaded end is inserted through a knockout into the box or enclosure and a second locknut installed on the threads to lock the connector to the enclosure. When properly installed the connector is rated as concrete-tight by Underwriters' Laboratories. Sizes for conduit $\frac{1}{2}$ in through 4 in are available. The larger sizes require two set-screws built into the connector to properly secure to conduit.

EMT set-screw coupling

EMT Set-Screw Coupling: The EMT set is constructed to fit over the outside of the conduit being joined, and has a molded ring at the inside center to assure that both pieces of conduit are properly inserted. Set-screw couplings are manufactured in $\frac{1}{2}$ in through 4 in conduit sizes. The larger sizes require the use of two set screws to correctly secure the conduit in the coupling.

EMT Compression Connectors: The compression connector consists of four parts:

- Body
- Locknut
- Split steel compression ring
- Compression nut

EMT compression connector

Both ends of the fitting are threaded with a hex-shaped nut molded in the center of the threads. To secure the fitting to the conduit, the compression nut is slipped over the conduit followed by the split steel compression ring. The compression nut is then tightened onto the threaded body of the fitting thus compressing the ring onto the conduit. The connector is locked into the box or enclosure with a locknut. The supplied locknut is then secured against the box or enclosure making a concrete tight connection in $\frac{1}{2}$ in through 2 in size. The compression connector is made in $\frac{1}{2}$ in through 4 in conduit sizes.

EMT compression coupling

EMT Compression Coupling: The compression coupling is designed to join two lengths of EMT conduit. It is made up of five pieces: the body threaded at both ends, two split steel compression rings and two compression nuts. The compression nut followed by a compression ring is fitted over the conduit and then the compression nut is tightened onto the fitting compressing the ring and securing the conduit. The procedure is repeated on the second piece of conduit making a concrete tight connection in $\frac{1}{2}$ in through 2 in sizes. The coupling is available up to the 4 in conduit size.

EMT combination coupling

EMT Combination Coupling: The combination coupling is designed to join EMT conduit to flexible metal conduit or armored cable. The EMT side of the coupling uses a set screw to affix it to the conduit. The flexible metal conduit or armored cable end utilizes a two screw strap for securing to the flex or cable. The coupling is only available as a $\frac{1}{2}$ in EMT size to $\frac{3}{8}$ in flex size.

EMT Malleable Iron Fittings

EMT combination coupling

EMT Combination Couplings: The malleable iron combination coupling for joining EMT to armored cable or flexible steel conduit is available in $\frac{1}{2}$ in EMT size to connect to $\frac{3}{8}$ in or $\frac{1}{2}$ in cable or flex.

EMT Steel Fittings: The steel compression connector is much lighter in weight than the die-cast fitting. The connector is a four part fitting consisting of the body (threaded on both ends), the compression ring, compression nut and locknut. The compression nut is slipped over the conduit followed by the compression ring. Tightening of the compression nut collapses the compression ring onto the conduit making a raintight connection in sizes $\frac{1}{2}$ in to 2 in and a concrete-tight connection up to the maximum 4 in size. The connector is attached to the box or other enclosure by means of a lock ring on the threads inserted through the knockout of the box. The supplied locknut is then tightened against the outside of the box to firmly lock the connector in place.

Insulated steel compression connector

Insulated Steel Compression Connector: Many of the steel connectors are available with an insulating liner built into the body of the connector to conform with the electrical code in many cities. The outward appearance of the fitting remains the same as the uninsulated connector. The insulated fitting is installed in the same manner described for the regular compression connector.

Steel compression coupling

Steel Compression Coupling: The steel compression coupling consists of the body, two split compression rings, and two tapered compression nuts. The compression nuts, followed by the compression rings are slipped over the ends of the conduit to be joined and the conduit then inserted into the coupling. Tightening of the compression nuts compresses the split ring onto the conduit thus forming a secure concrete-tight seal in $\frac{1}{2}$ in to 4 in sizes and a watertight seal in $\frac{1}{2}$ in to 2 in sizes.

Steel set-screw connector

Steel Set-Screw Connector: The steel set-screw connector is supplied in conduit sizes from $\frac{1}{2}$ in to 4 in. The connector has a body, lock ring, set screw and is threaded on one end. The threaded end is inserted through the knockout of the box and a separate locking ring installed on the inside of the box. The supplied lock ring is then tightened against the outside of the box to make a stable connection. Conduit is then inserted into the end of the connector and the set screw firmly tightened onto the conduit. The connection is rated concrete-tight under the NEC.

Steel set-screw coupling

Steel Set-Screw Coupling: The steel set-screw coupling is furnished with two set screws used to secure the coupling to the conduit. Conduit is simply inserted into the ends of the coupling and the set screws tightened to complete the connection. This fitting must be taped with electrical tape to be considered concrete-tight. Sizes available are $\frac{1}{2}$ in through 4 in.

Steel snap-in blank

Steel Snap-in Blanks: The steel snap-in blanks in sizes $\frac{1}{2}$ in through 2 in are used to cover or blank out unused holes in boxes or enclosures. The tabs are crimped and lock the blank into the hole thus effectively closing it.

Nylon insuliner sleeve

Nylon Insuliner Sleeve: The insuliner sleeve is split to allow easy insertion into the throat of a connector to replace a broken insulator or making a previously uninsulated fitting conform to the NEC. They are available in sizes from $\frac{1}{2}$ to 4 in.

Steel offset nipple

Steel Offset Nipple: The steel offset nipple is used to offset the axis of a raceway ¾ in and is offered in ½ in to 2 in sizes. The nipple is threaded on both ends and supplied with a lock washer on each end.

EMT Steel Combination Couplings: The EMT steel combination coupling is suitable for joining EMT conduit to flexible metal conduit or armored cable. The coupling is made in two sizes to fit ½ in EMT and either ⅜ in or ½ in flex or cable. The EMT end of the coupling is of the compression type utilizing a steel split compression ring and a tapered compression nut. The flex/cable end uses a two-hole clamp requiring two set screws to affix the flex/cable.

EMT steel combination coupling

Flex metal conduit connector

Flexible-Metal Conduit and Armored-Cable Connector: The connector attaches to the box or enclosure by inserting the threaded end into the knockout and applying a locknut on the interior of the box or enclosure. The locknut supplied with the connector is then tightened against the outside surface of the enclosure or box to secure the connector.

The flexible conduit is then inserted into the other end of the connector and secured in place by tightening the two screw strap-type clamps against the flex. The same type of connector is also used to connect non-metallic cable to a box or enclosure. The connector is only available in ½ in knockout size which will accommodate flexible-metal conduit and armored cable from a minimum of .440 in to a maximum of .610 in.

45° Angle Flexible-Metal Conduit Squeeze Connectors: The 45° angle connector is useful in situations where it is necessary for the flex or cable to arrive at the termination point at an angle from above or below the box or enclosure. The connector is available with either the two-screw bar-type clamp, or a one-screw compression-ring-type clamp to hold the conduit in the connector. The connector is affixed to the box or enclosure by use of the furnished locknut on the exterior and another locknut on the interior surface of the box. The connector is in sizes to fit ½ in and ¾ in knockouts.

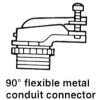

90° flexible metal conduit connector

90° Angle Flexible-Metal Conduit Connector: The 90° connector is used to connect the conduit or cable when it arrives at the termination point in a position making use of a right angle connector necessary. The connector is attached to the box or enclosure and the flex conduit or cable in the same manner as the 45° connector. The connector is manufactured in sizes to adapt to knockouts from ½ in to 2 in.

Flexbible metal conduit connector

Straight Clamp Flexible-Metal Conduit Connector: Straight clamp connectors allow the flex to be attached in a straight line to the box or enclosure. The two screw bar type clamp holds the conduit firmly in the connector and the body of the connector is attached to the enclosure by the use of the furnished locknut on the exterior and a second locknut on the interior. The straight connector is available in $\frac{1}{2}$ in, $\frac{3}{4}$ in and 1 in knockout sizes.

Liquidtight Flexible-Metal Conduit Fittings

Flexible-metal conduit is available which is encased in a liquidproof covering for use in installations subject to outdoor weather conditions or large amounts of moisture in interior installations. Special conduit fittings are required to maintain the liquidtight characteristics of installation.

Liquidtight Straight Flex Connector: The liquidtight connector consists of five parts:

Liquidtight flex connector

- Body
- Ridged-tapered compression type nut
- Nylon split gland ring
- Casehardened locknut
- Moistureproof sealing ring

The body has a raised hexagon nut-shaped section next to the threaded box side connector portion of the body. The sealing ring is placed between the outside surface of the box or enclosure and the raised section of the connector produces a liquidtight seal between the connector and the box surface. The split gland ring and the compression nut is placed over the outside of the flex conduit. Tightening the nut onto the threaded end of the connector produces the liquidtight coupling of the flex to the connector. The connector is manufactured in the insulated throat style and the non-insulated style. Sizes available are $\frac{1}{2}$ in through 4 in hub size.

45° liquidtight flex connector

Liquidtight 45° Angle Flex Connector: The 45° liquidtight connector is constructed of the same five main parts as the straight liquidtight connector. Installation follows the same as that of the straight connector. The 45° angle permits the installation of flex conduit where it is required to approach the box or enclosure from a 45° angle above or below. The connector is furnished in sizes $\frac{1}{2}$ in through 2 in. The connector is available in both the insulated and uninsulated versions.

90° angle flex connector

Liquidtight 90° Angle Flex Connector: Similar to the 45° angle connector requiring the same installation procedure, the 90° connector is available insulated up to the 3 in size. This connector is required where a sharp 90° turn is necessary to approach the box or enclosure. The 90° version is also available as non-insulated up to the 4 in size.

Liquidtight adapter

Liquidtight Adapter: The adapter is used for coupling liquidtight flex conduit to threaded rigid conduit. Constructed of three parts, the body with a hexagon-shaped center built in, a split nylon gland ring and a tapered compression nut. The adapter has a female thread to fit rigid conduit. The tapered compression nut and the nylon gland ring are fitted over the flex and tightened to secure them after the female end is fitted onto the rigid conduit. The adapter is furnished in $\frac{1}{2}$ in, $\frac{3}{4}$ in and 1 in rigid conduit sizes.

Liquidtight straight grounding connector

Liquidtight Straight Grounding Connector: The grounding connector has a grounding lug permanently attached to the tapered compression nut which allows a ground wire to be inserted and secured with a set screw. The fitting is used where a separate ground wire is required due to hazardous conditions. The connector is manufactured in $\frac{1}{2}$ in through 3 in insulated conduit sizes and only to the 2 in uninsulated. The installation procedures are the same as with the standard liquidtight fittings with the exception of the addition of the ground wire.

Liquidtight 45° grounding connector

Liquidtight 45° Grounding Connector: The 45° grounding connector is precisely the same as the standard 45° liquidtight connector with the addition of the grounding lug attached to the tapered compression nut. The grounding version of the 45° fitting is used in the same type of installation as the standard fitting. It is made in $\frac{1}{2}$ in, $\frac{3}{4}$ in and 1 in size.

Liquidtight 90° Grounding Connector: The grounding lug is the only difference in the 90° grounding connector and the standard 90° liquidtight connector. The installation is the same with the exception of the addition of the ground wire. Sizes $\frac{1}{2}$ in through 2 in are available.

Liquidtight 90° grounding connector

Chapter 18

Conduit Bodies

Conduit bodies or "condulets" are required to properly distribute branch-circuit wiring installations from the load center to the various outlets, receptacles, switches, etc. encountered in commercial and industrial electrical projects.

Condulets are installed at convenient locations in conduit systems to:

1. Act as pull outlets for conductors to be installed in the conduit system and/or pulled around bends.

2. Make 90° bends in conduit systems.

3. Provide openings for making splices and taps in conductors.

4. Provide taps for branch conduit runs.

5. Act as a mounting outlet for lighting fixtures and wiring devices.

6. Provide access to conductors for maintenance and future system changes.

7. Connect conduit sections.

The selection of the type of conduit or condulets is determined from the configuration of the conduit system and intended function of the outlet body. The conduit body size is determined by the size of conduit and wire that is used. In addition, consideration must be given the hub size or style; that is, either for rigid, threaded, threadless, EMT or PVC. The type of environment will determine the material finish of the condulet required.

The condulets are available constructed of malleable iron, or copper-free aluminum. All of them are classified as raintight when used with gaskets under the covers. The flat-back design permits greater wiring capacity and the internal volume and maximum wire is clearly marked on the casting in the malleable iron style. The malleable iron bodies may be obtained finished with zinc-plated aluminum enamel or hot dip and/or mechanically galvanized finish.

Conduit Bodies for Rigid and IMC Conduit

Type C conduit body

Conduit Body Type C: This is the most compact of the condulets and its over all size approaches the outside diameter of the conduit making a neat installation. Type C is a straight-through configuration used in line with the conduit and generally used as a splice point in the conduit run. Sizes range from $\frac{1}{2}$ in to 3 in conduit size.

Type E conduit body

Conduit Body Type E: Has only one hub and used at the termination of a conduit run with an appropriate outlet cover to connect to an additional fixture or motor installation. Type E is available only in $\frac{1}{2}$ in, $\frac{3}{4}$ in and 1 in sizes.

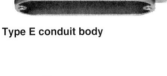

Type LB conduit body

Conduit Body Type LB: The L is indicative of the word "ell," while the B indicates "back." With the condulet opening facing the viewer, the 90° take off portion is to the back permitting a 90° turn in the opposite direction in a conduit run. The wire is first pulled into the body in sufficient length to reach the next splice or termination point. The wire is then inserted into the exit of the condulet and the excess length pulled to the next splice or termination. Type LB is manufactured in $\frac{1}{2}$ in through 4 in sizes. Sizes $1\frac{1}{4}$ in through 4 in are supplied with rollers built into each end of the wiring chamber to reduce friction when cable is pulled through the hubs. This eliminates cable damage while it makes installation relatively easy.

Type LR conduit body

Conduit Body Type LR: Type LR at first glance appears to be the same as the LB but the R stands for the 90° take-off hub being to the right when the opening of the condulet body faces the viewer in the vertical position. The LR is provided in all conduit sizes $\frac{1}{2}$ in through 4 in and in either malleable iron or aluminum.

Type LL conduit body

Conduit Body Type LL: With the conduit body opening facing the viewer, vertically, the 90° take-off hub goes to the left thus the designation L for "ell" and the second L for left. It is used where it is necessary for conduit run to continue at a right angle to the left. Available in conduit sizes $\frac{1}{2}$ in through 4 in.

Type T conduit body

Conduit Body Type T: Type T body has three hubs, one at each end in line with the conduit run, and one on the side at a right angle to the condulet opening. The body is used to allow a tap to be made in the wiring installation to be run off at a right angle to the main conduit run. It is furnished in $\frac{1}{2}$ in through 4 in conduit sizes.

Type TB conduit body

Conduit Body Type TB: Type TB is similar to the type T with the exception that the third hub extends to the back accounting for the B in the name. Used to allow a tap to extend at a 90° angle from the main conduit run to the back when the opening of the body is facing the viewer.

Type TA conduit body

Conduit Body Type TA: Type TA has four hubs, one at each end for the main conduit run and one at the back and one side of the opening for the tap runs. It is furnished in $\frac{1}{2}$ in through 2 in conduit sizes only.

Type X conduit body

Conduit Body Type X: Type X has four hubs, one at each end and one on each side of the condulet opening. The hubs on each side of the opening permit taps to extend to either side or up and down from the parallel run of conduit. The type X is available in $\frac{1}{2}$ in through 2 in conduit sizes.

Conduit Bodies for EMT

The conduit bodies for EMT conduit are constructed of copper-free aluminum. The hubs are equipped with slotted hex-head, steel-hardened cup point set screws to allow attachment to the nonthreaded EMT conduit. Sizes $\frac{1}{2}$ in through 2 in are die-cast aluminum and sizes $2\frac{1}{2}$ in through 4 in are sand cast aluminum. The conduit body configurations are the same for both rigid and EMT; that is, LB, T, X, etc. Please refer to the appropriate illustration under conduit bodies for rigid and IMC.

Conduit Body Type C: Type C for EMT provides access to conductors for pulling, splicing, and maintenance. The body has a hub at each end and is installed in line with the conduit. Type C sizes are $\frac{1}{2}$ in through 2 in. Uses set screws to attach to conduit.

Conduit Body Type LB: Has two hubs, one extending out the back of the body when the opening is facing the viewer and the other in line with the opening. This configuration permits the conduit run to be extended at a 90° angle away from the main conduit run to the back of the condulet. Available in $\frac{1}{2}$ in through 4 in sizes. Set screws tighten it to the conduit.

Conduit Body Type LL: Type LL has two hubs and one access opening. The main hub extends from the bottom of the body, the L or left hub extends from the left side of the body when the body opening is facing the viewer and the body is in a vertical position. Set screws attach the body to the conduit. It is available in conduit sizes $\frac{1}{2}$ in through 2 in.

Conduit Body Type LR: Type LR has two hubs in addition to the access opening. The main hub extends from the lower end of the body, the R or right hub extends from the right side of the body when the body opening is facing the viewer and the body is in a vertical position. Attaches to the main conduit with set screws and furnished in conduit sizes $\frac{1}{2}$ in through 2 in.

Conduit Body Type T: Conduit body type T has three hubs, two are in line with the primary conduit and one is in the center of the side of the body. The body attaches to the main conduit with set screws and is available in $\frac{1}{2}$ in through the 2 in conduit sizes. The T-type is used to make a tap on the wire or cable and extend the connecting wire to either the right or left of the primary conduit.

Covers and Gaskets for Conduit Bodies

Type BS Domed-Top Cover: The domed-top type cover is available in type BS manufactured of stamped steel, with steel screws and zinc plated. The type BS-Q is made of stamped copper-free aluminum, uses stainless

Type BS domed-top cover

steel screws and comes in a natural finish. The covers are made to fit all sizes conduit bodies from ½ in conduit size to 4 in. The attaching screws screw into pre-drilled and tapped holes in the conduit bodies.

Type BC Flat-top Design Cover: Cover is constructed of cast iron with stainless steel mounting screws. Standard sizes ½ in through 4 in fit all sizes of conduit bodies. The cover is hot dip galvanized to minimize corrosion.

Type BC flat-top cover

Type BCC clip-on cover

Type BCC Clip-On Covers: These are designed to only be used on malleable iron conduit bodies. Covers are attached to body with builtin clips locked in place by turning a stainless steel screw. The clips are made of zinc-plated steel. Manufactured to fit all malleable iron conduit bodies in sizes ½ in through 4 in.

Solid type sealing gasket

Solid Type Sealing Gasket: Made to fit over the access opening of all conduit bodies making them rain tight. Gasket is made of either velbuna or closed cell neoprene in sizes to fit bodies ½ in to 4 in.

Type SGNC cover gasket

Type SGNC Cover Gasket: Designed specifically to be used with clip-on conduit body covers. Standard sizes are ½ in through 4 in. The gasket has a perforation at each end to permit the insertion of the clips. It is made of neoprene and makes the conduit body rain tight.

Outlet Boxes for Rigid Conduit and IMC

The boxes are designated as type FS Shallow and type FD Deep with the type FD deep having approximately one third more internal volume than the type FS shallow. All of the FS and FD boxes are used to accommodate wiring devices and provide openings to splice, tap, or pull conductors. The hubs are taper threaded with integral bushings. Internal grounding screws are standard on all boxes and mounting lugs are standard on types FS, FD, FSC, FDC, FSA, and FDA.

The boxes are suitable for use in wet locations when used with gaskets under the covers and may also be used in concrete. Construction material is malleable or gray iron with zinc plating and aluminum enamel finish.

Single-gang FS and FD box

Single-Gang Type FS and FD: Suitable for a single device, has one hub in sizes ½ in , ¾ in and 1 in. Mounting lug on the upper left hand corner of the box allows mounting to wood, concrete or any suitable surface. The ½ in and ¾ in hub size box is generally large enough in the S or shallow version. The 1 in size hub generally calls for the use of the D or deep style. Type S has an inside volume of 21 cu in while type D has 28 cu in.

Single-gang type FSC and FDC box

Single-Gang Type FSC and EDC: Both the FSC and the FDC boxes have two hubs, one on each end of the box. The hubs are conduit sized ½ in, ¾ in and 1 in. These boxes are designed to be installed in line with the primary conduit run with the conduit extending on beyond the box. They are used for mounting a single wiring device, splicing of circuits, taps, or pulling boxes. The internal volume is 21 cu in in the S style and 28 cu in in the D style.

Single-gang type FSC and FDC box

Single-Gang Types FSL and FDL: The L in the type designation stands for "left" describing the direction of the hub located on the left side of the box when the main hub is on top and the opening is facing the viewer. The addition of the left hub permits a 90° turn to the left from the downward run of the primary conduit. Type FSL and FDL do not have integral mounting lugs but are designed to be supported by the conduit to which they are attached. They are furnished only in ½ in and ¾ in hub sizes. Internal volume is 21 cu in for the S type and 28 cu in for the D type.

Single Gang Type FSR and FDR: The R in the type designation stands for "right" describing the direction of the hub located on the right side of the box when the main hub is on top and the opening is facing the viewer. The addition of the left hub permits a 90° turn to the right from the downward run of the primary conduit. Type FSR and FDR do not have

Single-gang FSR and FDR box

intergral mounting lugs; rather, they are designed to be supported by the conduit to which they are attached. Furnished only in ½ in and ¾ in hub sizes the internal volume is 21 cu in for the S type and 28 cu in for the D type.

Single-Gang Type FSA and FDA: The FSA and FDA have only one hub which extends from the back of the box. Equipped with builtin mounting lugs, they are designed for installation on the end of a conduit run with the box opening facing in the same direction as the run. The hub is available only in ½ in and ¾ in conduit size. The box is used to mount a single wiring device, splice, tap, or as a pulling box. Internal volume is 21 cu in for the FSA and 28 cu in for the FDA.

Single-gang FSA and FDA box

Single-gang FSCC and FDCC box

Single-Gang Type FSCC and FDCC: Both the FSCC and FDCC have three hubs in either the ½ in or ¾ in sizes. Two of the hubs are on one end of the box with the third on the other end. The two hubs on the one end permit two runs of conduit to enter or leave the box allowing extension of circuits flowing into the box from the single hub on the other end. The box is used for mounting a single wiring device, splicing of circuits, junction box or pulling box. Internal volume is 21 cubic inches for FSCC and 28 cu in for the FDCC.

Single-Gang Type FSS and Type FDSS: These boxes have two hubs extending from one end of the box in either ½ in or ¾ in conduit size. The two hubs allow the application of two conduit runs for continuation of a circuit, splicing, tap or pulling. A single wiring device may also be mounted on the box. Internal volume is 21 cu in. Mounting lugs are not supplied.

Single-gang FSS and FDSS box

Single-gang FSCT and FDCT box

Single-Gang Type FSCT and FDCT: The three hubs extend one from the side and one from each end of the box forming a T configuration. This allows the circuits to continue in two directions opposite to that of the primary conduit run or in one other direction and a straight forward continuation of the primary. The box is used to mount a single-gang wiring device, splice, tap, or as pulling box. The box is supplied in ½ in, ¾ in and 1 in hub size. Internal volume is 21 cu in in the shallow version and 28 cu in in the deep version.

Single-gang FSX and FDX box

Single-Gang Type FSX and FDX: The FSX and FDX have four hubs all in the ½ in or ¾ in sizes. The hubs located on each side and each end permit the extension of circuits straight through and to the right and left of the box and primary circuit. The box is used to mount a single-gang wiring device, provide splice enclosure, permit a tap or as a pull box. Mounting lugs are not provided. Internal volume is 21 cu in in the S type and 28 cu in in the D type.

Two-Gang Type FS-2 and FD-2: The two gang FS-2 and FD-2 does not have mounting lugs and is supported by the conduit installed in the single hub. Hub sizes available are ½ in, ¾ in and 1 in. Internal volume FS-2 33.2 cubic inches, FD-2 46.0 cubic inches. Box is suitable for mounting two wiring devices' being used as a pull box or splicing a circuit.

Two-gang FS-2 and FD-2 box

Two-Gang Type ESC-2 and EOC-2: This box has two hubs, one on each end of the box, allowing the continuation of the circuit in line with the primary conduit. Hubs are furnished in ½ in, ¾ in, and 1 in sizes. Internal volume is 33.2 cu in in the shallow type and 46.0 cu in in the deep type. Box will accommodate installation of two single wiring devices, splicing, taps, and pull box.

Chapter 19

Outlet Boxes

Boxes are usually made of steel with a galvanized finish. For certain purposes, boxes made of an insulating material such as plastic or porcelain are used. NEC Article 370-3 states that nonmetallic boxes approved for the purpose may be used only with open wiring on insulators, concealed knob-and-tube work, nonmetallic-sheathed cable, and nonmetallic raceways.

The NEC establishes specifications for outlet boxes, junction boxes and switch boxes. When made of metal, they must be grounded and conform to the following:

- *Corrosion-resistant Metallic Boxes and Fittings:* Corrosion-resistant boxes and fittings shall be well galvanized, enameled, or otherwise be properly coded, inside and out, to prevent corrosion.

- *Thickness of Metal:* Sheet metal boxes and fittings, not over 100 cu in, shall be made from metal not less than 0.0625 in thick. Cast metal boxes shall have a wall thickness of not less than $\frac{1}{8}$ in thick, except that boxes of malleable iron shall have a wall thickness of not less than $\frac{3}{32}$ in.

- *Boxes Over 100 Cubic Inches:* Boxes of over 100 cu in shall be composed of metal and shall conform to the requirements for cabinets and cutout boxes, except that the covers may consist of single flat sheets secured to the box proper by screws or bolts instead of hinges. Boxes having covers of this form are for use only to enclose joints in conductors, or to facilitate the installation of wires and cables. They are not intended to enclose switches, cutouts, or other control devices.

Five commonly used outlet boxes are:

- Octagonal boxes with diameters of either $3\frac{1}{4}$ in or $3\frac{1}{2}$ in are referred to in the trade as 3-0 ("three-aught") boxes.
- Octagonal boxes with a diameter of 4 in are referred to as 4-0 boxes.
- Four-in square boxes, measuring 4 in on each side are often referred to as 4S or 4 square outlet boxes. They are also called "1900 boxes."
- A square box measuring $4\frac{11}{16}$ in on each side are referred to as 5S or five square boxes.
- Ceiling pan boxes are round-shaped and are either $3\frac{1}{2}$ in or 4 in in diameter.

Octagonal Boxes

3½ in octagon box

3½ in octagon box with NM cable clamps

3½ in Octagon Box #24141-½: The box is 1½ in deep with internal capacity of 11.8 cu in. Five ½ in conduit knockouts are provided with one in each of the four main sides and one in the back. Two threaded ears are located opposite each other on secondary surfaces of the box to permit the attachment of a cover with screws supplied with the box. The box is generally used for ceiling outlet installations.

3½ in Octagon Box #24151-N: The box is 1½ in deep with internal capacity of 11.8 cu in. The letter N stands for nonmetallic-sheathed cable and the box is supplied with two clamps arranged so that two runs of "Romex" type nonmetallic-sheathed cable may be secured under each clamp. Three ½ in knockouts, one in the back and one on each of two primary sides, are supplied. Four smaller knockouts to accommodate the cables are located under the clamps. Threaded ears for a cover attachment are standard.

Outlet Boxes 187

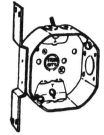

3½ in octagon box with mounting bracket

3½ In Octagon Box #24151-NV: The box is 1½ in deep with internal capacity of 11.8 cu in. The letter N signifies that the box is suitable for nonmetallic cable or "Romex" installation. The V stands for the mounting bracket permanently attached to the side of the box to permit the box to be mounted on a stud or other surface. Two ½ in knockouts are furnished, one in the back of the box and another on the side opposite the mounting bar. Four small knockouts under the supplied Romex cable clamps allow the cable to enter the box.

Round ceiling pan box

Round Ceiling Pan Box: This box is supplied in either 3½ in or 4 in in diameter. The ½ in deep pan boxes are generally used where space is limited and use of a deeper box is not practical or in a remodeling ceiling installation. There are four knockouts, two ½ in and two ¾ in on the pan side of the box. The two threaded ears allow attachment of a cover or device with screws.

4 in octagon box

4 In Octagon Box #54151-½: The 4 in octagon box has eight conduit knockouts located four in the rear surface of the box and one on each of four major sides of the rim. All of the knockouts are ½ in conduit size. The box is 1½ in deep and has an internal capacity of 15.8 cu in. Two ears located diagonal to each other on the outside rim are threaded to receive mounting screws for device or cover plate.

4 in octagon box with all ¾-in knockouts

4 In Octagon Box #54151-¾: This box is similar to the preceding box except that the knockouts are all ¾ in conduit size. The box is 1½ in deep with internal capacity of 15.8 cu in, and has two threaded ears to accept mounting screws for device or cover plate.

No 54151 4 in Octagon Box

4 In Octagon Box #54151-½ and ¾: This box is furnished with a mixture of half ½ in and half ¾ in knockouts. Box is 1½ in deep with internal capacity of 15.8 cu in. It is equipped with two threaded mounting ears to permit attachment of cover or device.

4 In Octagon Box #54151-CFB: This box has three knockouts — one in the rear surface and one on each of two side rim major surfaces. Pre-drilled holes adjacent to the knockouts permit the mounting of flex conduit or cable direct to the box while the fitted cable clamps with tightening screws included on the inside of the box permit the attachment of flex through the sides of the box. Its interior capacity is 15.8 cu in and is 1½ in deep.

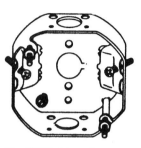

No. 54151 4 in octagon box with cable clamps

No. 54151-N 4 in octagon box

4 In Octagon Box #54151-N: This box is equipped with two nonmetallic-sheathed cable clamps, 1½ in deep. The internal capacity is 15.8 cu in and is furnished with three ½ in knockouts. Threaded mounting ears are for mounting device or cover plate.

4 In Octagon Box #54151-NL: This box is equipped with two nonmetallic-sheathed (Romex) cable clamps and three conduit knockouts. The knockouts are ½ in conduit size. The cable clamps eliminate the need for conduit connectors when nonmetallic-sheathed cable is used. The L in the nomenclature describes the L-shaped mounting bracket attached to the side of the box which allows the box to be mounted on a stud and flush with the wall's surface. Internal capacity is 15.8 cu in.

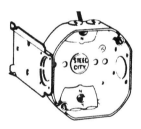

No. 54151-NL 4 in octagon box

No. 541514-NV in octagon box

4 In Octagon Box #54151-NV: It is supplied with two nonmetallic-sheathed cable clamps and three ½ in knockouts. The box is 1½ in deep with an internal capacity of 15.8 cu in. The V stands for V-shaped mounting bracket permanently attached to the box which permits mounting the box to a stud while maintaining a ½ in recess from the wall surface.

Outlet Boxes

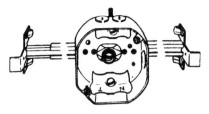

No 54151-NJ 4 in Octagon Box

4 In Octagon Box #54151-NJ: The N stands for nonmetallic-sheathed cable and the J stands for the adjustable bar joist hanger attached to the rear surface of the box. Constructed especially to be installed over ceiling joists the box is $1\frac{1}{2}$ in deep, has internal capacity of 15.8 cu in and is supplied with two nonmetallic-sheathed cable clamps with tightening screws.

No 54171 4 in Octagon Box

4 In Octagon Box #54171-$\frac{1}{2}$ and $\frac{3}{4}$: The box has a combination of $\frac{1}{2}$ and $\frac{3}{4}$ in knockouts located four in the rear surface and one on each of the major side rim surfaces. The $2\frac{1}{8}$ in depth results in a 22.5 cu in internal capacity. Eight conduit knockouts located one on each of four major rim sides and four in the rear allow variety of conduit entrances. Threaded ears opposing each other on the front of the rim permit mounting of device or cover plate.

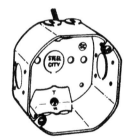

No 54171-NL 4 in Octagon Box

4 In Octagon Box #54171-NL: This box has two nonmetallic-sheathed cable clamps built in with tightening screws to secure cable. Threaded ears on front of rim provide ability to mount device, cover plate or extension ring. Two $\frac{1}{2}$ in conduit knockouts are provided with pre-drilled holes adjacent to them to allow use of flex connector. The furnished and attached L bracket permits mounting flush to the wall surface. The $2\frac{1}{8}$ in depth provides interior capacity of 22.5 cu in.

No 55151 4 in extension ring

4 In Octagon Box Extension Rings #55151-$\frac{1}{2}$ and $\frac{3}{4}$: The slots on the bottom of the open extension ring allow it to be attached to the screws on the ears provided on all of the 4 in boxes. The extension is $1\frac{1}{2}$ in deep, has four conduit knockouts, two $\frac{1}{2}$ in and two $\frac{3}{4}$ in and provides an additional 15.8 cu in of interior capacity to any of the 4 in octagon boxes.

Square Boxes

No. 52151-½ 4-in square box

4-In Square Box #52151-½, ¾, ½ and ¾: The standard 4 in square box is available with ½ in knockouts, ¾ knockouts, or combination of ½ and ¾ knockouts. The box is constructed of drawn steel, is 1½ in deep and has internal capacity of 21 cu in. A total of sixteen knockouts are distributed three to each of four sides and four in the rear surface of the box. Two threaded ears on opposing corners of the front of the box are for mounting cover plate or device.

No. 52151-½W 4-in square box

4-In Square Box #52151-½W, ¾W, ½ and ¾W: This welded construction box is supplied with ½ in, ¾ in, or a combination of ½ in and ¾ in knockouts. It is 1½ in deep but the welded construction gives it a 22.5 cu in internal capacity. The sixteen knockouts allow a wide variety of conduit entrance configurations. The knockouts are spaced 3½ in on center on each side of the 52151-½W, 2¾ in on center on the ¾W, and 2½ in and 1¾ in on center on the combination ½ and ¾W.

No. 52151-½ - ¾-EW 4-in square box

4-In Square Box #52151-½ and ¾ EW: This combination box has two ¾ in and two ½ in knockouts on the rear surface and one ¾ in and one ½ knockout on each of the four sides. The spacing of the rear surface knockouts permits conduit to enter the box on either 3½ in or 2¾ in centers. The side-surface knockouts permit conduit to enter on 2½ in and 1¾ in centers. The box is equipped with an H-mounting bracket, is of welded construction and has internal capacity of 22.5 cu in. The box is 1½ in deep.

No. 52151-MS-½ - ¾ 4-in square box

4-In Square Box #52151-MS-½ and ¾: The combination box has dual punching on the ¾ in knockouts which permits the knockout to be used as either a ¾ in or ½ in knockout. Two of the combination knockouts are supplied on the rear surface and one each on the side surfaces. The box is 1½ in deep, welded construction, has perfectly square corners, and an internal capacity of 22.5 cu in. It is supplied with a removable MS bracket which allows it to be mounted on the front surface of the stud. The MS box is four ounces heavier than the previous 4 in square boxes.

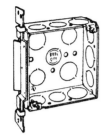

No. 52151-½-¾-V 4-in square box

4-In Square Box #52151-V-½ and ¾: The box is furnished with a flush mounted SV bracket to permit the box to be mounted flush with the wall surface and nailed or screwed flush to and directly to the supporting stud. The welded construction and square corner shaped result in 22.5 cu in of interior capacity. One $\frac{3}{4}$ in and two $\frac{1}{2}$ in knockouts are located on each side panel and two of each size knockouts are arranged on the rear panel of the box. The knockouts are arranged on $2\frac{1}{2}$ and $1\frac{3}{4}$ in centers on the side panels and on $3\frac{1}{2}$ and $2\frac{3}{4}$ in centers on the rear or bottom surface.

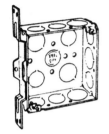

No. 52151-½-¾-CV 4-in square box

4-In Square Box #52151-CV-½ and ¾: The CV mounting bracket is permanently welded to the side of the box. The long shallow C shape of the bracket offsets the box from the supporting stud but is mounted flush to the front of the box and requires the box to be mounted recessed from the wall surface. Welded and square corner construction results in an interior capacity of 22.5 cu in. The box is $1\frac{1}{2}$ in deep and the knockouts are on $2\frac{1}{2}$ in and $1\frac{3}{4}$ in centers on the sides and $3\frac{1}{2}$ in and $2\frac{3}{4}$ in on the bottom of the box.

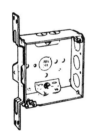

No. 52151-CVN 4-in square box

4-In Square Box #52151-CVN: The CVN box is furnished with two nonmetallic-sheathed cable clamps and tightening screws mounted on each of the opposing sides of the box. The cable clamps secure the nonmetallic-sheathed cable (Romex) when it is inserted through the special size knockouts provided. The shallow C-shaped mounting bracket is welded to one side of the box and requires that the box be mounted offset from the supporting stud and recessed from the surface of the wall. The $1\frac{1}{2}$ in depth and welded square corner construction provides 22.5 cu in of interior capacity. Three $\frac{1}{2}$ in knockouts are located on the side of the box opposite the mounting bracket. Threaded ears on opposing corners of the front of the box are for mounting a cover plate or wiring device.

No. 52151-X 4-in square box

4-In Square Box #52151-X: The X box is designed for use with metal-clad flex cable. The two special armored cable/metal-clad clamps are mounted opposing each other on the sides of the box with special knockouts to accommodate the metal-clad flex cable. Six $\frac{1}{2}$ in knockouts are located three on each of the other two sides of the box. A flat B-type mounting bracket is welded flush to the front of the box. The box is mounted recessed to the surface of the wall. The box is $1\frac{1}{2}$ in deep with an interior capacity of 2.5 cu in.

No. 52151-CVX 4-in square box

4-In Square Box #52151-CVX: The CVX box is furnished with a shallow C-shaped mounting bracket flush welded securely to the side of the box. The shape of the bracket requires the box to be mounted offset to the stud and recessed from the wall surface. Three $\frac{1}{2}$ in conduit knockouts provide access to the box on the side across from the mounting bracket. Special armored cable/metal clad clamps secure metal flex-type cable to the box. The $1\frac{1}{2}$ in deep, welded, square corner construction provides 22.5 cu in of interior capacity.

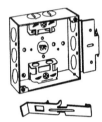

No. 52151-MSX 4-in square box

4-In Square Box #52151-MSX: The MSX box differs from the CVX box in the configuration of the mounting bracket. The MS bracket uses an adjustable clip to attach the box to the metal stud. Two armored cable/metal-clad clamps are located inside the box on opposing sides. Box is of welded construction, $1\frac{1}{2}$ in deep and has 22.5 cu in of interior room. Three $\frac{1}{2}$ in conduit knockouts are located opposite to mounting bracket side. The box is shipped in lots of 25.

No. 52171-½-¾ 4-in square box

4-In Square Box #52171-$\frac{1}{2}$, 4-In Square Box #52171-$\frac{3}{4}$, 4-In Square Box #52171-$\frac{1}{2}$ and $\frac{3}{4}$: All three of the standard #52171 boxes are $2\frac{1}{8}$ in deep and have 30.3 cu in interior capacity. They are of welded construction with square corners. The knockout sizes are designated by the -numbers. The boxes are not furnished with mounting brackets but must be either mounted onto the conduit or nailed directly to a mounting surface. This size box is most often used in commercial or industrial applications.

No. 52171-½-¾-E 4-in square box

4-In Square Box #52171-$\frac{1}{2}$ and $\frac{3}{4}$-E: The $\frac{3}{4}$ in knockouts are double punched to allow them to be used as $\frac{1}{2}$ in as required for the particular application. The $2\frac{1}{8}$ in depth, welded construction and square corners provide 30.3 cu in of interior capacity. Threaded ears on two opposite front corners of the box permit mounting of cover plate or wiring device. Mounting bracket is not provided and box must be supported by conduit or nailed directly to structure surface.

Outlet Boxes 193

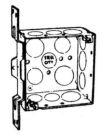

No. 52171-CV-½ 4-in square box

4-In Square Box #52171-CV-½ and ¾: Equipped with a shallow C-shaped, flush-mounted bracket, the box is designed to be mounted offset from the stud and recessed from the wall surface. The thirteen knockouts are arranged three to each of three sides and four in the rear or bottom surface of the box. The $2\frac{1}{8}$ in depth, welded construction and square corners provide 30.3 cu in of interior capacity. The knockouts are spaced on $2\frac{1}{2}$ in and $1\frac{3}{4}$ in centers, depending on the conduit size of the knockout on the sides, and on $3\frac{1}{2}$ in and $2\frac{3}{4}$ in centers, on the rear surface of the box.

No. 53151-½ extension ring

Extension Ring 53151-½ and ¾: Extension rings are used to increase the internal wiring capacity of the primary box. The 53151 ring is $1\frac{1}{2}$ in deep, constructed of drawn steel and adds 21.0 cu in to the interior capacity of the box. Twelve conduit knockouts in a combination of ½ in and ¾ in conduit sizes are provided three to each side of the ring. Threaded ears on the front of the ring allow the addition of a cover plate or wiring device. Similar ears on the rear of the plate are used to mount it to the box.

No. 72171-¾ square box

$4^{11}/_{16}$-In Square Boxes #72171-¾, $4^{11}/_{16}$-Square Boxes #72171-¾ and 1: The 72171 boxes are generally used in commercial or industrial applications. The $2\frac{1}{8}$ in depth and welded construction provides 42.0 cu in of usable interior capacity. Four threaded ears are located in the front of the box to allow mounting of a cover plate. The large, ¾ in and 1 in knockouts permit the use of larger conduit than the smaller boxes.

No. 73171-½ extension ring

Extension Ring #73171-½ and ¾: The #73171 extension ring is $2\frac{1}{8}$ in deep and doubles the interior capacity of the #72171 boxes by adding 42.0 cu in capacity. The ring has a rear surface with a large circular cutout which stabilizes the ring. Four threaded ears on the front of the ring allow an addition of a cover plate. The combination of ½ in and ¾ in knockouts on the four sides of the ring allow more conduit entrances to the large interior area.

Utility Boxes

No. 58351-½ utility box

Utility Box #58351-½: Generally referred to as an outlet box, the #58351 is designed to accommodate a single-wiring device such as duplex receptacle. The ten ½ in conduit knockouts are located three on each side, one on each end and two in the bottom of the box. Threaded ears on the front of the box are for the attachment of the wiring device. The box is 4 in long and $2\frac{1}{8}$ in wide by $1\frac{1}{2}$ in deep and has a 10.3 cu in capacity.

No. 58361-½ utility box

Utility Box #58361-½, Utility Box #58361-¾: The #58361 boxes are provided with either ½ in or ¾ in conduit knockouts. The $1\frac{7}{8}$ in depth increases the interior capacity to 13.0 cu in. Exterior dimensions of the box are 4 in by $2\frac{1}{8}$ in. These boxes are used in situations where more wiring capacity is required. The threaded mounting ears on the open front of the box permit the wiring device to be attached.

No. 58371-½ utility box

Utility Box #58371-½, Utility Box #58371-¾: The #588371 utility box is $2\frac{1}{8}$ in deep and has an internal capacity of 14.5 cu in. Supplied with either ½ in or ¾ in knockouts, the box offers solutions to many wiring situations requiring additional space for connections. Exterior dimensions of the box are 4 in by $2\frac{1}{8}$ in. Threaded ears on the front of the box allow mounting of a wiring device.

No. 58371-V-½ utility box

Utility Box #58371-V-½: The welded-on CV mounting bracket is recessed from the front of the box ½ in which allows the box to be mounted flush with the wall surface. Seven ½ in knockouts are located one each on the top and bottom surface, two on the rear surface, and three on the side of the box. The $2\frac{1}{8}$ in depth of the box gives it a 14.5 cu in internal capacity. Two threaded ears are provided to permit mounting of the wiring device.

Outlet Boxes

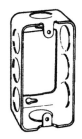

No. 59361-½ extension ring

Extension Ring #59361-½: The extension ring is 1⅞ in deep and provides an additional 13 cu in of capacity to any box it is mounted on. The ring has elongated slots in the back rim to fit over the mounting screws on the ears of the box. Two threaded ears are provided on the front of the ring for attachment of a wiring device. Eight ½ in conduit knockouts are located 3 to each side and one on each end of the ring.

No. 68371-½ utility box

Utility Box #68371-½: The box is 4⅛ in high by 2½ in wide by 2³⁄₁₆ in deep. The oversize box has 18.8 cu in of internal capacity and is used in situations requiring a larger number of conductors, connections or larger wire sizes. Ten ½- in knockouts are provided in the drawn steel box.

Gang Boxes and Partitions

Two Gang ½ and ¾, Three Gang ½ and ¾, Four Gang ½ and ¾: The multiple gang boxes are used in applications where several wiring devices are to be installed in a single location. All of them are 1⅝ in deep and have a combination of ½ in and ¾ in knockouts. Slots in the two sides of the box hold snap-in partitions in place to provide a separate compartment in the box for each device. The two-gang box is 6¹³⁄₁₆ in long, the three gang is 8⅝ in long and the four gang is 10⁷⁄₁₆ in long. The partitions have ears on each side which lock into the slots in the side of the box.

Multiple gang box and partition

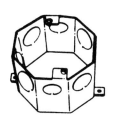

4-in octagon concrete box

4-In Octagon Concrete Box #54541-½ and ¾, 4-In Octagon Concrete Box #54561-½ and ¾: The concrete box is designed to be installed in a concrete slab or wall. Box is open at both ends, has mounting ears facing outward on the bottom, mounting ears facing inward on the top and open design requires it to be placed against a solid surface and then concrete poured around it. The combination knockouts are located one on each of the six sides. The #54541 is 2½ in deep with internal capacity of 29 cu in and the #54561 is 3½ in deep with internal capacity of 41.0 cu in.

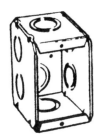

Tile wall box GW-125-C

Tile Wall Boxes

Tile Wall Box GW 125-C-½ and ¾
Tile Wall Box GW 225-C-½ and ¾
Tile Wall Box GW 325-C-½ and ¾
Tile Wall Box GW 425-C-½ and ¾: These tile wall boxes are designed to be installed in a tile faced wall flush with the surface. The 125 box is single gang, the 225 is double gang, the 325 is triple gang and the 425 is a four-gang box. All of the boxes have a continuous device mounting plate with threaded screw holes on the top and bottom in front of the box.

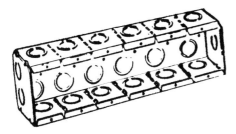

Tile wall box GW-425-C

The double punched concentric conduit knockouts may be used as either ½ in or ¾ in by simply knocking out the desired size. The 125 box is 1^{15}/$_{16}$ in long with 15.0 cu in capacity, the 225 is 3¾ in long with 31.6 cu in capacity, the 325 is 5^{9}/$_{16}$ in long with 47.4 cu in capacity and the 425 is 7⅜ in long with 64.0 cu in capacity. All of the boxes are 2½ in deep by 3¾ in high.

The tile wall boxes are also available in a deep style which is 3½ in deep and 3¾ in high. The length of the deep boxes is the same as that of the shallow box but the additional depth results in internal capacity of 22 cu in, 46.9 cu in, 71.0 cu in and 93.5 cu in respectively.

Adjustable Bar Hangers

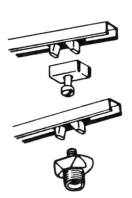

No. 6010-P adjustable bar hanger

Bar Hanger #6010-P: The 6010-P adjustable bar hanger is suitable for mounting appropriate boxes on ½ in, ¾ in and ⅞ in thick plaster walls. The snap-in block is threaded to accept the furnished stud which is inserted through the back of the box and then tightened to hold the box firmly. The bar length is adjustable from 10½ in to 18½ in and the end plate is ^{15}/$_{16}$ in deep.

No. 6010-AP adjustable bar hanger

Bar Hanger #6010-AP: The 6010-AP adjustable bar hanger is used to hang boxes on $\frac{1}{2}$ in, $\frac{3}{4}$ and $\frac{7}{8}$ thick plaster walls. The bar length is adjustable from $14\frac{1}{2}$ in to $26\frac{1}{2}$ in and the end plate is $\frac{15}{16}$ in deep. Box is attached to the bar hanger with a conduit-size threaded stud placed through a knockout and held in place with a locknut.

Grounding Devices

Every electrical circuit has a number of points at which it is vulnerable to breaks in the grounding path. The NEC devotes one entire article and a great number of additional sections as a guide to proper grounding. Effective grounding demands that the path to ground is *permanent* and *continuous* and that it has ample current carrying capacity and low impedance to permit all the overcurrent limiting devices in the circuit to work properly. The Occupational Safety and Health Act (OSHA) calls for equipment grounding in both new construction and existing installations and accepts four methods of grounding:

1. A supply cord carrying a grounding conductor
2. Metal conduit for direct wired equipment
3. Securing electrical equipment to the frame of a building's structure
4. A separate flexible wire or strap (by special permission)

Other grounding methods and devices will be discussed in another chapter but the grounding clip, grounding screws and pigtails are the accepted methods of providing grounding at the box.

Grounding clip

Grounding Clip: The grounding clip should be UL listed and approved. The clip is made of spring steel, zinc plated, and colored green. It accommodates No. 12 and No. 14 copper and No. 12 aluminum conductors in nonmetallic-sheathed cable. The wire is inserted under a built-in spring projection on the clip and the clip slipped over the edge of a box. Teeth on the clip hold it securely on the side of the box. The ground wire in any given circuit is attached to the clip on the box thus providing a continuous path to ground. Metal conduit attached to the box serves the same purpose.

Grounding Screw and Pigtail: The grounding screw and pigtail is supplied in three different configurations. GSC-12 is a 10-32 × $\frac{3}{8}$ slotted hexagon head washer face ground screw with green dye finish and a six-in solid copper No. 12 AWG insulated wire. GSB-12 is a 10-32 × $\frac{3}{8}$ slotted hexagon head washer face ground screw with green dye finish and a six-in solid copper No. 12 AWG bare wire. GSB-14 is a 10-32 × $\frac{3}{8}$ slotted hexagon head washer face ground screw with green dye finish and a six-inch solid copper No. 14 AWG bare wire. The ground screw is used to attach one end of the pigtail to the receptacle or other device. The other end of the pigtail is then inserted in the grounding clip to complete the ground path from device to box.

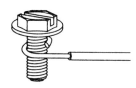

Grounding screw and pigtail

Cable protector

Cable Protector: The steel plate with attachment drive-in teeth protects nonmetallic-sheathed cables or raceway-type wires running through wood studs, joists, rafters or similar structural wood members from accidental nail, screw or drill penetration.

Box Covers

3$\frac{1}{2}$-In Round Box Cover: The covers have one slip over and turn slot and one straight slot which fit over the mounting ears on the 3$\frac{1}{2}$ in round or octagon boxes. The covers are required to close the box opening to protect the wiring inside the box. The cover is supplied in two styles—the #24-C-1 which is a flat blank cover and the 24-C-6 which is flat but has a $\frac{1}{2}$ knockout in the center of the cover.

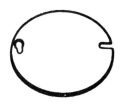

3$\frac{1}{2}$-in round box cover

4-In Octagon and Round Box Covers: The 4-in octagon and round covers are available in five different styles. The 54-C-1 is a flat blank cover, 54-C-6 is flat with $\frac{1}{2}$ in knockouts in the center, 54-C-3$\frac{1}{2}$ is center blanked, raised from the box $\frac{1}{2}$ in (supplies 3.0 additional cu in of interior capacity) with tapped ears on 2$\frac{3}{4}$ in centers, 54-C-3 is raised from the box front $\frac{5}{8}$ in (supplies additional 3.5 cu in of interior capacity) center blanked with tapped ears on 2$\frac{3}{4}$ in centers and the 5402-LR which is flat with the center blanked for a duplex receptacle.

4-in round and octagon box cover

4-in square box cover

4-In Square Box Covers: The standard 4-in square box covers are available in three different styles. The 52-C-1 is a flat cover with slots cut out from slipping over the mounting ears on the box. The 52-C-3 has blank center raised $5/8$ in above the box and tapped ears on $2\frac{3}{4}$ in centers and the 52-C-6 is a flat cover with $\frac{1}{2}$ in knockout in the center.

4-in square device cover

4-In Square Device Covers: The single-gang device cover has a mounting slot cut in each of four corners. The cover is supplied in five different styles. The #52-C-13 device mounting opening is raised $\frac{1}{2}$ in and has a 3.0 cu in capacity. The 52-C-14 opening is raised $\frac{3}{4}$ in with a 5.0 cu in capacity. The 52-C-14$\frac{5}{8}$ opening is raised $\frac{5}{8}$ in with a 4.0 cu in capacity. The 52-C-15 has a 1 in raised opening and a 7.0 cu in capacity. The 52-C-16 has a $1\frac{1}{4}$ in raised opening and a 8.3 cu in capacity.

4-in square two-device cover

4-In Square Two-Device Cover: The two-device cover has four slots, one on each corner for mounting to the box. The cover is supplied in different styles. The 52-C-17 has a $\frac{1}{2}$ in raised opening with 6.3 cu in capacity, the 52-C-18 has a $\frac{3}{4}$ in raised opening with 9.0 cu in capacity, and the 52-C-18$\frac{5}{8}$ which has a $\frac{5}{8}$ in raised opening with 7.3 cu in capacity.

4-in square cover for cut tile walls

4-In Square Cut Tile Wall Covers: The square corners of the raised device opening facilitate tile installation. The cover has a slot on each corner of the base to permit attachment to the mounting ears of the box. Supplied as the 52-C-49$\frac{1}{2}$ with $\frac{1}{2}$ in rise device opening and 3.5 cu in capacity, 52-C-49$\frac{3}{4}$ with $\frac{3}{4}$ in rise and 5.3 cu in capacity, 52-C-49-1 with 1 in rise and 7.0 cu in capacity, 52-C-49-1$\frac{1}{4}$ with a $1\frac{1}{4}$ in rise and 9.3 cu in capacity, the 52-C-50-1$\frac{1}{2}$ with $1\frac{1}{2}$ in rise and 1 cu in capacity to the 52-C-51-2 which has a 2 in rise device opening and 14.8 cu in capacity. The cover will fit flush to the surface on all tile installations. A two-device 4 in square cover is also available as the 52-C-52$\frac{3}{4}$ which has a $\frac{3}{4}$ in rise device opening and a capacity of 9.0 cu in.

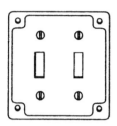

No. RS-5 4-in square surface cover

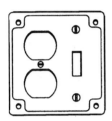

No. RS-2 4-in square surface cover

4-In Square Surface Covers: The surface covers are raised $\frac{1}{2}$ in and have a 5-cu in wiring capacity. The covers are available as:

- RS-2 — For one toggle switch and one duplex receptacle
- RS-5 — For two toggle switches
- RS-8 — For two duplex flush receptacles
- RS-9 — For one toggle switch
- RS-12 — For one duplex flush receptacle
- RS-16-CC — For one ground-fault receptacle
- RS-18-CC — For one ground-fault receptacle and one toggle switch
- RS-19-CC — For one ground-fault receptacle and one duplex receptacle

It is necessary to remove a portion of the ground-fault receptacle ear to mount receptacle to cover.

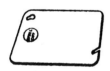

No. 72-C-1 flat blank cover

$4^{11}/_{16}$-In Square Covers: The cover is available in various configurations similar to the 4-in square covers.

- 72-C-1 — Flat blank cover
- 72-C-6 — Flat with a $\frac{1}{2}$ in knockout
- 72-C-13 — Single device $\frac{1}{2}$ in rise, 3.0 cu in capacity
- 72-C-14 — Single device $\frac{3}{4}$ in rise, 5.0 cu in capacity
- 72-C-14$\frac{5}{8}$ — Single device $\frac{5}{8}$ in rise, 4.0 cu in capacity
- 72-C-18$\frac{5}{8}$ — Two device $\frac{5}{8}$ in rise, 7.5 cu in capacity

No. 72-C-14 device cover

$4^{11}/_{16}$-In Square Surface Covers: The $4^{11}/_{16}$-surface covers are all $\frac{1}{2}$ in deep with a 7.5 cu in capacity and supplied as:

- RSL-5 — For two toggle switches
- RSL-8 — For two duplex flush receptacles
- RSL-9 — For one toggle switch
- RSL-12 — For one duplex flush receptacle

Outlet Boxes

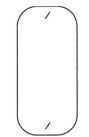

No. 58-C-1 utility box cover

4 × 2⅛-In Utility Box Covers: The flat utility box covers are available in the following configurations:

- 58-C-1—Blank
- 58-C-7—For duplex flush receptacle
- 58-C-30—For one toggle switch
- 58-C-16—For GFCI receptacle

No. 68-C-30 utility box cover

4⅛-In × 2½-In Utility Box Covers: These covers are all raised $3/16$ in above the face surface of the box and available as:

- 68-C-1—Blank
- 68-C-7—For duplex flush receptacle
- 68-C-30—For one toggle switch

Concrete box cover plate

Concrete Box Cover Plate (CBP): A round flat plate to cover the top of the concrete box. The plate has four knockouts spaced $3½$ in and $2½$ in on center. Mounting hole and mounting slot are provided to allow permanent attachment to the top of box.

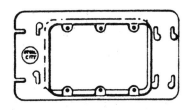

Gang box cover

Gang Box Covers: The covers are $4^{11}/_{16}$ in wide and raised $^{13}/_{16}$ in for plaster installation. Boxes are available as:

- 2-GC — Two gang, 7 in long, 8.5 cu in capacity
- 3-GC — Three gang, $8^{13}/_{16}$ in long, 13.5 cu in capacity
- 4-GC — Four gang, $10⅝$ in long, 18.3 cu in capacity

Switch Boxes

LCOW 2¼-in deep gangable switch box

LCOW 2¼-In Deep Gangable Switch Box: It is designed to be installed in old work, has beveled corners and comes with nonmetallic-sheathed cable (Romex) clamps. Box mounting ears are flush to the front of the box and two threaded switch mount ears are provided. One side of the box is removable to allow ganging with another box. Interior capacity of box is 10.5 cu in.

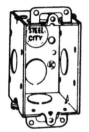

CDOW 2½-in deep gangable switch box

CDOW 2½-In Deep Gangable Switch Box: Designed for installation in remodeling-type work, the box slips into the wall opening and is held in place by the two mounting ears flush to the front opening. Five ½-in conduit knockouts are located one on each side and two in the rear or bottom of the box. It has a 12.5 cu in capacity. One side of the box may be removed to allow ganging with another box.

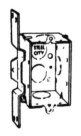

CDV 2½-in deep gangable switch box

CDV 2½-In Deep Gangable Switch Box: The CDV has five ½-in conduit knockouts located one on each of three sides and two in the bottom of the box. A "CV" bracket recessed ⅞ in from the front of box is nailed to a wall stud to provide mounting stability. Interior capacity is 12.5 cu in.

LXWOW 2½-in deep gangable switch box

LXWOW 2½-In Deep Gangable Switch Box: This is a combination box furnished with three ½-in conduit knockouts and two nonmetallic-sheathed cable clamps. Box is designed for installation in old work and has two mounting ears flush mounted to the front of the box. One side of the box is removable to facilitate ganging with another box. Threaded ears for mounting switch are provided. Interior capacity is 12½ cu in.

LXWOWC 2¼-in deep gangable switch box

LXWOWC 2½-In Deep Gangable Switch Box: Combination box has three ½-in conduit knockouts and two nonmetallic-sheathed cable clamps. Front mounting ears are flush with front of box. Interior capacity is 12.5 cu in. The box has two threaded ears, one on each side to accept support slips which make the box adaptable to any wall thickness up to ¾ in. This type of box is specifically designed to be used in old work and has one removable side to permit ganging with another box.

A-257 2½-in deep nongangable switch box

A-257 2½-In Deep Nongangable Switch Box: Box is equipped with two nonmetallic-sheathed cable clamps and two threaded switch mounting ears. Furnished nails for direct nailing to stud are slanted for easy drive in. The box is to be installed in new work. Capacity is 12.5 cu in.

LXOW 2½-in deep gangable switch box

LXOW 2½-In Deep Gangable Switch Box: Two armored cable/metal-clad clamps are supplied. Box has 12.5 cu in capacity and the two flush mounting ears are designed to be installed in remodeling or old work. Two ½-in conduit knockouts are located one to each side. One side of the box is removable to permit ganging.

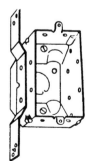

LVX 2½-in deep gangable switch box

LVX 2½-In Deep Gangable Switch Box: "CV" mounting bracket affixed to one side of box is recessed ⅞ in from the front of box. Two armored cable clamps with tightening screws are included. Interior capacity is 12.5 cu in and ½ in conduit knockout is on the other side of the box. The side of the box opposite the "CV" is removable to permit ganging with another box.

CW 2¾-in deep gangable switch box

CW ½ and ¾ Gangable Box 2¾-In Deep: Box is supplied with one ½ in or ¾ in knockout on each side and rear of the box. Mounting ears are flush with the front and two threaded device mounting ears are furnished. The 2¾ in depth provides 14.0 cu in of interior capacity. Gangable feature allows removing of one side.

CY ½ and CY ¾ Gangable Box 3½ In Deep: The boxes have either ½ in or ¾ in conduit knockouts located two on each side and one in the rear of the box. The vertical sides are removable to permit gang-type installation. The 3½ in depth provides 18.0 cu in of interior capacity. Mounting ears are flush with the front of the box and threaded device mounting ears are provided.

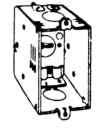

CXWOW 3½-in deep gangable switch box

CXWOW Gangable Box 3½-In Deep: The box is furnished with nonmetallic-sheathed cable clamps and tightening screws. Six ½-in conduit knockouts are located two on each side and one on each end. Two removable sides permit gang type installation and mounting ears are mounted flush with the front of the box. The 3½ in depth provides 18.0 cu in of interior capacity. Box is designed to be installed in old work.

CX Gangable Box 3½-In Deep: Armored cable/metal-clad clamps with tightening screws are installed in the box. Removable sides permit gang type installation and six ½-in conduit knockouts are provided. Mounting ears are flush with front of box and threaded device mounting ears are furnished. Box has 18.0 cu in of interior capacity.

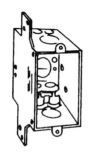

CXV 3½-in deep nongangable switch box

CXV Nongangable Box 3½-In Deep: The "CV" mounting bracket is recessed ⅞ in from the front of the box to permit the box to be mounted flush with the wall surface. The 3½-in depth provides interior capacity of 18.0 cu in. Armored cable/metal-clad clamps with tightening screws are pre-installed and the ½-in conduit knockouts allow conduit to be attached to the box.

Switch box supports

820-D Switch Box Supports: The supports are supplied in sets of two and used for mounting old work switch boxes in all types of wall materials.

SBEX Switch Box Extension: The switch box extension is used to increase the interior usable capacity of the switch box by 3.5 cu in.

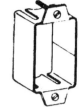

Switch box extension

Old Work Switch Box Support Clips: The support clips are designed to be mounted on the side of the box and then adjusted to hold the box against any thickness of mounting surface up to $\frac{3}{4}$ in.

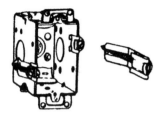

Old work switch box with support clips

Floor Boxes, Plates and Service Fittings

68-D floor box

68-D Floor Box: The floor box is constructed of steel and designed to be buried in the concrete floor. Four adjusting bolts are furnished and installed in the base of the box and permit the box to be adjusted up and down a total of $2\frac{1}{2}$ in before the concrete is poured and $\frac{1}{2}$ in after the pour. Dimensions of the box are $4\frac{1}{4}$ in × $4\frac{11}{16}$ in × $3\frac{5}{8}$ in. Four $\frac{3}{4}$-in conduit knockouts are provided on the sides and three $\frac{1}{2}$ in knockouts are located in the bottom of the box. The 68-D is used to install either a duplex or single receptacle depending on the cover plate used.

601-I floor box

601-I Floor Box: The 601-I floor box is rectangular and measures $4\frac{1}{4}$ in × $5\frac{1}{8}$ in × $3\frac{9}{16}$ in and is adjustable $2\frac{1}{2}$ in before concrete is poured and $\frac{15}{16}$ in after the pour. Four 1-in conduit knockouts are provided on the sides and two $\frac{3}{4}$ in knockouts are on the bottom. Box is designed to use a P60 cover plate for either a single or duplex receptacle.

  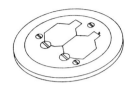

P60 Cover Plate   P60-CP Cover Plate   P60CACP Duplex

Carpet and cover plates for Nos. 68 and 601 floor boxes

Carpet and Cover Plates For #68 and #601 Floor Boxes: Flat cover plates are available with threaded center plugs of ½ in, ¾ in, 1 in, 1¼ in, 2 in and the P60-CP which accepts the duplex carpet floor plate #P60CACP. The center plugs are used to install various service fittings for power, communications, and data.

665 series floor box

664 Series Floor Box: The adjustable box is rectangular in shape and measures 7 in × 4¾ in and adjustable 2½ in up or down before concrete is poured and ½ in after the pour. The two gang box includes one duplex receptacle face plate but does not include the receptacle. Knockouts provided are: four ½ in, two ½ in - ¾ in concentric style, and two ¾ in - 1 in.

665 series floor box

Carpet Plate for #664 Floor Box: The carpet plate for the 664 style floor box completely covers the duplex receptacle plate with a fold down lid. The cover plate is secured to the underlying box with screws. Plate is available colored grey, brown, or beige. The additional duplex receptacle face plate must be purchased separately and the duplex receptacle is not included in the price of the box or plate.

Outlet Boxes

641 and 642 Series Floor Boxes: The 641 is a one-gang water-tight box adjustable $2\frac{1}{2}$ in before concrete pour and $\frac{1}{2}$ in after the pour. The 642 series box is a two-gang box adjustable $2\frac{1}{2}$ in before the pour and $\frac{1}{2}$ in after the pour. Four conduit plugs are available on the 641 box and six are provided on the 642 box.

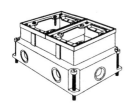

Cover Plates for 641 and 642 Floor Boxes

Cover Plate P64-$\frac{1}{2}$-2 and P64-$\frac{3}{4}$-2: The standard plate is available with either a $\frac{1}{2}$ in or $\frac{3}{4}$ in threaded center plug. The plugs are for mounting service fittings for power, communications and/or data.

P64P-CP cover plate

Carpet Cover Plate P64: This plate is available in three different configurations:

P64-DS cover plate

- P64P-CP — a single receptacle cover with screw top
- P64-DS — a duplex receptacle plate with lids
- P64-DU — a plate with screw-in plugs for mounting service fittings

641 and 642 series floor boxes

Service Fitting for Power, Communications and Data

SFH-40-RG service fitting

Service Fitting #SFH-40-RG: This low-profile fitting is designed to accept a duplex receptacle and is mounted to the floor box with a brass nipple in the center plug of the floor-box cover. The fitting is finished in brushed aluminum. Dimensions are $4\frac{3}{8}$ in wide × 3 in deep × $2\frac{5}{8}$ in high.

SFH-50-2RG service fitting

Service Fitting #SFH-50-2RG: This brushed-aluminum fitting features interchangeble face plates to accept various types of plug-in devices required for communications or data-processing equipment. This fitting is 5 in wide, $3\frac{3}{8}$ in deep, and 3 in high. It is mounted to a floor box with a nipple in the center plug of the floor-box cover.

Service Fitting #SFL-10: A low-profile, brushed-aluminum fitting adaptable to various outlets required for power, communications, or data transmission. The fitting is $4\frac{3}{8}$ in wide, $3\frac{1}{8}$ in deep, and $2\frac{5}{8}$ in high.

SFL-10 service fitting

Challenger Reinforced Phenolic Switch Boxes

Single-Gang, Nail-On box: The single-gang, nail-on, reinforced hard plastic boxes are available in the following five styles:

SINGLE-GANG NAIL-ON PHENOLIC SWITCH BOXES					
Mfr. Number	Size cu. in.	Mounting	Dimensions in Inches		
			Length	Width	Depth
2030	16.0	Bottom nails	$3\frac{9}{16}$	$2\frac{1}{4}$	$2\frac{13}{16}$
2000	18.0	Bottom nails	$3\frac{11}{16}$	$2\frac{1}{4}$	$3\frac{1}{8}$
1240	18.0	Recessed angle nails	$3\frac{11}{16}$	$2\frac{1}{4}$	$3\frac{5}{16}$
1250	20.3	Recessed angle nails	$3\frac{1}{8}$	$3\frac{3}{8}$	$3\frac{5}{16}$
1050	21.0	Compound angle nails	$3\frac{5}{8}$	$2\frac{5}{16}$	$3\frac{1}{2}$

Single-gang, nail-on switch box

Single Gang, Snap-In Bracket: The single gang, snap-in bracket box is designed with mounting ears attached to the top and bottom of the box front to prevent the box from sliding past the surface of the wall. The No. 8-gauge, snap-in brackets are compressed while the box is inserted through the wall opening; the brackets then expand to secure the box in the opening. It is supplied with nonmetallic-sheathed cable (Romex) clamps and tightening screws. The model #7020-8 provides 10 cu in inside capacity and is $3^{15}/_{16}$ in long $\times$ $2^{3}/_{16}$ in wide $\times$ $2^{1}/_{4}$ in deep. The model #7010-8 provides 12.5 cu in capacity and is $3^{1}/_{4}$ in $\times$ $2^{3}/_{16}$ in $\times$ $2^{11}/_{16}$ in.

Single-Gang, Offset Bracket: The offset bracket box is available with the large flat bracket recessed from the front of the box $1/_{2}$ in, $1/_{4}$ in and $3/_{8}$ in to accommodate installation in corresponding wall thicknesses. The bracket is suitable for mounting to wood or steel studs. The box has outside dimensions of $3^{5}/_{8}$ in $\times$ $2^{1}/_{4}$ in $\times$ $2^{13}/_{16}$ in and internal capacity of 16.0 cu in.

Two-Gang Nail-On Box: Two gang nail-on boxes are available in three styles. The #1032C is $3^{5}/_{8}$ in $\times$ $4^{1}/_{16}$ in $\times$ $2^{11}/_{16}$ with interior capacity of 27.5 cu in and equipped with nonmetallic-sheathed cable clamps. The #1052 is $3^{5}/_{8} \times 4^{1}/_{16} \times 3^{1}/_{8}$ with interior capacity of 32.5 cu in. Number 1052C $3^{5}/_{8}$ in $\times$ $4^{1}/_{16} \times 3^{1}/_{8}$ and has interior capacity of 32.5 cu in. It is equipped with nonmetallic-sheathed cable clamps.

Two-Gang Bracket Box: The hard plastic #6062-4 two-gang bracket box is supplied with two #4 swing-out brackets which are folded against the box while the box is being inserted through the cutout in the wall surface. The brackets then swing out behind the sheet rock or plaster of the wall and prevent the box from moving forward. The two sets of plaster ears mounted on the front top and bottom of the box prevent it from sliding through the cutout. Box is $3^{3}/_{16}$ in $\times$ $4^{3}/_{16} \times 2^{1}/_{2}$ with interior capacity of 25 cu in. Two sets of nonmetallic-sheathed cable clamps are furnished.

Three-Gang Nail-On Box: The #1043-C three-gang box is furnished complete with compound angled nails and three sets of nonmetallic-sheathed cable-screw clamps. Box is $3^{5}/_{8}$ in $\times$ $5^{7}/_{8}$ in $\times$ $2^{3}/_{4}$ and has interior capacity of 42.9 cu in.

4-In Square Box: The #4000-N 4-in square box has a special "N" metal nail-on mounting bracket affixed to the rear edge of the box. Outside dimensions are $4\frac{5}{16}$ in × $4\frac{5}{16}$ in × $2\frac{3}{8}$ in with interior capacity of 30.0 cu in. Three nonmetallic-sheathed cable clamps are provided.

4-In Square Box Covers and Plaster Rings: The #4041 flat cover is used when the 4-in square box is utilized as a junction or pull box. The #4042-12 single-gang plaster ring has a $\frac{1}{2}$ in raised opening to fit through the wall opening and allow the mounting of a single device.

$3\frac{1}{2}$-In Round Outlet Box: The $3\frac{1}{2}$-in round box is available in three styles. The 3190-C is $3\frac{1}{4}$ in deep, has a 22.8 cu in interior capacity and has two compound angled nails for mounting. Screw-type nonmetallic-sheathed cable clamps are furnished and the box is flanged on the front edge to prevent passing through the wall opening. No. 3090N is $2\frac{1}{8}$ deep with capacity of 13.5 cu in and has a "N" style nail-on metal bracket attached to the back surface of the box. The 3080-9 is $2\frac{1}{8}$ in deep, has a capacity of 14.0 cu in, and is equipped with plaster ears and #9 snap-in mounting bracket. The same style boxes are also available in a 4-in round version under the numbers 4170 and 4070. The 4170 uses compound angled side nails for mounting and the 4070 is equipped with the "N" style metal nail-on bracket.

Nonmetallic Switch and Outlet Boxes (Soft Plastic)

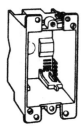

Single-gang 7887 NM box

#7887 Single-Gang Box: The box is furnished with four removable cable knockouts and designed specifically to be used in remodeling or old work. Plaster ears are molded flush to the front of top and bottom surfaces of the box and molded retainers on each end of the box hold it against the inside surface of the wall. Threaded device mounting ears are located between the plaster ears. The box is $2\frac{13}{16}$ in deep with internal wiring capacity of 15.0 cu in.

#7302 Single-Gang Box: The nail-on style box has captive nails mounted at an angle to permit easy installation. Four $\frac{1}{2}$-in removable cable knockouts are located two at each end of box. It is $2\frac{7}{8}$ in deep with capacity of 18.0 cu in. Device-mounting screw holes are on the inside top and bottom of box. The nail holding brackets are set back $\frac{1}{2}$ in from front of box.

Single-gang 7302 NM box

Outlet Boxes 211

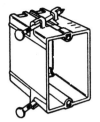

Single-gang 7820 NM box

#7820 Single-Gang Box: The nail-on style box has captive nails mounted at an angle to permit easy installation. Box is $3\frac{3}{16}$ in deep and has usable inside wiring capacity of 20.3 cu in. Four removable cable knockouts are placed two at each end of box. Threaded-device mounting holes are on the inside top and bottom of box. The nail-holding brackets are set back $\frac{5}{8}$ in from the front of box.

#7883 Single-Gang Box: The side mounting bracket is set back $\frac{5}{8}$ in from the front and molded directly to the box. Four removable cable knockouts are located two at each end. The box is $3\frac{3}{16}$ in deep with wiring capacity of 20.3 cu in.

Single-gang 7883 NM box

$3\frac{1}{2}$-In Diameter Ceiling Box: The #7119 $3\frac{1}{2}$ in round ceiling box is $2\frac{5}{16}$ in deep with interior capacity of 16.0 cu in. The furnished mounting nails are fixed at an angle to provide easy driving into the stud and set back $\frac{1}{2}$ in from front of box. Four integral clamps that push-in and lock cable in place are included.

7119 NM ceiling box

4-In Diameter Ceiling Box: The #7823 4-in round ceiling box is $2\frac{1}{2}$ in deep with interior capacity of 23.5 cu in. The mounting nails are set back $\frac{5}{8}$ from the front of the box and set at an easy driving angle. A ground plate is included and four integral clamps that push-in and lock the cable in place are standard.

7823 NM ceiling box

Two-Gang PVC Switch Box #7488: The box has plaster ears and threaded-device mounting ears on the top and bottom at the front of the box, no mounting bracket or nails are furnished. Eight integral, no-tie cable clamps are furnished. It is $2\frac{13}{16}$ in deep with 25.0 cu in interior capacity. This box was designed to be used in old work.

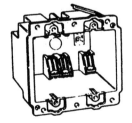

Two-gang 7488 PVC switch box

Two-Gang PVC Switch Box #7835: The 3 in deep two-gang box has 34.3 cu in capacity. The four threaded-device mounting ears are on the inside surface of the front of the box and eight integral,

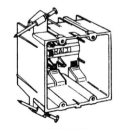

Two-gang 7835 PVC switch box

no-tie cable clamps are provided. Eight cable knockouts permit separate circuit wiring for each device. Two mounting nails are provided.

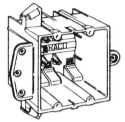

Two-gang 7834-B PVC box

Two-Gang PVC Switch Box #7834-B: The face-mount bracket allows the box to be nailed to the front of the stud and the $\frac{1}{2}$ in setback of the bracket lets the box extend through a $\frac{1}{2}$ in wall to the surface. The 3 in deep box has 34.3 cu in capacity and is supplied with eight integral, no-tie cable clamps and eight cable knockouts.

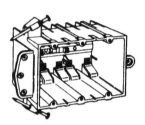

Three-gang 7846 PVC box

Three-Gang PVC Switch Box #7846: The box has combination of face-mount bracket and angled nails to secure the large size box. The $2^{11}/_{16}$ in depth results in 45.0 cu in of internal wiring capacity. Bracket is set back $\frac{1}{2}$ in to permit mounting of the box flush to the wall surface. Twelve integral, no-tie cable clamps and twelve cable knockouts are included.

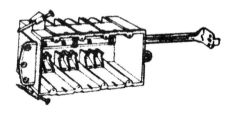

Four-gang 7645 PVC box

Four-Gang PVC Switch Box #7645: The box is $2^{5}/_{8}$ in deep and has 58.0 cu in of internal capacity. Fourteen integral, no-tie cable clamps and 14 cable knockouts are included. The large size of the box requires a face mounting bracket, captive angled nails and a long side bracket which nails to another stud.

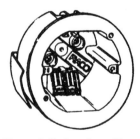

Nonmetallic box for ceiling fans

Nonmetallic Box for Ceiling Fans: The box is 4 in in diameter, $2^{5}/_{16}$ in deep and has 12.5 cu in wiring capacity. Four integral wiring clamps and four cable knockouts are supplied. The box is UL listed for support of ceiling fans up to 35 lbs.

Boxes, Covers and Accessories (Weatherproof)

Aluminum cast weatherproof boxes are offered in an enameled finish in gray or marine green. An independent testing laboratory proved by testing with a corrosive salt spray that the marine green finish lasted up to 5 times longer than the standard gray finish.

There are several manufacturers of the cast weatherproof boxes. As has happened with other tools and materials in the electrical industry, one manufacturer's name has become synonymous with the particular product. In the case of the weatherproof outdoor junction box, the name "Bell Box" has become the commonly used name.

Standard cast weatherproof boxes are available in one-gang, two-gang and three-gang sizes. In all sizes there is always one hub out of the back of the box with additional hubs located only on the top and bottom of the box. If hubs are required on all four sides of the box, the letter "S" is added to the catalog number of the particular sizes of boxes ordered. In the trade the outlet hubs are often referred to as outlet holes. The hubs are female threaded and the use of male threaded connectors or plugs is required for proper installation of the box.

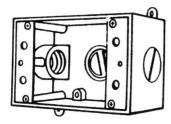

Three-hub, single-gange IH3-1 WP box

Three-Hub, Single-Gang Box #IH3-1 and IH3-2: The three-hub, single-gang box is supplied with $\frac{1}{2}$ in hubs or $\frac{3}{4}$ in hubs located on each on top, bottom, and back of the box. The box measures $2\frac{3}{4}$ in × $4\frac{1}{2}$ in × 2 in. Reducing bushings may be used, if required, to adapt the $\frac{3}{4}$ in hubs to $\frac{1}{2}$ in conduit size.

Three-Hub, Single-Gang Box #IH3-1-LM and IH3-2-LM: The three-hub, single-gang box is supplied with either $\frac{1}{2}$ in hubs or $\frac{3}{4}$ in hubs located one each on the top, bottom and back and has two mounting lugs cast onto the box. The box is $2\frac{3}{4}$ in × $4\frac{1}{2}$ in × 2 in. If necessary, reducing bushings may be used to reduce the $\frac{3}{4}$ in hubs to $\frac{1}{2}$ in.

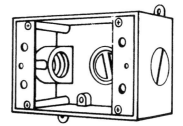

Three-hub, single-gang, IH3-1-LM WP box

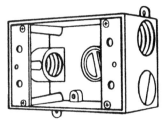

Four-hub, single-gang IH4-1 WP box

Four-Hub, Single-Gang Box #IH4-1, IH4-2, IH4-1-LM, and IH4-2-LM: The #4 in the model number refers to the four hubs located two on the top, one on the bottom and one in the back of the box. The -1 refers to ½ in hubs, -2 refers to ¾ in hubs and the LM specifies mounting lugs. Box is 2¾ in × 4½ in × 2 in.

Five-Hub, Single-Gang Box #IH5-1, IH5-2, IH5-1-LM, &IH5-2-LM: The #5 in the model number refers to the 5 hubs located two on the top, two on the bottom and one on the back of the box. The -1 indicates ½ in conduit hubs, the -2 indicates ¾ in hubs and the LM refers to the boxes with mounting lugs. The box is 2¾ in × 4½ in × 2 in.

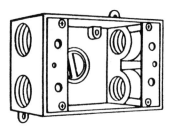

Four-hub, single-gang IH4-1 WP box

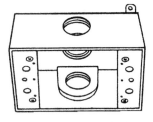

Five-hub, single-gang IH5S2-1 WP box

Five-Hub, Single-Gang Box #IH5S2-1-LM and IH5S2-2-LM: The "S" in the model number refers to the one hub located in each side of the box. In addition there is a hub on the top, on the bottom and on the back of the box. The -1 or -2 indicates either ½ in or ¾ in hubs and the LM refers to the mounting lugs. The box is 2¾ in × 4½ in × 2 in.

Three-Hub, Two-Gang Box #2IH3-1 and 2IH3-2: The first #2 in the model number refers to the two-gang type of box, the IH refers to weatherproof cast box, the #3 indicates the number of hubs and -1 or -2 specifies ½ in or ¾ in hubs respectively. Due to size and weight of box, mounting lugs are provided. Hubs are located one each on top and bottom and one in the back of box.

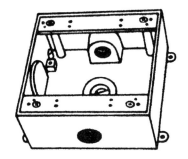

Three-hub, two-gang 21H3-1 WP box

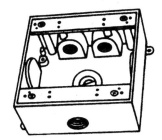

Four-hub, two-gang 21H4-1 WP box

Four-Hub, Two-Gang Box #2IH4-1 and 2IH4-2: The two-gang box is furnished with two hubs in the top, one hub in the bottom, and one hub in the back. The box is available with either $\frac{1}{2}$ in or $\frac{3}{4}$ in threaded hubs. Two cast-on mounting lugs are standard.

Four-Hub, Two-Gang Box #2IH4S-1 and #2IH4S-2: The hubs are located one each on the top, bottom, back and one side of the box. It is available with either $\frac{1}{2}$ in or $\frac{3}{4}$ in threaded hubs. Two cast-on mounting lugs are standard.

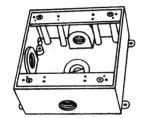

Four-hub, two-gang 21H4S-1 WP box

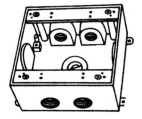

Five-hub, single-gang 21H5-1 WP box

Five-Hub, Two-Gang Box #2IH5-1 and #2IH5-2: The five hubs are arranged two on the top, two on the bottom and one in the back of the box. It is supplied with either $\frac{1}{2}$ in or $\frac{3}{4}$ in threaded hubs. Cast-on mounting lugs are standard.

Five-Hub, Two-Gang Box #2IH5S-2: Five $\frac{3}{4}$ in hubs are located two in the top, one in the bottom, one in the back and one on one side of the box. Two mounting lugs molded on opposing corners of the back of the box assure stable installation.

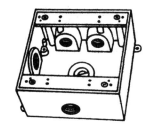

Five-hub, two-gang 21H5S-2 WP box

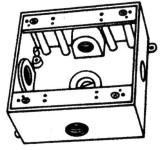

Five-hub, two-gang 2IH5S2-1 WP box

Five-Hub, Two-Gang Box #2IH5S2-1 and #2IH5S2-2: The -1 indicates $\frac{1}{2}$ in hubs and the -2 indicates $\frac{3}{4}$ in hubs, all of which are located one each on the top, bottom, both sides and the back of the box. Two mounting lugs are molded on opposing corners of the back of the box.

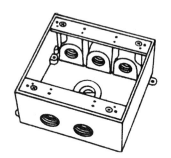

Six-hub, two-gang 2IH6-1 WP box

Six-Hub, Two-Gang Box #2IH6-1 and 2IH6-2: The six $\frac{1}{2}$ in or $\frac{3}{4}$ in hubs are placed three on the top, two on the bottom and one on the rear of the box. Furnished molded mounting lugs are on opposing back corners of the box.

Seven-Hub, Two-Gang Box #2IH7-1 and 2IH7-2: The box is supplied with either $\frac{1}{2}$ in or $\frac{3}{4}$ in hubs located three on top, three on the bottom and one on the back. Device mounting plates at the top and bottom front permit installing two devices. Standard molded mounting lugs are on opposing corners at the rear of the box.

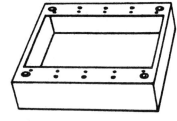

Seven-hub, two-gang 21H7-1 WP box

Seven-hub, two-gang 21H7S2-1 WP box

Seven-Hub, Two-Gang Box #2IH7S2-1 and 2IH7S2-2: One hub in each of two sides, two hubs in top and bottom and one hub in rear of box permit multiple circuits and box is available with either $\frac{1}{2}$ in or $\frac{3}{4}$ in hubs. Threaded device mounting plates at top and bottom front are available for two devices. Molded mounting lugs are on opposing corners at the rear of box.

Extension ring 2IHE

Extension Ring #2IHE: The extension ring comes complete with gasket and mounting screws. Use of the extension ring increases the capacity of the box by 50 percent. No hubs are provided in the extension ring.

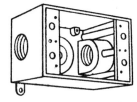

Three-hub, two-gang 2IHD3-1 WP box

Three-Hub, 1-Gang Box #IHD3-1, IHD3-2, and IHD3-3: The "D" in the model number stands for "deep" and indicates that the box is $2\frac{5}{8}$ in deep instead of the standard 2 in. The three hubs are located one on the top, on the bottom, and in the rear of the box. All three hubs are supplied in the same size either $\frac{1}{2}$ in, $\frac{3}{4}$ in or 1 in. Two molded-on mounting lugs are standard.

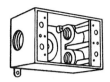

Four-hub, 1-gang IHD4-1 box

Four-Hub, 1-Gang Box #IHD4-1 and IHD4-2: The $2\frac{5}{8}$ in deep box has four hubs located 2 on top, 1 on the bottom, and 1 on the back with two standard molded mounting lugs. The box is available either $\frac{1}{2}$- or $\frac{3}{4}$- in hubs.

Five-Hub, Two-Gang Box #2IHD5-2 and 2IHD5-3: The 5-hub, $2\frac{5}{8}$ in deep box is only available with $\frac{3}{4}$ in or 1 in hubs. The hubs are placed 2 on the top, 2 on the bottom and 1 in the rear of the box. Threaded-device mounting plates are an integral part of the box. Molded mounting lugs on the opposing corners of the rear provide firm installation.

Five-hub, 2-gang 2IHD5-2 box

7-hub, 3-gang 3IHD7-2 box

Seven-Hub, 3-Gang Box #3IHD7-2: The box has three hubs on the top, three hubs on the bottom and one hub in the rear. Only $\frac{3}{4}$ in hubs are available. Mounting lugs are on opposing corners at the rear of the box. A switch or receptacle is mounted to threaded internal molded lugs.

Weatherproof Box Covers

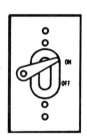

Single-gang switch cover

Single-Gang Switch Cover #CCT: The cover is designed to accept a switch packed with spacers. The switch is not included with the cover.

Single-Gang Switch Cover #CCT-3: The cover is furnished with 3-way switch 5A, 250V or 10A, 125V.

Single-Gang Power Outlet Cover #CCPO: This power outlet cover is for 30A and 50A receptacles 2.125 in in diameter, box-mount type. The cover mounts directly to front of box.

Two-Gang Duplex Receptacle Cover #2CCD and Two-Gang Single Receptacle Cover #2CCS: The duplex cover accepts two duplex receptacles and the single receptacle cover accepts two single receptacles. Receptacles are mounted directly to the box and the cover then installed over them.

Combination Single- and Duplex-Receptacle Cover #2CCSD and Combination GFCI and Single-Receptacle Cover #2CCSG: The #2CCSD combination cover accepts a single receptacle and a duplex receptacle. The #2CCSG is designed to accept a ground fault interrupter receptacle and a single receptacle. Both receptacles are mounted to the box and the cover installed over them.

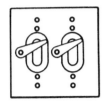

Double switch cover

Double Switch Cover #2CCT: The two-gang double switch cover offers weatherproof protection to a pair of switches which are supplied separately. Cover is supplied with spacers which operate the standard single-pole switches when the cover handle is moved. Switches are mounted to the box and the cover installed to cover them.

Switch and Single-Receptacle Cover, Two Gang #2CCTS: The 2CCTS accepts one standard single-pole switch and one single receptacle. The switch portion of the cover is marked ON and OFF. Both the switch and the single receptacle are installed on the box and the cover then installed over them to provide weather protection.

Two-Gang Duplex Receptacle Cover #2CCD-FS and 2CCD-FSL: The FS indicates the flush mount feature of the cover and the FSL represents the flush mount plus lock feature of the optional cover. The duplex receptacles are mounted to the box and cover installed over them.

Switch and Duplex-Receptacle Cover, Two Gang #2CCTD: This is designed to provide weatherproof quality for a standard switch and a duplex receptacle. The cover is supplied with spacer for the switch and has the ON and OFF switch position marked.

Blank Cover, Two Gang #2CCB: The blank cover is designed to be used when the weatherproof box serves as a junction box without a wiring device or is in a standby mode.

Horizontal Single-Gang #CCD and Vertical Single-Gang #CCDV Device Mount Covers: Both the CCD and the CCDV are designed to accommodate one duplex receptacle mounted on a single-gang box. The addition of the "V" specifies the vertical-mount version. Both covers have upward opening lids.

Horizontal Single-Gang #CCG and Vertical Single-Gang #CCGV Device Mount Covers: Both covers have upward opening lids and are designed to cover a single ground fault interrupter (GFI) receptacle. The receptacle is mounted directly to the box and the cover installed over it.

Horizontal Single-Gang #CCS and Vertical Single-Gang #CCSV Device Mount Covers: This cover is designed to provide weatherproof protection for one single-style receptacle mounted on the horizontal under the #CCS and vertical under the #CCSV. The receptacle is mounted to the box and cover installed over it.

Blank Cover, Single Gang #CCB: The blank cover is used to cover a box being used as a junction box or in a standby mode for later installation of a wiring device The cover provides weatherproof protection for the taps or splices enclosed.

Standard Lampholder Covers

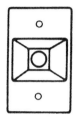
Lampholder cover

The lampholder covers are available in three styles — LC-11, LC-21, and LC-31. The LC-11 has one 1/2-in tapped hole to accommodate installation of one lampholder device. The LC-21 has two $\frac{1}{2}$ in tapped holes to accept two lampholder devices, and the LC-31 has three $\frac{1}{2}$ in tapped holes to accept three lampholder devices.

Outdoor Lampholders With $\frac{1}{2}$ in Threaded Arm

Economy Style #L-80: The economy lampholder has the standard $\frac{1}{2}$ in threaded arm for mounting to the cover. The lampholder is silver and designed to accept a 75- to 150-W Par 38 or R40 lamp. The weatherproofing gasket is designed to fit around the base portion of the lamp.

Outdoor lampholder

Utility Style #L-80-D: The utility-style lampholder is similar in appearance to the L-80 but has a weatherproofing gasket to be installed high in the neck of the holder. It will accept 75- to 150-W Par 38 or R40 lamps. The holder is silver-finished.

Decorative Style #L-80-C: The #L-80-C is equipped with the external-type gasket fitting on the top of the neck of the holder and is chrome-plated. The holder is suitable for 75- to 150-W Par 38 and R40 lamps.

Standard Lampholder #L-300: This lampholder is UL approved to accept 75- to 150-W Par 38 lamps and up to 300-W R40 lamps. The finish is silver and the threaded mounting arm is $\frac{1}{2}$ in.

Deep-Shield Lampholder #L-800: The deep-shield lampholder is designed for the lamp to be recessed from the open end of the lampholder thus shielding the lamp from damage. The recessed mounting prevents side light. The holder is silver-finished and suitable for 75- to 150-W Par 38 and R40 lamps.

Weatherproof outlet box

Weatherproof Outlet Box #S-47 and S-48: The S-47 box is 4 in in diameter with four $\frac{1}{2}$-in threaded hubs located in the side of the box and one $\frac{1}{2}$-in hub in the back. The S-48 box is 4 in in diameter with four $\frac{3}{4}$-in hubs located in the side of the box and one $\frac{3}{4}$-in hub in the back. The hubs will accept standard conduit connectors or lampholders.

Outlet Box Cover #S-1 and S-3: The 4-in round covers are available in the S-1 one-hub configuration and the S-3 three-hub configuration. The S-3 is supplied with a one-hub plug.

Weatherproof outlet box cover

Chapter 20

Pull and Junction Boxes

Pull and junction boxes provide the access points for pulling and feeding conductors into the raceway. They can be used effectively as T, cross, offset and right-angle junctions for single or paralleled groups of rigid steel, aluminum, PBC or EMT conduit. Their use is required in conduit runs where there would otherwise be a greater number of bends or offsets between outlets or fittings than is permitted.

The larger the box, the easier it is to install or position the conductors, make taps or install cable supports. Field experience has proved it is false economy to skip on the size or number of junction pull boxes. However, they should not be used where they are not needed, as an excessive number introduces unnecessary handling and splicing of conductors.

To insure sufficient working space in pull and junction boxes, Section 370-28 of the NEC specifies minimum dimensions for boxes with raceways containing #4 and larger conductors. In the smaller box sizes, Section 370-16 governs the dimensions.

In straight pulls, where the conduit enters and leaves on opposite sides of the box, the length of the box must be at least 8 times the nominal diameter of the largest raceway. For example, if the conduit size to be installed is 4 in, then $8 \times 4 = 32$ and 32 in is the minimum required length of the junction or pull box. Good practice suggests a minimum length of $12 \times$ the diameter to allow some excess capacity and efficient installation. Width and depth of the boxes are governed by size of conduit and number of conductors involved. In determining the correct size box to use the dimensions of locknuts and bushings must also be taken into consideration.

Width of an outlet or junction box must be at least 6 times the trade diameter of the largest conduit, but it is recommended that 10 times the diameter of the largest conduit be used. Depth of the box should be governed by the locknut dimensions of the conduits involved.

Angle pulls are those where conductors enter one side and leave from other than the opposite side. Minimum NEC dimensions or a box used for angle pulls are determined in a different manner than from the straight pull-type junction box. Reference to the NEC should be made to determine proper size box to coincide with the particular installation condition.

Pull and junction boxes come in two basic designs, either National Electrical Manufacturers Association (NEMA) 1 for indoor installations or NEMA 3 for outdoor installations. On the indoor-type box, two forms of covers are available, either surface or flush. The flush cover is used with the J-box installed in a wall and the outer edge of it is flush with the wall surface. The surface cover is used when the box is installed on the surface of the wall.

Conductor Support Boxes

To permit installation of suitable conductor supports in vertical raceways, conductor support boxes must be included to permit the installation of supports where the weight of the cable would place an excessive strain on the conductor terminals. This is particularly important in high-rise buildings where heavy feeder conductors are involved.

When conduit-type conductor supports with split taper are used, hard fiber bushings are available that fit over the end of the conduit risers. The size of the support box is determined by the rules governing pull boxes. Where a box is to be used as combined splice and support box, there should be ample space left above the support units to make the splices.

Wireway or Gutter

Wireway or gutter, as it is more familiarly known in the trade, is a form of raceway. In addition to being used as a form of conduit, it is quite often used as a pull box or junction box.

Gutter is a square-type conduit usually offered in sizes 3 in × 3 in, 4 in × 4 in, 4 in × 6 in, 6 in × 6 in, 8 in × 8 in, 10 in × 10 in, and 12 in × 12 in. It is also available on special order in sizes 6 × 8, 8 × 10, and 10 × 12. The wireway or gutter is furnished in lengths of 12 in, 18 in, 24 in, 36 in, 48 in, 60 in, 72 in and 120 in.

Basic gutter design is U-shaped and is available with either screw cover, flush cover or hinged cover. Lengths of gutter generally come equipped with the screw-type cover but with open ends. The reason for this is that quite often one, two or more sections of gutter are required to be spliced together. Additional accessories for wireways include such items as end flanges, wireway Ts, elbows, crosses and 45 degree elbows.

In addition to the standard form of gutter a screw cover rain-tight gutter is available in dimensions from 3×3 through 8×8 and lengths of 72 in. The rain-tight gutter can be furnished with or without various sizes of threaded hubs. The rain-tight gutter is constructed of one piece with solid ends. All standard gutter is supplied with knockouts on both sides and in the end.

NEMA Type Enclosures For Electrical Equipment (1000 V Maximum)

Nonclassified Location Enclosures

Type 1 Enclosures: Type 1 enclosures are intended for indoor use primarily to provide a degree of protection against limited amounts of falling dirt in locations where unusual service conditions do not exist.

Type 2 Enclosures: Type 2 enclosures are intended for indoor use primarily to provide a degree of protection against limited amounts of falling water and dirt. They are not intended to provide protection against conditions such as internal condensation.

Type 3 Enclosures: Type 3 enclosures are intended for outdoor use primarily to provide a degree of protection against rain, sleet, wind-blown dust, and remain undamaged by ice forming on the enclosure. They are not intended to provide protection against conditions such as internal condensation or internal icing.

Type 3R Enclosures: Type 3R enclosures are intended for outdoor use primarily to provide a degree of protection against rain and sleet, and to be undamaged by the formation of ice on the enclosure. They are not intended to provide protection against conditions such as internal condensation or icing.

Type 3S Enclosures: Type 3S enclosures are intended for outdoor use primarily to provide a degree of protection against rain, sleet, and wind-blown dust, and to provide for operation of external mechanisms when ice laden. They are not intended to provide protection against conditions such as internal condensation or internal icing.

Type 4 Enclosures: Type 4 enclosures are intended for indoor or outdoor use primarily to provide a degree of protection against wind-blown dust and rain, splashing water, hose-directed water, and to be undamaged by the formation of ice on the enclosure. They are not intended to provide protection against conditions such as internal condensation or internal icing.

Type 4X Enclosures: Type 4X enclosures are intended for indoor or outdoor use primarily to provide a degree of protection against corrosion, wind-blown dust and rain, splashing water, hose-directed water, and to be undamaged by the formation of ice on the enclosure. They are not intended to provide protection against conditions such as internal condensation or icing.

Type 5 Enclosures: Type 5 enclosures are intended for indoor use primarily to provide a degree of protection against settling airborne dust, falling dirt, and dripping non-corrosive liquids. They are not intended to provide protection against such conditions as internal condensation.

Type 6 Enclosures: Type 6 enclosures are intended for indoor or outdoor use primarily to provide a degree of protection against hose-directed water and the entry of water during temporary submersion at a limited depth. They are to remain undamaged by the formation of ice on the enclosure. They are not intended to provide protection against conditions such as internal condensation or internal icing.

Type 6P Enclosures: Type 6P enclosures are intended for indoor or outdoor use primarily to provide a degree of protection against hose-directed water and the entry of water during prolonged submersion at a limited depth. They are to be undamaged by the formation of ice on the enclosure. They are not intended to provide protection against conditions such as internal condensation or icing.

Type 12 Enclosures: Type 12 enclosures are intended for indoor use primarily to provide a degree of protection against circulating dust, falling dirt, and dripping non-corrosive liquids. They are not intended to provide protection against such conditions as internal condensation.

Type 12K Enclosures: Type 12 K enclosures are intended for indoor use primarily to provide a degree of protection against circulating dust, falling dirt, and dripping non-corrosive liquids. The knockouts shall be provided only in the top or bottom walls, or both. The enclosures are not intended to provide protection against such conditions as internal condensation.

Type 13 Enclosures: Type 13 enclosures are intended for indoor use primarily to provide a degree of protection against lint, dust, spraying of water, oil and non-corrosive coolant. They are not intended to provide protection against conditions such as internal condensation.

Classified Location Enclosures

Type 7 Enclosures: Type 7 enclosures are for indoor use in locations classified as Class I, Groups A, B, C, or D, as defined in the NEC.

Type 8 Enclosures: Type 8 enclosures are for indoor or outdoor use in locations classified as Class I, Groups A, B, C, or D, as defined in the NEC.

Type 9 Enclosures: Type 9 enclosures are intended for indoor use in locations classified as Class II, Groups E, F, and G, as defined in the NEC.

Type 10 Enclosures (MSHA): Type 10 enclosures shall be capable of meeting the requirements of the Mine Safety and Health Administration, 30 C.F.R., Part 18.

Screw-Cover Pull Boxes

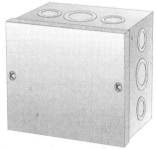

Gray-enamel Type SC surface pull box

Gray Enamel-Type-SC Surface: The screw cover (SC) box is designed to be surface mounted on the face of the wall. Knockouts are supplied in various numbers depending upon the outside dimensions of the box. It is also available without knockouts. The smaller boxes have three knockouts located on each of four sides for a total of twelve. The large boxes may have as many as six concentric knockouts on each side for a total of 24. The knockouts range from 1 in down to $\frac{1}{2}$ in conduit size. Inside dimensions (height, width, and depth) of the boxes with knockouts range in inches from $4 \times 4 \times 4$ to $24 \times 24 \times 8$. The boxes without knockouts are available in sizes from $6 \times 6 \times 4$ to $30 \times 36 \times 8$.

Galvanized Type SC surface pull box

Galvanized-Type SC Surface: The galvanized SC box is supplied only without knockouts. The available inside dimensions vary from those of the gray enamel box. The galvanized box is sized from $4 \times 4 \times 4$ to $36 \times 36 \times 12$.

Hinged-Cover Pull Boxes

Type A gray enamel hinged junction box

Type A Gray Enamel: This box offers easy accessibility in many wiring projects. It is available in sizes ranging from $4\frac{1}{2} \times 5 \times 3$ to $12 \times 12 \times 6$ with at least twelve concentric knockouts, and more in the larger sizes. All type A boxes are hinged on the height dimension.

Gasketed Screw-Cover Boxes

Moisture-proof seam-welded box

Moisture-proof, Seam Welded Box: It is designed for outdoor or indoor use and meets requirements for NEMA 3 or 4. The box is fabricated from heavier than code gauge bonderized paint-grip galvanized steel. The seams are continuously welded. The cover contains a neoprene gasket to make it moisture proof. The cover screws thread into sealed wells to prevent liquids, dust etc., from entering the box. No holes or knockouts are provided. Available in sizes $6 \times 6 \times 4$ to $12 \times 12 \times 6$.

Transformer Cabinets

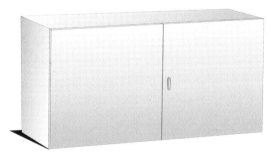

Double-door transformer cabinet

Double-Door Transformer Cabinet: All of the double-door cabinets are 10 in deep and either 30 in wide × 30 in high, 36 in wide × 30 in high or 36 in wide × 36 in high. Cabinets are UL listed and contain meter seal provision. Cabinets are finished with gray enamel and are NEMA Type 1.

Pull and Junction Boxes

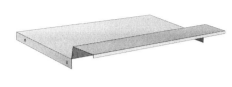

Protective weather hood

Protective Weather Hood: The hood is designed to be installed in the field as a shoe-box cover style. The dripproof front and flanges on the sides and rear offer protection from falling dirt, rain, snow and other elements. The hood is attached with self-tapping screws, finished in gray enamel and fits the double-door transformer cabinets. Sizes are 30 in × 10 in and 36 in × 10 in.

Weatherproof Rain-tight Gutter

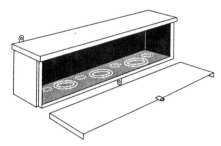

NEMA 3R screw-cover gutter

NEMA 3R Screw Cover Gutter: Gutter is UL listed, finished in gray enamel and available in inside dimension sizes 4 × 4 × 12 in long to 8 × 8 × 48 in long. Lengths ranging from 12 in to 48 in in increments of 12 in are available in all sizes of the gutter.

Oil-tight Pushbutton Enclosures

PB standard enclosure

PB Standard Enclosure: The PB standard is supplied in the following sizes:

SIZES OF STANDARD PUSHBUTTON ENCLOSURES			
Type	Width	Height	Depth
PB-1	3½ in	3¼ in	2¾ in
PB-2	5¾ in	3¼ in	2¾ in
PB-3	8 in	3¼ in	2¾ in
PB-4	10¼ in	3¼ in	2¾ in

SIZES OF STANDARD PUSHBUTTON ENCLOSURES *(Cont.)*			
Type	Width	Height	Depth
PB-5	12½ in	3¼ in	2¾ in
PB-6	9½ in	6¼ in	3 in

The two narrow enclosures are designed to accept four pushbutton controllers while the four larger size enclosures accept six controllers. The PB standard enclosure is also available in an extra-deep version with the catalog No. PBXD-1, -2, and -3. A stainless steel enclosure for corrosive locations is available in the same configuration as the PB standard.

PBC sloping PB enclosure

"PBC" Sloping Front PB Enclosure: The sloping front enclosure accommodates 16 pushbutton controllers and is generally used in heavy industrial applications. The enclosure is available in three sizes; PBC-8 which is $10\frac{3}{8} \times 12 \times 6\frac{3}{4}$, PBC-16 is $14\frac{1}{4} \times 12 \times 9$, and the PBC-20 is $14\frac{1}{4} \times 14\frac{1}{4} \times 9$.

NEMA 4 enclosure

NEMA 4 Enclosure: The NEMA 4 enclosure and junction box is finished in gray polyester powder inside and outside and has mounting flanges located on the back, top and bottom. The enclosure is furnished in sizes ranging from 16 in to 36 in high × 12 in to 30 in wide × 6 in to 10 in deep. The cover is full-hinged and can be locked with the supplied hasp. The enclosure is also manufactured in a stainless steel version.

NEMA 12 steel panel enclosure

NEMA 12 Steel Panel Enclosure: The wall mount enclosure offers maximum protection from dust, dirt, oil and water. It is finished with a gray prime exterior and a baked white enamel interior. Enclosure has a single full-hinged door and lock and is constructed of 14 gauge metal body and 12 gauge sub-panel. Thirty three sizes are manufactured ranging from 16 to 42 in high × 12 to 36 in wide × 6 to 16 in deep.

Two-door, floor-mounted NEMA 12 enclosure

Two-Door Floor-Mounted NEMA 12 Enclosure: The floor-mounted enclosure is finished with a gray prime exterior and baked white enamel interior. The four floor stands are 12 in high and two lifting rings are attached to the top surface to allow power lifting into place at the installation site. The double door is equipped with a latching device. The enclosure is manufactured in a range of sizes fro 60 to 72 in high × 48 to 72 in wide × 10 to 12 in deep.

Unflanged Junction or Pull Boxes

Types YS Cast-Iron Box and Cover and YS-A Cast-Aluminum Box and Cover: The YS boxes are designed to be surface mounted, and for raintight rating, external mounting means must be provided. Special factory-supplied mounting lugs may be specified. The box is dust-tight, raintight, watertight and rated as NEMA 4 suitable for general purpose outdoor installation. The finish is hot-dip galvanized. A neoprene gasket is provided to seal the cover which is attached with stainless-steel screws. Type YS boxes have a post in each corner for which allowances must be made when conduit entrances are to be located close to a corner. Enclosure comes in 28 sizes from 4 to 18 in long × 4 to 18 in wide × 2 to 12 in deep.

Type YS box and cover

Flanged and Overlapping Cover Boxes

Types YL Cast-Iron Box and Cover and YL-A Aluminum Box and Cover: The type YL box has an overlapping cover and is listed by UL as NEMA Type 4. Box is suitable for outdoor use, or where subject to rain, dripping or splashing water. The cast-iron box finish is hot-dip galvanized, the aluminum box is natural finish and both use stainless-steel cover screws. A neoprene gasket is provided to seal the cover. Special mounting lugs are available to facilitate surface mounting and preserve the watertight or raintight rating. Type YL boxes have a post in each inside corner for which allowances must be made when conduit entrances are to be located close to a corner. Boxes are manufactured in varying wall thicknesses from $\frac{3}{16}$ to $\frac{7}{16}$ in. Internal dimensions run from 4 to 25 in long × 4 to 24 in wide × 3 to 12 in deep.

Type YL box with overlapping cover

Type YF Cast-Iron Box for Surface Mounting: The YF box is suitable for use indoors or outdoors where they would be subjected to splashing water, seepage of water, falling or hose-directed water and severe external condensation (watertight NEMA 4). The box will withstand occasional submersion under a 6-ft head of water for 30 minutes (NEMA 6 requirements). It is finished in hot-dip galvanized and has a neoprene cover gasket. The flange is ground before galvanizing and the cover is attached with furnished stainless-steel screws. Wall thickness varies from $\frac{5}{32}$ to $\frac{5}{16}$ in and inside dimensions range from 4 to 18 in long × 4 to 12 in wide × 4 to 6 in deep.

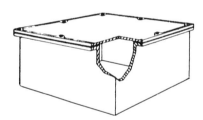

Type YF box with flanged cover

Hinged Cover Box

Type YW Cast-Iron Box for Surface Mounting: The type YW hinged-cover boxes make ideal enclosures when equipment within the box has to be inspected frequently and easy access is required. The hinges are adjusted at the factory for proper gasket pressure, but can be readjusted in the field. Box is supplied with a neoprene gasket and stainless steel hardware. It is UL listed as NEMA 4. Size range is 6 to 24 in long × 4 to 24 in wide × 4 to 12 in deep.

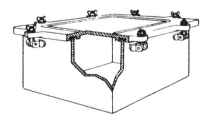

Type YW box with hinged cover

Pull and Junction Boxes

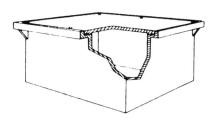

Type YR box with flanged recess cover

Type YR Cast-Iron Box for Flush Mounting: The type YR box is outside flanged with a recessed cover. It is suitable for outdoor use where subject to rain, dripping or splashing of water. It is rated as watertight NEMA 4. Designed especially for flush mounting in walls or floors or can be used for surface mounting using mounting lugs. Furnished with a plain cover, but can be supplied with steel checkered-plate covers suitable for foot traffic. Cover is sealed with supplied neoprene gasket and stainless-steel cover screws. Finish is hot-dipped galvanized. Sizes available are 4 to 24 in long × 4 to 18 in wide × 4 to 10 in deep.

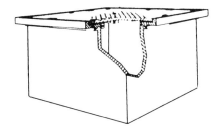

Type YT sidewalk box

Type YT Checkered-Cover Sidewalk Boxes: Constructed of cast iron, the boxes are specially designed to be mounted in sidewalks and other flat-concrete surfaces. Checkered covers are made to withstand pedestrian traffic. Flanges and covers are interchangeable to permit replacement without disturbing the box or conduit system. Heavy checkered-steel covers can be furnished to accommodate vehicular traffic. Cross-ribbed checkered cover has pry-bar slots. Box is furnished with a hot-dipped galvanized finish, a neoprene gasket and stainless-steel cover screws. Sizes range from 6 in × 6 in × 4 in to 18 in × 18 in × 12 in.

Motor Starter Enclosures

Enclosures with Class 419 Manual Motor Starters: Enclosures for single-phase manual motor starters are available in two styles — shallow or deep — to meet any plant installation requirement. The units are explosionproof, dust-ignition proof and weather resistant. They may be used to install a wide variety of electrical wiring devices; and come complete with covers and with a $\frac{3}{4}$ in hub opening on the top and bottom. They are available either with or without manual motor starter or tumbler switch. The complete units (enclosure with starter) are ideal for manual starting of small ac or dc motors in locations made hazardous by the presence of flammable vapors, gases or highly-combustible dusts. Enclosure is constructed of copper-free cast aluminum, has a gray vinyl finish to resist corrosion, cast in hubs, interchangeable heater units, and stainless-steel

Motor starter enclosure

cover bolts. The enclosures comply with NEMA 7 – Class 1, Group C, D; NEMA 9 – Class II, Group E, F, G; UL Standard 698. The starter is equipped with an easily-readable operator handle with provisions for padlocking in the ON or OFF position. When ordering, it is necessary to specify Class number and enclosure type. Also specify starter horsepower, number of poles, full-load amps and voltage.

Chapter 21

Wire and Cable

Sizes and Types of Conductors

A conductor is a substance that carries electric current. An insulator is a substance that does not carry electric current. Since no conductor is perfect, and because any conductor has at least a small amount of resistance, it is better to define a conductor as a substance with a very low resistance. No insulator is perfect and because no insulator has an infinite resistance, it is better to define an insulator as a substance with a very high resistance. Therefore, conductors, resistors and insulators are basically resistive substances. However, they are classified into different groups, because a practical conductor has extremely low resistance, a load resistor has a moderate resistance, and a good insulator has extremely high resistance.

Wires are used to conduct electric power from the point where it is generated to the point where it is used. A conductor may be wire, bus bar or any other form of metal suitable for carrying current. All wires, therefore, are conductors but not all conductors are wires. Copper bus bars, for example, are conductors but are not referred to as wires. Because all wire has resistance, this prevents an unlimited flow of current and causes voltage drop. Therefore, for any given load, you must select the size wire that causes only a reasonable voltage drop.

Several factors determine how much current a conductor can carry safely and economically in a given installation. These include conductor size (large wires carrying more current); type of insulation (heat resistant types permit higher ratings); length of circuit (voltage drop is greater on long runs; number of conductors in a raceway (lower ampacity allowed as number of conductors increases); and the ambient air temperature (additional derating because of high temperatures).

Wires used as electrical conductors are generally made of copper, however, aluminum wire is also used particularly in residential type of house wiring. Although silver is an excellent conductor of electricity, it is seldom used because of its high cost. Most wire is round, although square and rectangular forms are also used in some applications. The basic description of a round wire is its diameter. This means the diameter of wire itself, disregarding any insulation that might be used to cover the wire. The diameter of a wire can be measured accurately with a micrometer.

Circular Mils

To understand the meanings of different sizes of wire, you must understand something about the scheme used in numbering these sizes. The units used are "mil" and "circular mil." A mil is $\frac{1}{1000}$ in (0.001 in). A circular mil (its abbreviation is c.m. or cmil) is the area of a circle 1 mil in diameter. Thus a wire that is 0.001 in or 1 mil in diameter is said to have a cross-sectional area of 1 circular mil. Since the areas of two circles are always proportional to the squares of their diameters, it follows that the cross-sectional area of a wire 0.003 in or 3 mils in diameter is 9 circular mils; it follows that of 0.100 in or 100 mils in diameter is 10,000 circular mils, etc. The cross-sectional area of any round wire in circular mils is equivalent to the diameter of the copper only in mils or thousandths of an inch, squared or multiplied by itself.

Electricians use the word mil to describe the diameter of a wire. If the micrometer shows that a wire has a diameter of 0.001 inches, we say that its diameter is 1 mil. Electricians also use the term "square mil" to describe the cross-sectional area of a square or rectangular wire. If a square wire is 1 mil on a side, it is said to have a cross-sectional area of 1 square mil. It follows that if a square wire is 2 mils on a side it has a cross-sectional area of 4 square mils. Again, if a rectangular conductor is 3 mils wide and 2 mils thick, it has a cross-sectional area of 6 square mils.

A comparison is made between a circular mil and a square mil because if a square conductor is to be placed by a round conductor, or vice versa, we must use the same cross-sectional area to obtain the same conductivity. Therefore, electricians are required to know how to change circular mils into square mils and vice versa.

Electricians usually check wire sizes with an American Standard Wire Gauge (AWG). This gauge is also called a Brown & Sharpe gauge. For brevity the terms AWG or B&S are used. This gauge is not the same as that used for steel wires or used for non-electrical purposes — for example, fence wires.

Number 14 wire, which is the size most commonly used for ordinary house wiring, has a copper conductor 0.064 in or 64 mils in diameter. Wires smaller than this are numbers 16, 18, 20, and so on. Number 40 has a diameter of approximately 0.003 in, as small as a hair, many still finer sizes are made. Sizes larger that number 14 are numbers 12, 10, 8, and etc. It should be noted that the larger the number, the smaller the diameter of the wire.

In this way, sizes proceed until number 0 is reached; the next sizes are numbers 00, 000, and finally, 0000 which is almost $\frac{1}{2}$ in in diameter. Numbers 0, 00, 000, and 0000 are usually designated as 1/0, 2/0, 3/0 and 4/0. As still heavier sizes are reached, they no longer are designated by a numerical size but simply by their cross-sectional areas in circular mils, beginning with the 250,000th circular mils (250 kcmil) and up to the largest recognized standard size of 2 million circular mils (2000 kcmil).

Wire sizes 18 and 16 are used chiefly for flexible cords, for signal systems and for similar purposes, where relatively small amperages are involved. Number 14 to 4/0 are used in ordinary residential wiring and, of course, in industrial work where the still heavier sizes are also used. Number 14 is the smallest wire size permitted by the NEC for ordinary wiring. The even sizes of wire, such as numbers 18, 16, 14, 12, 10, 8, etc., are commonly used. The odd sizes such as 15, 13, 11, and 9 (with the exception of numbers 3 and 1) are seldom used in wiring installations. The odd sizes, however, are commonly used in the form of magnetic wire for manufacturing motors, transformers, etc., for which purposes even fractional sizes such as number $15\frac{1}{2}$ are not at all uncommon.

One way to compare wire sizes is to remember that any wire which is 3 sizes larger than another will have a cross-sectional area exactly twice that of the other. For example, number 11 has an area exactly twice that of number 14. Number 3 wire has an area exactly twice that of number 6. Any wire that is six sizes heavier than another has exactly twice the diameter, four times the area, of the other. Another example is number 6 wire has exactly twice the diameter and four times the area of number 12.

A rule-of-thumb for determining required wire sizes for any given installation is that a given wire size assures a given ampacity. For example, 14 gauge wire has an ampacity of 15 amperes. This is true only if (1) a copper conductor is used; (2) the insulation is rated for 60 or 70 degrees centigrade; (3) there are not more than three conductors in a raceway, conduit, or wireway, (4) the ambient temperature is not more than 30 C. (or 86 degrees F.).

The allowable wire size cannot be determined simply from the known load. When any of the four conditions above do not apply, the allowable wire size must be determined from the NEC Tables 310-16 through 310-19. For large circuits, a larger conductor size may be required because of limits to allowable voltage drop than would otherwise be required for branch circuits.

Voltage Drop

Since a wire has resistance, there is a voltage drop from the source end to load end of the line. For example, if 117 V are applied to the source end, a line drop of 20 V would reduce the voltage at the load to 97 V. This might be an excessive voltage drop. For example, an electric light bulb that normally operates at 117 V will be dimmed if operated at 97 V. All electric motors require a specified voltage level to operate properly; lower levels will cause the motor to run hot and may burn them out. In practical situations, a reasonable compromise is determined by the electrician so a line drop is tolerable without incurring undue costs due to the use of needlessly large wire.

In selecting wire size for a particular installation, electricians are bound by the NEC which establishes the allowable current capacity for insulated wires. NEC Tables 310-16 through 310-19 list the maximum current that is permitted to flow in various sizes of wire. Rubber insulated wire tends to heat up more than varnished cambric insulated wire, and therefore, less current is allowed in rubber-insulated wire. Other insulating materials permit more rapid escape of heat and thus are allowed to carry more current.

Types of Wire

The National Electrical Code recognizes many different types of wire that may be used in wiring buildings.

Wires normally used in ordinary wiring are suitable for use at any voltage up to and including 600 V, with the exception of most flexible cords and fixture wires used only for the interior wiring of fixtures, which are suitable only for use up to 300 V.

Aluminum Wire

Aluminum wire is being used to an ever increasing extent. Although aluminum is lighter and easier to bend than copper, a number 14 aluminum wire has greater resistance than an equal length of number 14 copper wire. Therefore, a larger diameter aluminum wire must be used for a given load or current demand. Generally aluminum wire two sizes larger than the required size of copper wire is required to handle a specific load.

There is a trend toward using aluminum in place of copper; the trend accelerates when there is a shortage of copper. Using aluminum wire does introduce new problems not common to using copper.

Aluminum is next to impossible to solder by methods available to most contractors. In practice, this has resulted in aluminum being used mostly in heavy sizes and where long runs are the rule.

Formerly, using aluminum wire in connection with brass or copper terminals led to electrolytic action, and in turn, this led to corroded, high resistance joints and heating. In addition the aluminum tended to "flow" so that connections at terminals that were originally tight later became loose, leading to higher resistance and heating. Much progress has been made in solving these problems, so that at the present time considerable quantities of aluminum are being used in ordinary house and residential construction.

Switches, receptacles and similar devices may be used with aluminum wire only as they are specifically listed by UL for that purpose. Among the requirements is that the device be so constructed that the wire can be wrapped at least ¾ of a turn around the terminal screw, and the device contacts must be made of an alloy that is compatible with both aluminum and copper wire. If a device has push-in terminals without screws, aluminum wire must not be used.

Stranded Wires

A stranded wire consists of a group of wires which are usually twisted to form a metallic string. Stranding improves the flexibility of a wire. When common sizes of wire are used for ordinary wiring purposes, there is usually no reason why the copper conductor should not be one single solid conductor. However, where considerable flexibility is needed, as in flexible cord, the conductor, instead of being one solid wire, consists of a great many strands of fine wire twisted together. The number assigned to such a conductor is determined by the total cross-sectional area of all these individual strands added together. Building wires number 8 and larger are stranded; solid wires are too stiff to be practical. Stranding of each size has been entirely standardized so it is not necessary to specify the size of the individual strands.

Ampacity of Wire

The rated ampacity of a conductor is based on the size of wire, type of insulation, ambient temperature and the number of conductors in a raceway.

Copper is not harmed by heat; insulation is harmed by heat. If insulation is overheated, it is harmed in various ways, dependent on the degree of overheating and the kind of insulation. Some kinds melt, some harden, some burn. In any event, insulation loses its usefulness if overheated, leading to breakdowns and fires.

The ampacities specified in various tables for any particular kind and size of wire is the amperage it can carry without increasing the temperature of its insulation beyond the danger point. The rated ampacity of each kind and size of wire is based on the assumption that the wire is installed where the ambient temperature is 30 degrees Centigrade or 86 degrees Fahrenheit, the normal temperature in an area while there is no current flowing in the wires. When current does flow, heat is created and the temperature will increase above the normal ambient. NEC Table 310-13 shows the maximum temperature that the insulation of each kind of wire is permitted to reach. That temperature will be reached when the wire is carrying its full rated ampacity in a room where the temperature is 30 degrees C or 86 degrees F.

Identifying Kinds of Wire

It is practically impossible to determine the type of wire just by looking at it. There is no difference in appearance between types of T, TH, THW and several other types. The NEC requires that all wires and cables (except special types) be plainly identified by printing the type designation on the surface of the wire or cable at regular intervals. Other markings required are the size, the voltage limitations and the manufacturer's name or trademark. Other markings not required but usually shown include the month and year of manufacture. The abbreviation UL (Underwriters' Laboratories) may be shown. If the conductor is aluminum, the word aluminum or its abbreviation must be shown on the wire.

In addition to the markings on the wire, the tag on the coil or reel must show the type, size, voltage limitation, manufacturer's name, the month and year of manufacture, the designation for aluminum if the conductor is aluminum, the words "NEC Standard" or an abbreviation, and any other markings if they do not confuse or mislead.

Insulation Types

Most ordinary types of plastic insulated wire is what the code calls Type T. Some manufacturers no longer make the ordinary Type T and have replaced it with Type TW or THW. THW wire is similar to Type TW but withstands a greater degree of heat, and consequently it has a higher ampacity rating than TW. If the insulation has a "W" imprinted on it this indicates that it may be used in dry or wet locations. If the identifying mark includes an "H" the wire can be safely used to withstand temperatures higher than 140 degrees F. If the insulation identifier has two "H"s it can be used to operate in still higher temperatures. Many wires currently being used are marked THHN or THWN which indicates that they have a final extruded layer of nylon which is exceedingly tough mechanically, besides having excellent insulating properties. The construction of THHN and THWN type insulation results in a reduction in overall diameter of the wire compared to previous types, especially in the smaller sizes, which permits the use of smaller conduit than that required for other type of insulation.

Classes of Insulation

Substances used as insulators in practical electrical work are classified into four groups:

- *Class A Insulation:* Consists of (1) cotton, silk, paper, material similar to paper when impregnated or immersed in an insulating liquid; (2) molded or laminated materials with cellulose filler, phenolic resins or similar resins, (3) films or sheets of cellulose acetate or similar cellulose products; and (4) varnishes or enamel applied to conductors.

- *Class B Insulation:* Consists of mica, or fiberglass both with a binder.

- *Class C Insulation:* Consists of porcelain, glass, quartz or other similar materials.

- *Class D Insulation:* Consists of cotton, silk, paper or similar materials not impregnated or immersed in insulating liquid.

Wire and Cable

Type MC cable

Type MC Interlocked 5 kV nonshielded armor cable

Type MC 15 kV interlocked armor cable

Interlocked Armor-Type MC 600 Volt Cable: This cable type may have either three or four conductors, Class 8 stranded copper, insulated with heat and moisture resistant cross-linked polyethylene (Type XHHW) phase identified, cabled together with suitable fillers and bare copper ground conductor (two grounds/four conductor). The cable core is covered with binder tape, aluminum or galvanized steel interlocked armor, with flame and sunlight resistant black P9C jacket. Jacket is available under the armor and in colors. Application is as 600 V. Type MC cable is rated at 90 degrees C in wet or dry locations; for aerial installations, or when used in metal racks, trays, troughs, cable trays, or direct buried, for power and control circuits not exceeding 600 V in manufacturing and processing plants, substations and generating stations. When installed as specified by the National Electrical Code, cables meet the requirements of OSHA. This cable may be used in NEC Class I and II, Division 2 and Class III, Division 1 and 2 hazardous locations. It is available in sizes AWG or kcmil 14 through 750 and ampacities 20 to 428 amperes.

Interlocked Armor-Type MC 5kV Nonshielded Cable: This cable consists of Class B stranded copper, extruded conductor shield, insulated with heat and moisture resistant cross-linked polyethylene (Type XLP) or ethylene propylene rubber (Type EPR). Three conductors are cabled together with suitable fillers and bare copper ground conductor. Cable core is covered with binder tape, aluminum or galvanized steel interlocked armor, with flame and sunlight resistant yellow P9C jacket. Jacket is available under armor and in other colors. Application is as armored Type MV-90 cable for installation aerially or in racks, trays, troughs, cable trays, or direct buried for power circuits not exceeding 5000 V in manufacturing and processing plants, substations and generating stations. Available in sizes AWG or kcmil to 750 and ampacity at 40 degrees C of 52 through 525 amperes.

Interlocked Armor-Type MC 15 kV 100 Percent and 133 Percent Insulation Level: This cable type has three conductors, Class B strand, annealed copper, extruded semiconducting XLP shield, heat and moisture resistant cross-linked polyethylene, Type XLP or ethylene propylene rubber, Type EPR, 100X or 133X insulation level, semiconducting insulation shield, and copper shield overall PVC jacket. It is cabled together with suitable fillers and bare copper ground conductor. Cable core is covered with binder tape, aluminum or galvanized steel interlocked armor, with flame and sunlight resistant red PVC jacket. Jacket available under armor and in other colors. Application is as armored Type M9-9O cable for installation aerially or in rack, tray, trough, cable trays, or direct buried; for poser circuits not exceeding 15,000 V in manufacturing and processing plants, substations, and generating stations. Sizes available are AWG or kcmil 2 through 750 with ampacities of 145 to 570 at 40 degrees C.

Bare copper stranded wire

Bare Copper Wire — Solid or Stranded: Bare copper, solid or stranded wire is available in tempers hard, medium-hard, or soft. Stranded conductors are concentrically stranded. Some sizes are available in tinned copper. Application is to be used on insulators for overhead distribution circuits or for grounding conductors. Solid bare copper is available in AWG sizes 14 through 8. Stranded bare copper sizes are AWG or kcmil through 1000. Sizes 1/0 through 1000 may be supplied in any required length.

Type THW PVC-insulated wire

Type THW PVC-Insulated Building Wire: This wire type is rated for a maximum of 600 V. It is PVC insulated, with an annealed uncoated copper conductor. The PVC insulation is surface printed. This type wire is available in many colors and is heat, moisture and oil resistant. It is used as a general-purpose wire for lighting and power in residential, commercial, and industrial buildings. The maximum conductor temperature is 75 degrees C in wet or dry locations. For circuits not exceeding 600 V. Available in sizes AWG or kcmil 14 through l000, with ampacities according to size. The ampacity listed in the NEC is for not more than three conductors in raceway, 75 degree C conductor temperature and 30 degree C ambient in wet or dry locations. Wire is shipped on 500 ft spools or reels. Fourteen through 10 gauge solid and stranded also available on 2500 ft spools. Eight through l000 kcmil also available on 1000 ft reels.

Building Wire Type THHN/THWN, TFFN 600 V PVC Insulation, Nylon Jacket: Type THHN,THWN or MTW meets or exceeds all applicable ASTM specifications, UL standard 83, UL standard 1063 (MTW), Federal Specification J-C-30B, and requirements of the National Electrical Code. It is an annealed uncoated copper conductor, PVC insulation, nylon jacket, surface printed. Used for general-purpose wiring in accordance with the NEC, maximum conductor temperature of 90 degree C in dry locations and

THHN/THWN 600 V building wire

75 degrees C in wet locations, 600 V, for installation in conduit or other recognized raceway, Also used for wiring of machine tools (stranded), appliances, and control circuits not exceeding 600 V. Sizes available are AWG or kcmil from 14 through 1000 with variable ampacity of 15 to 615 A being based on not more than three single-insulated conductors in raceway or in free air with an ambient temperature of 30 degrees C (86 degrees F). Stock colors available are black, white, red, blue, green, yellow, orange, brown, purple, pink, and gray.

SHHW-2 600-V SLP insulation

Building Wire XHHW-2 600 V XLP Insulation: This is an annealed copper conductor, thermosetting chemically cross-linked polyethylene insulation, surface printed; high heat and moisture resistant. It is primarily used in conduit or other recognized raceways for services, feeders, and branch-circuit wiring, as specified in the NEC. XHHW-2 conductors may be used in wet or dry locations at temperatures not to exceed 90 degrees C. Voltage rating for XHHW conductors is 600 V. Used as general-purpose conductor for lighting and power in residential, commercial, and industrial buildings. It has a maximum conductor temperature of 90 degrees C. This wire is suitable for use in low-leakage circuits requiring a dielectric constant of 3.5 or less, such as isolated circuits supplying anesthetizing locations per NEC Article 517-160. It is also suitable for use as low leakage inductive (loop) vehicle detector wire in accordance with state and municipal requirements. Available in sizes AWG or kcmil 14 through 1000 with ampacity, based on not more than three conductors in raceway in free air with an ambient temperature of 30 degrees C (86 degrees F.) and a maximum conductor temperature of 90 degrees C in wet or dry locations of 15 to 615 A.

XLP-USE-2 or RHW-2 XLP insulation

Building Wire XLP-USE-2 or RHH or RHW-2 XLP Insulation: An annealed copper conductor, thermosetting chemically cross-linked polyethylene insulation, surface printed. It is used for lighting and power applications in accordance with the NEC and for other general purpose wiring applications. Suitable for use in circuits not exceeding 600 V at conductor temperatures not exceeding 90 degrees C in wet or dry locations. May be installed in raceway, duct, direct burial and aerial installations. Available in sizes AWG or kcmil of 12 through 1000. Ampacity ranges from 20 to 615 A, depending on size of wire and there not being more than three conductors in raceway in free air with an ambient temperature of 30 degrees C (86 degrees F) and maximum conductor temperature of 90 degrees C.

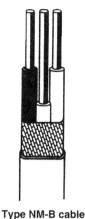

Type NM-B cable

NM-B, 600-V, 90-Degree C Nonmetallic-Sheathed Cable: Type NMB cable is manufactured as two or three conductor cable, with or without a ground wire. Copper conductors are annealed (soft) copper. Conductor insulation is 90 degree C-rated polyvinyl chloride (PVC), nylon jacketed, color coded for identification purposes. NM-B cable is primarily used in residential wiring as branch circuits for outlets, switches, and other loads. NM-B cable may be run in air voids of masonry block or tile walls where such walls are not subject to excessive moisture or dampness. Available in sizes AWG 14 through 2 with ampacity ratings of 15 to 95 amperes, depending on temperature and size.

Type UF cable

UF-8 600-V, Underground Feeder and Branch-Circuit Cable: Type UF-B cable is manufactured as a two or three conductor cable with or without a ground wire. Copper conductors are annealed (soft) copper. Phase conductors are polyvinyl chloride (PVC) insulated, nylon jacketed, color coded for identification purposes. Type UF-B cable is generally used as feeder to outside post lamps, pumps and other loads or apparatus. UF-B cable may be used underground, including direct burial. Multiple conductor UF-B cable may be used for interior branch circuit wiring. Highly resistant to acids, alkalis, corrosive fumes, chemicals, lubricants and ground water. It is non-corroding, sunlight-resistant and will not support combustion.

Type SEU cable

SEU 600-V, Service Entrance Cable Type SE: Style U cable is constructed with type XHHW conductors. Copper XHHW conductors are annealed (soft) copper. Cable assembly plus a concentrically applied neutral and reinforcement tape are jacketed with a sunlight resistant grey polyvinyl chloride (PVC). Type SE, Style U service entrance cable is used to convey power from the service drop to the meter base and from the meter base to the distribution panel board. It may also be used for certain branch circuits as permitted by the National Electrical Code. SEU phase conductors are suitable for use at temperatures not to exceed 90 degree C in dry locations, 75 degrees C in wet locations. Voltage rating for SE, style U cable is 600 V. Cable is supplied in AWG/kcmil sizes g to 4/0 with ampacity ratings of 50 through 260 depending on size of cable and temperature.

Type SE cable

SER 600-V, Service Entrance Cable, Type SE: Style SER cable is constructed with XHHW conductors. Copper conductors are annealed (soft) copper. Cable assemble plus fiberglass reinforcement tape are jacketed with a sunlight resistant grey polyvinyl chloride (PVC). Available as three conductor (two insulated phase conductors, insulated neutral) or four-conductor (two insulated phase conductors, insulated neutral, bare equipment ground).

Type SE, Style SER service entrance cable is primarily used as panel feeders in multifamily dwellings, but it may be used in all applications where Type SE cable is permitted. Type SER cable may be used in wet locations at temperatures not to exceed 75 degrees C, 90 degrees C in dry locations. Voltage rating is 600 V. Sizes available are AWG 8 through 3/0 and ampacity dependent on wire size and temperature requirements and wet or dry conditions of 50 to 225. Cable has 7 strands in the smaller (8 to 2) sizes and 19 strands in sizes 1 through 4/0.

MC 600-V, 90-Degree C Dry Metal-Clad Cables, THHN Conductors: Type MC cable is an assembly of two or more insulated conductors with an insulated ground wire. The assembled conductors have an overall moisture resistant tape and are enclosed in a galvanized steel interlocked cladding or aluminum cladding. Type MC cable is ideal in commercial, industrial and utility applications. Suitable for wiring systems in manufacturing and processing plants, as secondary feeders in industrial and commercial distribution systems and for supplying power to station auxiliaries in power stations and substations. This type wire is available in AWG sizes 14, 12, and 10; all wire is solid.

Type MC, 600-V metal-clad cable

ACTHH/BX 600-V, 90-Degree C, Dry Metal-Clad Cables THHN Conductors: ACTHH/BX is a 600 volt 90 galvanized steel armored cable, employing Type THHN thermoplastic conductors with each conductor individually wrapped in paper. ACTHH/BX is supplied with an aluminum bond wire and is also available in aluminum armor. Cable is ideal for commercial, industrial and multifamily residential branch circuit and feeder wiring, up to 600 V. It offers reliability for 120/208V systems and for higher voltage systems such as 480/277V. Available in AWG sizes 14 through 10 with solid conductors and sizes 8 through 2 with 7 strands.

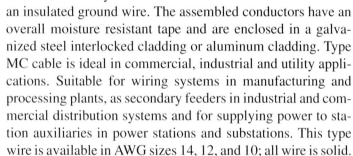

Type ACTHH/BX 600-V metal-clad cable

Tray Cable

Type TC THHN, 600-V cable

Type TC, THHN Conductors, PVC Jacket, 600-V cable: Individual conductors of stranded uncoated copper Type TFFN (16 AWG), Type THWN or THHN (14 to 10 AWG), color-coded, two conductors flat, three or more conductors twisted. PVC jacket overall, surface printed. Also available in sizes 8 through 1000 kcmil in three and four conductor, with/without ground. Cable is used as flame-retardant multiconductor control, signal or power cables rated 600 V, 90 degrees C in dry locations and 75 degrees C in wet locations. Specifically approved for installation in cable trays per Article 340 of the NEC. Also approved for use in Class 1 remote-control and signaling circuits per Article 726-11 (b) of the NEC. Type TC cable is suitable for use in Class 1 and ll, Division 2 hazardous locations. Cables may be installed in open air, in ducts or conduits, in tray or trough, and are suitable for direct burial. Cable is available with seven strands in sizes AWG 16, 14, 12 and 10.

Type TC XHHW 600-V cable

Type TC,.XHHW Insulation, PVC Jacket,600-V Cable: Stranded uncoated copper conductors, 30 mils flame-retardant cross-linked polyethylene insulation (for sizes 16, 14, 12 and 10), color coded, two conductors flat, three or more conductors twisted with suitable fillers where necessary to make round, cable tape, PVC jacket overall, surface printed. Also available in EPR insulation with CPE jacket. Cable is used as superior flame-retardant multi-conductor control, signal or power cable rated 600 V, 90 degree C in wet or dry locations. Specifically approved for installation in cable tray per Article 340 of the NEC. Also approved for use in Class 1 remote-control and signaling circuits per Article 725-11 (b) of the NEC. Type TC cable is suitable for use in Class 1 and 11, Division 2 hazardous locations. Cables may be installed in open air, in ducts or conduits, in tray or trough, and are suitable for direct burial.

Flexible Cable

Type SOW-A/SO Neoprene Jacketed portable Cord: Conductors sizes are 18 through 2 AWG fully-annealed stranded bare copper per ASTM B-174. Insulation: premium grade color coded 90 degree EPDM. Jacket: black neoprene, temperature range -40 degrees C to +90 degrees C, voltage rating 600 V. Jacket Marking: Size/Conductors Type SOW-A 90 degrees C Type SO 90 degrees C (Sizes 18-10 AWG), size/conductors Type SO 90 degrees C (sizes 8-2 AWG). Jacket is also marked with P number and MSHA approval. Type SOW-A and type SO cord is used for portable tools

Type SOW-A/SO portable cord

and equipment, portable appliances, small motors and associated machinery. Available with 2, 3, or 4 conductors in AWG sizes 18 through 10 and with up to 5 conductors in the larger sizes.

Neoprene Jacketed Type SOW-A/SO Portable Control Conductors: 18 through 16 AWG fully annealed stranded bare copper per ASTM B-174. Insulation premium grade color coded 90 degree C EPDM. Jacket: black neoprene, temperature range -40 degrees C to +90 degrees. voltage rating 600V. Jacket is also marked with P number and MSHA approval. Cable is used on: tools, cranes, hoists, track systems Heavy industrial, processing, construction equipment and other electrical equipment. Number of stranded conductors varies from 5 to 64.

Type SOW-A/SO portable control cable

Type SJTO and STO Oil-Resistant, Thermoplastic Jacket, 60-degree C, UL Listed, Portable Cord Conductors: 18 through 10 AWG fully annealed stranded bare copper per ASTM B-174. Insulation: premium grade color coded PVC. Jacket: thermoplastic (PVC), gray or yellow, temperature range -20 degrees C to + 60 degrees C. Voltage rating 300 V Type SJTQ, 600 V Type STO. Jacket marking: Size/Conductors TYPE SJTO (or STO). Jacket is also marked with P number and MSHA approval. Cable is suitable for use on portable motors, floor maintenance equipment, hospital equipment, sound equipment, washing machines, portable lights, lamps and similar equipment. Conductor sizes 10 through 18.

Type SJTO and STO oil-resistant cable

Thermostat Cable: UL Type CL2 is a multiconductor thermoplastic insulated cable, solid-bare copper conductor with a PVC jacket overall. It resists combustion and shorting to ground and between conductors, for low-voltage applications. The cable is used in low-voltage applications such as thermostat controls, heating and air conditioning installations, burglar alarms, and intercom systems.

Thermostat cable

Locomotive and drilling-rig cable

Locomotive and Drilling Rig Cable: This cable is designed specifically for the wiring of diesel electric locomotives and rail car equipment. It is also recommended for high-flexibility uses such as motor or generator leads, battery lead wires, jumper leads and offshore drilling rig applications. The cable conforms to ICEA-NEMA standards, ICE publication S-68-516. It also meets the IEEE flame test at 70,000 BTU/hr input.

It can be UL Listed as shipboard cable and printed and labeled as such. It is suitable for use in wiring diesel electric locomotives, oil and gas drilling rigs, mining and earth moving equipment, general shipyard use, motor leads, apparatus leads, and heavy-duty flexing applications.

Welding Cable Neoprene Class K-30 AWG: The annealed copper conductor conforms to ASTM Specification B-172 and is specially stranded to achieve high flexibility. Covering the conductor, a Kraft crepe paper enhances the flexibility and prevents discoloration of the conductor. A specially compounded or NBR/PVC neoprene jacket covers the conductor. Welding cable is used for connections from electrode holder and clamp to arc welder, bus, welding box or transformer in shipyards, mines and all hazardous locations. It is also suitable for portable lighting systems in the entertainment industry or any portable lighting systems. It is available in sizes 4/0 through 6 AWG.

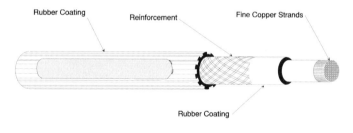

Welding cable

Apparatus and motor-lead wire

Apparatus and Motor Lead Wire SRML and SRK-ML: This is a flexible-stranded tinned copper conductor, SRML-silicone rubber insulation, fiberglass braid jacket. It has a SRK-ML silicone rubber insulation, aramid fiber braid jacket. Both constructions are saturated with flame-, heat- and moisture-resistant finish. The cable is used for leads to motor, transformers, or other electrical equipment where hazardous and high temperature conditions exist requiring flexible heat-resistant conductors.

TGGT 600-V, 250-degree C cable

Appliance and Industrial Lead Wire TGGT and MG 600V: This is a stranded nickel-coated copper conductor, PTFE Teflon tapes covered by wrapped fiberglass yarns. The overall braid of fiberglass yarn is impregnated with a PTFE saturant. The standard color is tan, however, other colors can be furnished. It is used for internal wiring of commercial, industrial and household ovens, cooking and drying equipment. It is ideal for hot spots, severe and critical locations in industrial processing such as iron and steel fabrication, glass plants and petrol chemical refineries. It has lead wires for band and strip heaters. It is available in sizes AWG 22 to 6 and stranded according to size from 7 to 133 strands.

Type MG is flexible stranded, solid nickel or 27 percent nickel-coated copper insulated with glass reinforced mica tapes, jacketed with an overall fiberglass braid impregnated with a high temperature flame, and has a heat and moisture resistant finish. The standard color is brown but other colors are available upon request. Type MG is for internal wiring of domestic and commercial ovens and cooking appliances and similar high temperature environments. It is also ideal for use in wiring electric heaters and for lead and equipment wiring in iron and steel mills as well as glass plants and cement kilns. It is available in sizes 22 through 6 AWG with stranding from 7 to 133 depending on size.

HG 600-V, 538-degree C cable

High-Voltage Power Cable

MV 90-XLP, 5000 V nonshielded, nonjacketed

MV 90 Dry-XLP, 5000V Nonshielded, Nonjacketed: This is an annealed uncoated copper conductor, conductor shield, XLP black thermosetting chemically cross-linked polyethylene insulation per ICEA Pub. No. S-66-524 and NEMA pub. No. WC7. The cable is used where NEC jurisdiction applies; as 5000V nonshielded power cable, Type MV-90 Dry, for use at conductor temperatures not exceeding 90 degree C in dry locations. The cables meet requirements of Article 310-6 of the NEC. Size range is AWG or kcmil 8 through 1000 with 7 to 61 strands depending on size.

MV 90, Dry, SLP/PVC 5000 V shielded cable

MV 90 Dry-XLP/PVC 5000V Shielded, 100 and 133 Percent Insulation Levels: This is an annealed copper conductor, conductor shield, XLP thermosetting chemically cross-linked polyethylene insulation per ICEA Publication number 5-66-524 and NEMA Publication number WC7. It has a PolyKote, semiconducting tape, #22 AWG metallic wire shielding tape, black polyvinyl chloride jacket overall. The cable is used as 5000V shielded power cable, Type MV-90 for general-purpose applications, is in accordance with the NEC at conductor temperatures not exceeding 90 degrees C in wet or dry locations for normal operation, 130 degrees C for emergency overload, and 250 degrees C for short-circuit conditions. It is suitable for installation in conduit, trough, ducts, aerial and direct burial applications. Available in sizes AWG/kcmil 8 to 1000 and number of strands from 7 to 61, depending on size.

MV-90 wet or dry, SLP/PVC, 15,000 V shielded cable

MV-90 Wet or Dry-XLP/PVC, 15,000V Shielded, 100 Percent Insulation Level: It is an annealed copper conductor, extruded conductor shield, XLP thermosetting chemically cross-linked polyethylene insulation per ICEA Pub. No. 5-66-524 and NEMA Pub. No. WC7. It is covered overall with PolyKote, semiconducting tape, #22 AWG metallic wire shielding tape, black polyvinyl chloride jacket. The cable is used as medium voltage MV-90 power cable for use in main feeder, distribution and branch circuits in industrial, commercial and electric utility installations. The cables may be used in wet or dry locations in circuits not exceeding 15,000V. 100 percent insulation level, at conductor temperatures not exceeding 90 degrees C for normal, 130 degrees C for emergency overload and 250 degrees C for short-circuit conditions. It is suitable for installation in conduit, trough, ducts, aerial and direct burial applications. It is available in sizes AWG/kcmil 2 through 1000 and with number of strands varying according to size from 7 to 61.

MV-90 Wet or Dry-XLP/PVC 15,000V Shielded, 133% Insulation Level: It is an annealed copper conductor, extruded conductor shield, XLP thermosetting chemically cross-linked polyethylene insulation per ICEA Pub. No. 5-66-524 and NEMA Pub. No. WC7. It is covered overall with PolyKote semiconducting tape, #22 AWG metallic wire shielding tape, black polyvinyl chloride jacket. It is used as medium-voltage MV-90 power cable for use in main feeder, distribution and branch circuits in industrial, commercial and electric utility installations. The cable may be used in wet or dry

MV-90 wet or dry, SLP/PVC, 15,000 V shielded cable

locations in circuits not exceeding l5,000V, 133% percent insulation level, at conductor temperatures not exceeding 90 degrees C for normal, 130 degrees C for emergency overload and 250 degrees C for short-circuit conditions. It is suitable for installation in conduit, trough, ducts, aerial and direct burial applications. It is available in sizes AWG/kcmil 2 through 1000 and with 7 to 61 strands, depending on size.

MV-90 Wet or Dry, EPS/Hypalon 5,000V Nonshielded: It is an annealed copper conductor, extruded conductor shield, ethylene-propylene-rubber insulation per ICEA Pub. No. S-68-516 and NEMA Pub. No. WC8, discharge and moisture resistant Hypalon jacket. It is used where NEC jurisdiction applies; as 5000V nonshielded power cable, Type MV-90, for use at conductor temperatures not exceeding 90 degrees C in wet or dry locations. The cables meet requirements of Article 310-6 of the NEC. It is available in sizes AWG/kcmil 6 through 1000 and supplied according to size in strands of 7 through 61.

MV-90, wet or dry, XLP/PVC 15 kV shielded cable

MV-90 Wet or Dry, EPR/PVC 5,000/8,000V, Shielded: It is an annealed copper conductor, extruded conductor shield, ethylene-propylene-rubber insulation per ICEA Pub, No. 5-68-516 and NEMA Pub. No. WC8. It has an extruded insulation shield, five mil copper shielding tape, and black polyvinyl chloride jacket. It is used as medium voltage MV-90 power cable for use in main feeder, distribution and branch circuits in industrial, commercial and electric utility installations. The cables may be used in wet or dry locations in circuits not exceeding 5,000V 133 percent insulation level or 8,000V 100 percent insulation level, at conductor temperatures not exceeding 90 degree C for normal,

MV-90, wet or dry, XLP/PVC 5 – 8 kV shielded cable

130 degrees C for emergency overload and 250 degrees C for short-circuit conditions. It is suitable for installation in conduit, tray, trough, ducts, aerial and direct burial applications. Sizes available AWG/kcmil 6 through 1000 with the conductors stranded 7 to 612 depending on cable size.

MV-90, wet or dry, EPR/PVC 15 kV cable

MV-90 Wet or Dry EPR/PVC 15,000V, Shielded, 133 Percent Insulation Level: It is an annealed copper conductor, extruded conductor shield, ethylene-propylene-rubber insulation per ICEA Pub. No. 5-68-516 and NEMA Pub. No. WC8. It has an extruded insulation shield, five mil copper shielding tap, black polyvinyl chloride jacket. I is used as medium voltage MV-90 power cable for use in main feeder, distribution and branch circuits in industrial, commercial and electric utility installations. The cables may be used in wet or dry locations in circuits not exceeding 15,000V 133 percent insulation level, at conductor temperatures not exceeding 90 degrees C for normal, 130 degrees C for emergency overload and 250 degrees C for short-circuit conditions. It is suitable for installation in conduit, tray trough, ducts, aerial and direct burial applications. AWG/kcmil sizes available are 2 through 100 and with stranded conductors varying from 7 to 61.

Airport lighting cable

Airport Lighting Cable FAA-L-824 Type C XLP Insulation 5,000V: The 5,000V annealed uncoated stranded copper conductor with separator tape, XLP insulation, and surface printed is used for airport lighting and control circuits. It has a 90 degree C conductor temperature in wet or dry conditions. The single conductor is Type C and available in AWG sizes 8, 6 and 4 with 7 strands.

Hook-up Wire

Type SIS switchboard wire

Type SIS Switchboard Wire: The tinned stranded copper conductors are insulated with a 90 degree C rated chemically cross-linked polyethylene compound that makes small diameters and light weights possible. It also provides greater tensile strength and resistance to flame, fungus, tears and abrasion. It is UL listed to be used in panelboards, switchboards and control apparatus. For use at 600 V 90 degree C maximum operation in dry locations in accordance with the NEC. AWG sizes 18 through 4 with 16 to 133 strands according to the size are available.

Machine tool and appliance wire

Type MTW/AWM/TEW Machine Tool and Appliance Thermoplastic Equipment Wire: Type MTW/AWM/TEW wire is thermoplastic-insulated and is designed to meet the requirements for machine tool and control applications. This wire is heat, moisture, and oil resistant, UL rated for machine tool wire and appliance wiring material, and C.S.A. rated for thermoplastic equipment wire. Voltage ratings for all three applications are 600 V. It is used for appliances, machine tools, and building applications. It is available in sizes AWG 18 through 2 with stranded conductors 16, 19 and 26. Colors available are black, white, red, blue, green, orange, yellow, brown, purple, gray, pink, and tan.

Mining Cable

Flat portable power cable Type W, G, or GGC

Flat-Portable Power Cable Type W, G, or GGC 90-degree C Conductors: Number 8 through 4/0 AWG are fully-annealed stranded bare copper per ASTM B-172. Insulation is premium grade color coded 90 degree C EPOM. Jacket 90 degree C black, neoprene also available, temperature range-40 degrees C to + 90 degrees C, voltage rating for two conductors is 600V, three conductors -600/200V. Jacket marking: Sizes, 8-1: (SIZE)/2 TYPE (W or G) 600V 90 degrees C MSHA. The cable is used for power supply cable for mobile or portable power equipment such as shuttle cars, coal cutters and other applications where subject to extreme mechanical abuse. Resists oil, acid, alkalies, heat, flame, moisture and most chemicals.

Type W is available in sizes AWG 8 through 4/0 with 2 conductors and 259 strands. Type G is available in sizes AWG 8 through 4/0 with 2 conductors and 259 strands, with a ground conductor AWG sizes 8 through 1/0 depending on cable size. Type GGC is available in sizes 6 through 4/0 with 3 conductors and 259 strands, with two ground conductor sizes AWG 10 through 3 depending on cable size.

Type W round 600/2000-V cable

Type W Round 90-Degree C 600/2000V Conductors: Number 8 through 500 kcmil is fully annealed stranded bare copper per ASTM B-172. Insulation: premium grade color coded 90 degree C EPDM. Jacket: black neoprene, temperature - 40 degree C to + 90 degrees C, voltage rating 600/2000V. Jacket Marking: (#Conductors)/(Size) Type W 600/2000V 90 degrees C MSHA. They are used as heavy-duty service power supply cable, ac systems (grounded and ungrounded), mobile and portable electrical equipment, mining equipment, motor and battery leads. Two conductor cables are used on dc or ac single-phase systems where grounding is not required. Three conductor cable are used on three-phase ac systems where no grounding is required or on dc systems with one conductor for grounding. Four conductor cables are used on two- or three-phase ac systems with one conductor used for grounding. Five conductor cable are used in applications where separating the system neutral from the frame ground is required. Available as 2, 3, 4 or 5 conductor cable in sizes AWG/kcmil 8 through 500 with number of strands varying according to size of cable from 1133 to 1235. To determine ampacities per NEC refer to the latest edition.

Type G or GGC Round 90-Degree C 600/2000V Conductors: Number 8 through 500 kcmil are fully annealed bare copper per ASTM-172. Insulation: premium grade color coded 90 degree C EPDM. Jacket: 90 degree C Super Vu-Tron, black, neoprene, temperature range from 40 degrees C to + 90 degrees C, voltage rating 600/2000V. Jacket Marking:

Type G or GGC round 600/2000-V cable

(Type G or GGC) Sizes 8-1: (Size)/3 Type G or GGC 600/2000V 90 degree C MSHA. Sizes 1/0 and up: three conductor (Size) Type G or GGC 600/2000V 90 degrees C MSHA. They are used in heavy-duty service as power supply cable, and mobile and portable electrical equipment. Also for shuttle cars, coal cutters, drills and shovels where grounding is required for portable equipment. Two conductor - use with single phase ac systems where grounding is required. Three and four conductor -use on three-phase ac systems where grounding is required. The two-conductor cable is available as Type G in sizes AWG 8 through 4/0 with 133 through 259 strands and two grounding conductors ranging in sizes from AWG #10 through AWG #1. The three-conductor cable is available in Type G or GGC and sizes AWG/kcmil 8 through 500 with 133 through 1235 strands and two grounding conductors ranging in size from AWG#10 through 2/0. The four-conductor cable is available as Type G in sizes AWG 8 through 4/0 with 133 through 259 strands depending on cable size and four-ground conductors in AWG sizes from #12 through #4.

Type G or GGC round 600/2000-V cable

Type MPF-GC Portable Power Cable 60 Degree C 5-15kV Conductors: Number 6 through 500 AWG/kcmil are fully annealed copper per ASTM B-33, stranded per ASTM B-8. Insulation: premium grade cross-linked polyethylene (XLP) per ICEA S-66-524. Jacket: PVC, black, temperature range from 10 degree C to +60 degree C, voltage rating 5 - 15kV. Jacket Marking: 3/C, gauge, voltage, MPF-GC and P number to indicate full compliance with Pennsylvania Bureau of Mines Safety Code. The cable is used in bore holes, shafts, horizontal runs, and aerial suspension. Circuits rated 5, 8, 10 and 15kV. They are also used in other mining applications. The 5kV cable is available in sizes AWG 6 through 4/0 with 7 through 19 strands and two-grounding conductor sizes AWG #10 through #1. The 8kV cable is available in sizes AWG #6 through 4/0 with 7 through 19 strands and two-grounding conductor sizes AWG #10 to #1 depending on cable sizes. The 15kV cable is made in AWG sizes 2 through 4/0, stranded 7 to 19 depending on sizes with two grounding conductors, sizes 6 through 1.

Type SHD-GC and Type W magnet crane cable

Types SHD-GC and Magnet Crane Cable Type W 90 Degrees C Conductors: AWG/kcmil 6 through 500 are fully annealed tin-coated copper per ASTM B-33, stranded per ASTM B172. Insulation: premium grade color coded EPDM, black neoprene jacket, temperature ranges from 40 degrees C to + 90 degrees C. Voltage rating 2001-5000V (SHD-GC) (600V-2kV also available as special order). Jacket Marking: (SHD-GC): Sizes 6-1: Manufacturer's name (Size)/3 type SHD-GC . . . Volt sizes 1/0 and up: Manufacturer's name 3 Cond. (Size) Type SHD-GC . . . V (Magnet

Crane Cable): (Size) 2 Type W 600/2000V 90 degree C MSHA Crane Cable. The SHD-GC is used for emergency or temporary power, underground ac mining, dry and moist locations, severest working conditions requiring portable shielded cables, and magnet crane cable applications. Type 2001-5000V available in sizes AWG/kcmil 6 through 500 with three conductors stranded 259 to 427 and two ground wires size AWG 8 through 4/0. Type 90 degree C Magnet Crane Cable Type W 2 is available in sizes AWG 8 through 2 with two conductors stranded 133 through 259.

Special Cable Fabrications

Preassembled aerial cable (XHHW)

Preassembled Aerial Cable Cross-Linked Polyethylene Insulation (XHHW) 600V: This specification covers three conductor preassembled, nonshielded aerial cables. Cross-linked polyethylene (XLP) insulated conductors are cabled and bound to a supporting messenger. All cables manufactured under this specification are suitable for 600V operation. These cables comply in all respects with ICEA and NEMA standards. Three or four conductors, Class B strand, bare annealed copper, insulated with heat and moisture resistant cross-linked polyethylene. The conductors shall be cabled with a left-hand lay and a messenger of 30 percent EHS copperweld shall be laid parallel and bound with a flat-bare copper binding strap. The cables are also available in three-conductor 5000V shielded or unshielded construction and three-conductor, 15,000V shielded construction. The 15kV cable is available in three and four conductor, with 100 percent or 133 percent insulation level. Cross-linked polyethylene (XLP) insulated ozone-resistant cables are recommended for use in the transmission and distribution of electrical energy, indoors and outdoors, for industrial and electrical utility aerial power cable systems. This cable may be used at continuous conductor temperature of 90 degrees C, 130 degrees C emergency overload, and at 250 degrees C short-circuit conditions. Where NEC jurisdiction applies the cables are listed as 90 degrees C in dry locations and 75 degrees C in wet locations.

Utility Cables

ASCR Transmission/Distribution Cable Aluminum Conductor, Steel Reinforced: This cable is constructed of aluminum alloy 1350-H19 wires concentrically stranded about a steel core. The core wire for ACSR is available with Class A, B, or C galvanizing; "aluminized" aluminum coated (AZ); or aluminum clad (AW). Additional corrosion protection is available through the application of grease to the core or infusion of the complete

ASCR transmission distribution cable

cable with grease. It is used as bare overhead transmission cable and as primary and secondary distribution cable. ACSR offers optimal strength for line design.

Variable steel core stranding enables desired strength to be achieved without sacrificing ampacity. Cable is available in sizes AWG/kcmil 6 through 336.4 with stranded conductors numbering 6 through 26, depending on size, and ampacity of 105 through 529.

Triplex service-drop cable

Triplex Service Drop Aluminum Cross-Linked Polyethylene Insulated: These conductors are concentrically stranded, compressed 1350-H19 aluminum and insulated with either polyethylene or cross-linked polyethylene. The neutral messenger is concentrically stranded 6201, AAC or ACST. The cable is also available in duplex and quadruplex construction. It is used to supply power, usually from a pole-mounted transformer, to the user' service head where connection to the service entrance cable is made. It is to be used at a voltage of 600V or less phase-to-phase and at conductor temperatures not to exceed 75 degrees C, for polyethylene insulated conductors or 90 degrees C,and for vulcanized interlinked polyethylene insulated conductors. Sizes available AWG/kcmil 6 through 336.4 with number of strands of the phase conductor varying from 1 to 19 strands and the bare neutral messenger in the same sizes as the phase conductor stranded as 6 in sizes 6 through 4/0 and 18 in size 336.4.

Secondary UD triplex 600-V cable

Secondary UD Triplex 600V: The conductors are concentrically stranded, compressed 1350-H26 aluminum, single-pass insulated with an inner layer of vulcanized interlinked polyethylene, cross-linked polyethylene, and covered with a bonded layer of insulation grade cross-linked high density polyethylene. The neutrals are triple yellow extruded stripe. Black or solid yellow neutrals are available. Conductors are durably surface printed for identification. Two-phase conductors and one neutral conductor are cabled together with a right-hand lay to produce the triplex cable configuration. Cables are also available in single, duplex, and quadruplex construction. The cable is also available paralleled. It is used for secondary distribution and underground service at 600V or less, either direct burial or in ducts. Especially suited for applications requiring superior resistance to mechanical damage. Rated 90 degrees C continuous operation, 130 degrees C emergency overload and 250 degrees C short circuit. It is tested in accordance with AIEE Paper 59-27, "Procedure L-260."

15 kV primary UD cable

15kV Primary UD Cable Aluminum or Copper Up to 35kV: The phase conductor is concentrically stranded, compressed soft copper or 1350-H19 aluminum alloy. The cable is composed of the conductor, covered by a semi-conducting cross-linked polyethylene strand shield, a cross-linked polyethylene primary insulation per ICEA Pub. No. 66-524 and NEMA Pub. NO. WC7, and a semiconducting cross-linked polyethylene insulation shield. Conductors are available with either 100 percent or 133 percent insulation levels. A concentric neutral of bare copper wire is applied over the insulation shield. It is available in 25kV, 28kV and 35kV ratings, and also in 15kV water impervious cable. Predominantly used for primary underground distribution; also suitable for direct burial in wet or dry locations, or for installation in conduit, troughs or trays. It is to be used at 15,000V or less and at conductor temperatures not to exceed 90 degrees for normal operation. Aluminum conductor cable available in AWG/kcmil sizes 2 through 1000 with stranded conductors numbering 7 through 61 and a neutral consisting of 10 to 20 wires in AWG sizes 14 to 10 and the stranded shield made up of 20 to 30 wires. The copper conductor cable available in AWG/kcmil sizes 2 through 750 with stranded conductors numbering 7 through 61 and a neutral consisting of 16 through 29 wires depending on cable size. The stranded shield is composed of 20 to 30 wires.

Data Communication Cables

Telco Data Cable Non-Plenum: This cable is available in AWG 22 and 24 with solid, bare-copper conductors. Insulation is high density polyethylene, color coded pair 1 red/green, pair 2 yellow/black. Jacket is flame retardant PVC colored either beige or gray. Cable is suitable for inside use as residential protector to subscriber's phone. Packaging is 1000 ft in boxes. Conforms to NEC CMX.

Plenum and nonplenum data cable

Telco Data Cable Plenum: This cable is available in AWG 22 and 24 with solid-bare copper conductors. The insulation is fluoropolymer plenum-approved resins, color coded pair 1 black/white, pair 2 red/green. The jacket is fluoropolymer plenum-approved resins. It is suitable for inside use as protector phone and general-purpose plenum. Packaging is 1000 ft in boxes or on spools or reels. Conforms to NEC CMP.

General-purpose and riser nonplenum cable, UTP levels 1 or 2

General-Purpose Non-Plenum UTP-Level 1 or Level 2: This cable is furnished in only AWG #24 with solid-bare copper conductors arranged in two or three pairs. The insulation is high-density polyethylene colored gray. The jacket is of flame-retardant PVC gray in color. The cable is used for inside wiring and cable assemblies. Packaged 1000 ft in boxes. Conforms to NEC CM.

Riser Cable Non-Plenum UTP-Level 2: These conductors are AWG #24 solid-bare copper in 3, 4, 6, 12, 25, 50 or 100 pairs. The insulation is semi-rigid PVC gray in color. The jacket is flame-retardant gray PVC. The cable application is for inside wiring, cable assemblies, UTP performance level 2, IBM Type 3, CMR and CM circuits. Packaging is 1000 ft in boxes Conforms to NEC CMR.

Alvyn-sheathed, nonplenum, riser, Type AR cable

Alvyn Sheathed Non-Plenum Riser Type AR: These conductors are size AWG 24 or 26 solid-annealed copper. General characteristics are 100-ohm nominal impedance, low attenuation, with excellent crosstalk capabilities. Insulation is inner layer of expanded polyethylene and outer layer of PVC. Alvyn sheath is surfaced marked for identification, corrugated polymer-coated aluminum shield adhered to jacket. The jacket is flame-retardant PVC. The cable is used for vertical runs in riser and general horizontal wiring. It is available on reels. The AWG #24 is manufactured in 100 to 1800 pairs while the #26 is available in 300 to 3600 pairs. Conforms to NEC CMR.

Plenum UTP, levels 1 or 2 voice cable

Plenum UTP - Level 1 or Level 2 Voice Cable: The conductors are #24 AWG solid-bare copper insulated with plenum-approved resins and jacket of fluoropolymer plenum-approved resins. The number of pairs available are 2, 3, 4, 6, 12 and 25. The cable is suitable for installation without conduit in air plenums for voice applications RS 232. Packed 1000 ft in boxes or on spools or reels. NEC CMP.

Foil Shield Plenum STP-Level 2: It is constructed of four pairs of solid-AWG #24 bare copper conductors and is insulated with NEC plenum-rated insulation. The shield is overall aluminum/polyester foil with drain wire. The jacket is NEC plenum rated. Application: for installation without conduit in air plenums, for digital voice and data transmission. Packaging is 1000 ft on reels. Complies with NEC CMP.

Foil-shield plenum STP cable

Cross-connect cable

Cross-Connect Cable: The conductors are AWG #24 solid-bare copper and the cable is available with 1, 2 or 3 pairs with semi-rigid PVC insulation. The pairs are color coded Blu/Wht-Wht/Blu, Blu/Yel-Yel/Blu, Blu/Red-Red/Blu, Org/Wht-Wht/Org, Grn/Wht-Wht/Grn, Blk/Wht-Wht/Blk, Red/Wht-Wht/Red. The cable is used for cross-connecting station equipment and PBX terminals. Packaging is 1000 ft on reels.

Telecommunications Interconnect Silver Satin Line Cord: The conductors are AWG #26 tinned stranded copper with thermoplastic insulation. The jacket is gray, silver satin PVC. The cable is used as interconnect cable on phone sets, FAX equipment, as patch cords and security-system wiring. It is available with 4, 6, or 8 pairs in a flat configuration and with 2, 3, or 4 pairs on the round style. The color code for the pairs is Blk/Red, Grn/Yel, Brn/Gry, Gry/Org, Org/Blu. The cable is packaged 1000 ft on reels.

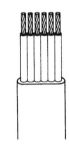

Telecommunications innerconnect, silver-satin, line cord.

Computer interconnect DTP line-cord data cable

Computer Interconnect DTP Line-Cord Data Twisted Pairs: The conductors are AWG #24 tinned stranded copper with thermoplastic insulation available in desired color codes. The jacket is black PVC. It is manufactured with 2, 3, or 4 pairs shielded or unshielded and is used for computer and peripherals wiring, patch cords and security systems. It is packaged in 1000 ft on reels.

Chapter 22

Electrical Connectors and Terminations

Wire-Nut Connectors

Wire Nuts: The wire nut has become an industry standard for connecting two or more conductors, replacing the old method of soldering and taping the connection. The wire-nut connector is available in five colors which indicate size of the connector and applicable size and number of conductors it may be used on. The connectors are UL listed as pressure-type wire connectors and comply with Federal Specification W-S-610D, are UL 486C listed (#ED5238), CSA C 222.2 #188 certified (#LR9706) and rated at 105 degrees C and 221 degrees F to handle all branch circuit and fixture splicing applications. The ribbed caps offer a positive grip and the fixed square-wire spring creates its own threads as it draws the conductors all the way into the connector, creating a secure pressure grip that will not relax with time. The gray wire nut is designed to handle 300 V and connect AWG No. 22 through No. 14 conductors with a minimum of one No. 20, one No. 22 and a maximum of 2 No. 16. The blue nut handles 300 V and connects AWG No. 22 through No. 14 conductors with a minimum of three No. 20 and a maximum of three No. 16. Orange nuts will handle 600 V and AWG No. 2 through No. 14 conductors with a minimum of three No. 20 and a maximum of three No. 16 and one No. 18. Yellow nuts are rated at 600 V and four AWG No. 18 through No. 10 conductors with a minimum of one No. 14, one No. 18 and maximum of four No. 14 and one No. 18. The red wire nut is rated at 600 V and AWG No. 18 through No. 8

Various sizes of wire-nut connectors

conductors a minimum of two No. 14 and a maximum of two No. 10 and two No. 12.

Twister wire-nut connectors

Twister Wire-Nut Connectors: The "Twister" connectors have a ribbed cap and built-in swept wings that provide extra torquing power for secure installation. A "nut driver" which fits the ribbed cap may also be used to further tighten the connector onto the conductors. Two available sizes are identified by the tan and gray colors. The tan nut will handle three AWG No. 22, three No. 14, three No. 12 or four No. 12. The large gray nut will handle three No. 14, four No. 12, two No. 8 stranded or two No. 10 with four No. 12.

Set-screw wire connectors

Set-Screw Wire Connectors: Set-screw connectors eliminate the need to restrip wires when rewiring. The tightened set screw prevents loose connections caused by vibration and sleeve-set screw configuration permits visual inspection of a splice without disturbing it. To make a splice, the wires are inserted into the sleeve, the set screw tightened, and then the flame-retardant shell is screwed onto the sleeve. Three sizes are available: model 10 is UL listed and CSA certified for a maximum of 300 V, and will accommodate AWG No. 24 through No. 16 with a minimum of two No. 24 or a maximum of one No. 16 with two No. 18 conductors. Model 11 is rated for 600 V, and will accommodate AWG No. 22 through No. 16 with a minimum of three No. 20 and a maximum of three No. 16 conductors. Model 22 is rated at 600 V and accommodates AWG No. 16 through No. 10 with a minimum of two No. 14 and a maximum of two No. 10 with one No. 12. Temperature rating is 150 degrees C maximum.

Electrical Connectors and Terminations

Crimp connectors

Crimp Connectors: Zinc-plated steel crimps are designed for making secure, permanent pressure-type wire connections on a wide variety of UL listed and CSA certified branch circuit and fixture wiring applications. The connectors may be crimped and installed by using the crimping nest on electrician's side cutter pliers or with a regular crimping tool. The connectors are available in three sizes, all rated at 600 V. The small size will connect AWG No. 18 through No. 10 with a minimum of one No. 14 and one No. 18 or a maximum of two No. 10 and 2 No. 14. Medium size will connect AWG No. 18 through No. 8 with a minimum of three No. 12 or a maximum of four No. 10. Large connectors will connect AWG No. 18 through No. 4 with a minimum of one No. 14 and one No. 16 or a maximum of two No. 8 and one No. 6 connectors are also available in a solid-copper version.

Crimp Terminals and Splices: The tin-plated copper or brass terminals provide a long, corrosion-free life. Terminals have either brazed or butted seams which prevent splitting and serrations on the crimp portion grip wires tight to reduce pullouts and electrical resistance. The terminals are clearly marked with wire and stud ranges and the funnel-shaped entry insulators guide wires into the crimp area and eliminate strand turn back. Beveled barrel openings ease wire entry. Insulated terminals have color-coded insulation: red for wire sizes No. 22-18 AWG, blue for No. 16-14 AWG, and yellow for No. 12-10 AWG. Wire range is stamped on the metal surface for easy interpretation.

Bare ring terminal

Bare Ring Terminals: Suitable only for copper to copper (Cu/Cu) terminations, the bare ring terminal is available in seven sizes for wire ranges: AWG No. 22-18, No. 16-14, No. 12-10, No. 8, No. 6, No. 4, and No. 2. Acceptable stud sizes vary from size 4 to $\frac{3}{8}$ in according to wire size of the terminal.

Vinyl Insulated Terminal: This terminal has an expanded entry to permit easy entry of wire and is suitable for Cu/Cu terminations only. Sizes available are for wire ranges: AWG No. 22-18, No. 16-14, No. 12-10. Acceptable stud sizes vary from size 4 to bolt size $\frac{3}{8}$ in. The terminal is also available with nylon insulation. It is rated for 600 V maximum and 90 degrees C maximum.

Vinyl insulated terminal

Bare spade terminal

Bare Spade Terminal: The spade terminal is manufactured in three sizes with wire ranges of AWG No. 22-18, No. 16-14, and No. 12-10. The spade for wire sizes AWG No. 22 through 14 may be attached to studs sizes 6, 8, 10 and the spade for AWG No. 12-10 accommodates stud sizes 6, 8, 10 and $\frac{1}{4}$ in.

Vinyl or Nylon Insulated Spade Terminal: This terminal has an expanded entry to permit easy entry of wire, and is suitable for Cu/Cu terminations only. Sizes available are for wire ranges: AWG No. 22-28, No. 16-14, and No. 112-10. The spade for wire sizes AWG No. 22 through 14 may be fitted to stud sizes 6, 8, or 10 while the spade for AWG No. 12-10 is suitable for sizes 6, 8, 10 and $\frac{1}{4}$ in.

Vinyl or nylon insulated spade terminal

Vinyl or nylon insulated snap-spade terminal

Vinyl or Nylon Insulated Snap-Spade Terminal: The snap-spade terminal has a small protrusion on one side of the blade that allows the terminal to be "snapped" onto the stud. The vinyl insulated terminal is rated for 600 V maximum and 90 degrees C. The nylon insulated terminal is rated for 300 V maximum and 10 degrees C. Both terminals are available in sizes by wire ranges AWG No. 22-18, No. 16-14, and No. 12-10. The two smaller size terminals will attach to stud sizes 6, 8, and 10. The AWG No. 12-10 terminal is suitable for sizes 8, 10 and 14.

Bare Butt Splice: The uninsulated butt splice is designed to make a splice in two wires inserted in opposite ends of the splice, butted against each other and secured by crimping onto each wire. The splice is offered in seven sizes to fit wire ranges AWG No. 2-18, No. 16-14, No. 12-10, No. 8, No. 6, No. 4, and No. 2. The splice resembles a small seamless metal tube.

Bare butt splice

Vinyl insulated butt splice

Vinyl Insulated Butt Splice: The vinyl insulated butt slice has an expanded entry on each end to make insertion of wire easy without fold over of strands. Wire is inserted into each end of the splice, butted against the opposite wire and secured by crimping midway to each end from the middle. Splice is manufactured in three sizes to fit wire in the range of AWG No. 22-18, No. 16-14, and No. 12-10.

Bare parallel splice

Bare Parallel Splice: The bare parallel splice is a seamless metal tube designed to splice two wires inserted parallel to each other and then secured by crimping the splice firmly down onto the internal wires. Splice is made in seven sizes to accommodate wire in ranges AWG No. 22-18, No. 16-14, No. 12-10, No. 8, No. 6, No. 4, and No. 2.

Vinyl Insulated Parallel Splice: The vinyl insulated splice is furnished with expanded ends to permit the easy insertion of wire and prevent folding over of strands when stranded wire is used. Two wires are inserted parallel to each other and then secured by crimping the splice firmly down onto the internal wires. Splice is offered in three sizes to splice wire AWG No. 22-18, No. 16-14, and No. 12-10.

Vinyl insulated parallel splice

Crimp Disconnects

Bare female disconnect

Bare Female Disconnect: These are of butted seam construction in the wire holding tube with a striated tab folded over on the sides to make the female disconnect portion. Conductor is inserted into the tube and the tube crimped to secure the conductor. The female tab is applied over a male spade to complete the circuit. The female tab dimensions will vary according to the wire size and intended usage from .11 in × .02 in to .25 in × .032 in. Applicable wire ranges are AWG No. 22-18, No. 16-14, No. 14-12 and No. 12-10. Disconnect is for use on copper only.

Insulated female disconnect

Vinyl and/or Nylon Insulated Female Disconnect: The vinyl and nylon insulated disconnect is available in three sizes to cover wire ranges AWG No. 22-18, No. 16-14, and No. 12-10. The nylon insulated style has metal insulation support. The female tab size varies according to wire size and intended use with ranges from .11 in × .02 in to .25 × .032 in. Disconnect is for use on copper connections only. The insulated disconnects have an expanded entry to provide easy insertion of conductor. Also available as fully insulated style.

Bare male disconnect

Bare Male Disconnect: It is designed for use on copper only and to accommodate wire in the range of AWG No. 22-18, No. 16-14, and No. 12-10. The flat male spade is drilled in the center to provide a lock when applied to the female connector. The male tab size is standard at .25 in × .032 in.

Vinyl male disconnect

Vinyl Male Disconnect: The vinyl insulated male disconnect has an expanded entry formed of vinyl to permit the easy insertion of conductor and avoid foldback of strands on stranded conductors. Insulated disconnects are for use with Cu/Cu only and will accommodate conductors AWG No. 16-14 and No. 12-10. The insulated disconnect is also available in a fully insulated version with insulation covering both the tube and the blade.

Mechanical Lug Connectors (Ideal)

Dual-Rated Aluminum/Copper (Al/Cu) Single-Barrel Connectors: The chamfered wire entry allows an easier and more reliable connection. The lugs are UL listed at 600 V and are acceptable for use up through 2,000 V, and may be used up through 8,000 V if installed in accordance with Section 310-6 of the NEC. The electro tin plating and high strength aluminum alloy construction resists corrosion, reduces contact resistance and gives maximum conductivity. The L-style lugs are available as:

L-STYLE LUG CONNECTORS					
Model Number	Wire Range	Dimensions (in)			Stud Size
		Length	Width	Height	
LA-OR	6 – 14 AWG	1 1/16	3/8	1/2	1/4 in
LA-6	6 – 14 AWG	1 1/16	1/2	1/2	1/4 in
LA-2	2 – 14 AWG	1 15/32	1/2	9/16	1/4 in
LA-1/0	1/0 – 14 AWG	1 15/32	5/8	25/32	1/4 in
LA-2/0	2/0 – 14 AWG	1 15/32	5/8	25/32	1/4 in
LA-250	250kcmil – 6 AWG	2.0	1.0	1 1/8	5/16 in
LA-300	300kcmil – 6 AWG	2.0	1.0	1 1/8	5/16 in
LA-350	350kcmil – 6 AWG	2 1/4	1 1/8	1 1/4	3/8 in
LA-500	500kcmil – 4 AWG	2 13/16	1 1/2	1 9/16	3/8 in

L-STYLE LUG CONNECTORS *(Cont.)*				
Model Number	Wire Range	Dimensions (in)			Stud Size
		Length	Width	Height	
LA-600	600kcmil – 2 AWG	3 3/16	1 1/2	1 9/16	3/8 in
LA-800	800 – 300kcmil	3 1/2	1 3/4	1 5/16	5/8 in
LA-1000	1000 – 350kcmil	3 1/2	1 3/4	1 5/16	5/8 in

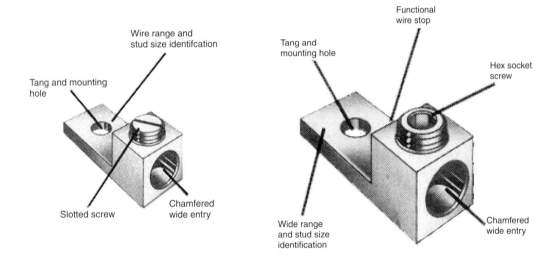

L-model single-barrel connectors

Dual Rated (Al/Cu) Panelboard Connectors: These have chamfered entry for faster and more reliable connections. Lugs are UL listed at 600 V and are acceptable for use up through 2,000 V if installed in accordance with Section 310-6 of the NEC. Lugs meet ANSI C119-4 requirements and are constructed of electro tin plated high-strength aluminum alloy. Multiple barrels permit connecting two, three, or four conductors to the same panel busbar.

Panelboard connectors are available as follows:

Model Number	Wire Range	Dimensions (Inches)			Stud Size
		Length	Width	Height	
PV2-600	600kcmil – 2AWG	$4^{29}/_{32}$	$1^{1}/_{2}$	3	$^{3}/_{8}$
PV2-750	750kcmil – 1/0AWG	$4^{29}/_{32}$	$1^{9}/_{16}$	3	$^{3}/_{8}$
PV3-600	600kcmil – 2AWG	$4^{29}/_{32}$	$2^{1}/_{2}$	3	$^{3}/_{8}$
PV3-750	750kcmil – 3/0AWG	$4^{29}/_{32}$	$2^{27}/_{32}$	3	$^{3}/_{8}$
PV4-600	600kcmil – 2AWG	$4^{29}/_{32}$	$2^{1}/_{2}$	3	$^{3}/_{8}$
PV4-750	750kcmil – 3/0AWG	$4^{29}/_{32}$	$2^{27}/_{32}$	3	$^{3}/_{8}$

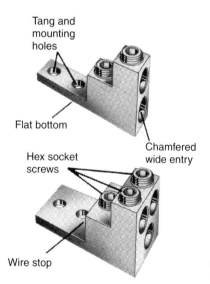

PV2 and PV3 panelboard connectors

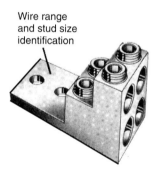

PV4 panelboard connector

Dual-Rated (Al/Cu) Splicer Reducers: The splicer reducers are used to splice two cables of different sizes. They are UL 486B listed AL9CU, constructed of electro tin plated high-strength aluminum alloy. They are rated at 600 V by UL and acceptable for use up through 2,000 V. They may be used up through 8,000 V if installed in accordance with Section 310-6 of the NEC. Splicer reducers are available as:

SPLICER REDUCERS

Model Number	Wire Range	Dimensions (Inches)			Stud Size
		Length	Width	Height	
SR-2	2 – 14 AWG	1 3/8	1/2	9/16	5/8
SR-1/0	1/0 – 14 AWG	1 29/32	3/4	3/4	7/8
SR-250	250kcmil – 6AWG	3 15/16	1	1 1/8	1 15/16
SR-350	350kcmil – 6AWG	4 3/16	1 1/8	1 3/16	2 1/16
SR-500	500kcmil – 3/0AWG	5	1 3/8	1 1/2	2 7/16
SR-750	750kcmil – 250kcmil	6 1/4	1 5/8	1 3/4	3 1/16
SR-1000	500 – 1000kcmil	8	1 3/4	2	3 15/16

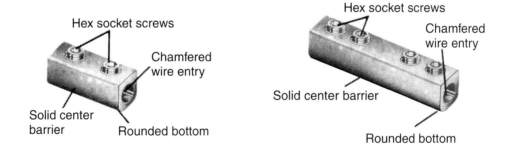

SR splicer reducers

Dual-Rated (Al/Cu) Two-Barrel Connectors: Two-barrel connectors allow the connecting of two conductors to the same panel busbar. The chamfered wire entries permit easier entry and more secure connections. Connectors are constructed of electro tin plated high-strength aluminum alloy which resist corrosion, reduces contact resistance and gives maximum conductivity. UL 486B listed AL9CU and rated UL 600 V but acceptable for use up to 2,000 V. They may be used up through 8,000 V if installed in accordance with Section 310-6 of the NEC. UL 486B listed for dual-rated connectors and 486A listed for copper connectors. They are available as:

Model Number	Wire Range	Dimensions (Inches)			Stud Size
		Length	Width	Height	
L2A1-1/0	1/0 – 14AWG(2)	$1^{15}/_{32}$	$1^{1}/_{8}$	$^{25}/_{32}$	$^{1}/_{4}$
L2A1-2/0	2/0 – 14AWG(2)	$1^{15}/_{32}$	$1^{1}/_{4}$	$^{25}/_{32}$	$^{1}/_{4}$
L2R1-250	250kcmil – 6AWG(2)	$2^{9}/_{16}$	$1^{5}/_{8}$	$1^{3}/_{16}$	$^{3}/_{8}$
L2A1-350	350kcmil – 6AWG(2)	$2^{7}/_{8}$	$1^{15}/_{16}$	$1^{1}/_{4}$	$^{1}/_{2}$
L2A1-600	600kcmil – 2AWG(2)	$3^{3}/_{16}$	$2^{13}/_{32}$	$1^{9}/_{16}$	$^{1}/_{2}$
L2A1-800	800 – 300kcmil(2)	$3^{1}/_{2}$	$3^{1}/_{2}$	$1^{15}/_{16}$	$^{5}/_{8}$
L2A1-1000	1000 – 500kcmil(2)	$3^{1}/_{2}$	$3^{1}/_{2}$	$1^{15}/_{16}$	$^{5}/_{8}$
L2D2-350	350kcmil – 6AWG(2)	$4^{1}/_{4}$	$2^{15}/_{16}$	$1^{3}/_{8}$	$^{1}/_{2}$
L2D2-600	600kcmil – 2AWG(2)	$5^{5}/_{16}$	$2^{3}/_{4}$	$1^{1}/_{2}$	$^{1}/_{2}$
L2D2-800	800 – 300kcmil(2)	$6^{3}/_{16}$	$3^{1}/_{2}$	$1^{7}/_{8}$	$^{1}/_{2}$
L2D2-1000	1000 – 500kcmil(2)	$6^{3}/_{16}$	$3^{1}/_{2}$	$1^{7}/_{8}$	$^{1}/_{2}$

DUAL-RATED TWO-BARREL CONNECTORS

Dual-Rated (Al/Cu) Switch-Gear Connectors: All of the SG models are UL486B listed AL9CU and constructed of electro tin plated high-strength aluminum alloy. The chamfered wire entry permits easy insertion of the conductor and assures a better connection. The connectors are UL listed at 600 V, are acceptable for use up through 2,000 V, and may be used up through 8,000 V if installed in accordance with Section 310-6 of the NEC. They are also listed 486A for use with copper conductors.

Electrical Connectors and Terminations

SWITCHGEAR CONNECTORS					
Model Number	Wire Range	Dimensions (Inches)			Stud Size
		Length	Width	Height	
SG-250	250kcmil-3/0AWG	3	1	$1\frac{3}{16}$	$\frac{3}{8}$
SG-350	350kcmil-4AWG	$4\frac{11}{16}$	$1\frac{1}{4}$	$1\frac{9}{16}$	$\frac{1}{2}$
SG-500	500-400kcmil	$4\frac{11}{16}$	$1\frac{1}{4}$	$1\frac{9}{16}$	$\frac{1}{2}$
SG-800	800-300kcmil	$6\frac{3}{16}$	$1\frac{1}{2}$	$1\frac{7}{8}$	$\frac{1}{2}$
SG-1000	1000-350kcmil	$6\frac{3}{16}$	$1\frac{5}{8}$	$1\frac{7}{8}$	$\frac{1}{2}$

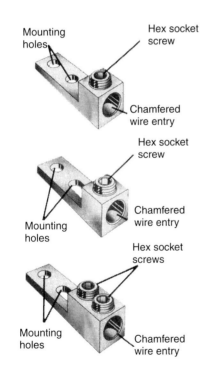

Single-conductor switchgear connectors

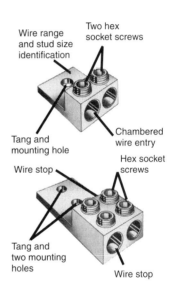

Two-conductor switchgear connectors

Dual-Rated (Al/Cu) Connectors With Multiple Stud Holes per NEMA for Spade-Type Transformers and Busbars: The connectors are made of clear-coated high-strength aluminum alloy. They are UL rated at 600 V and may be used through 2,000 V. They are acceptable for use through 8,000 V if installed in accordance with Section 310-6 of the NEC.

SPADE-TYPE TRANSFORMER AND BUSBAR CONNECTORS

Model Number	Wire Range	Number of Conductors	Dimensions (Inches)		
			Length	Width	Height
L1D2-250	250kcmil – 6AWG	1	$5 \frac{5}{16}$	$1 \frac{1}{4}$	$1 \frac{1}{2}$
L2D4-250	250kcmil – 6AWG	2	$5 \frac{5}{16}$	3	$1 \frac{1}{2}$
L3D4-250	250kcmil – 6AWG	3	$5 \frac{5}{16}$	3	$1 \frac{1}{2}$
L4D6-250	250kcmil – 6AWG	4	$5 \frac{5}{16}$	$4 \frac{5}{8}$	$1 \frac{1}{2}$
L6D8-250	250kcmil – 6AWG	6	$5 \frac{5}{16}$	$6 \frac{5}{32}$	$1 \frac{1}{2}$
L8D8-250	250kcmil – 6AWG	8	$5 \frac{5}{16}$	$8 \frac{1}{4}$	$1 \frac{1}{2}$
L1D2-350	350kcmil – 6AWG	1	$5 \frac{5}{16}$	$1 \frac{1}{4}$	$1 \frac{1}{2}$
L2D4-350	350kcmil – 6AWG	2	$5 \frac{5}{16}$	3	$1 \frac{1}{2}$
L3D4-350	350kcmil – 6AWG	3	$5 \frac{5}{16}$	$3 \frac{1}{2}$	$1 \frac{1}{2}$
L4D6-350	350kcmil – 6AWG	4	$5 \frac{5}{16}$	$4 \frac{23}{32}$	$1 \frac{1}{2}$
L6D8-350	350kcmil – 6AWG	6	$5 \frac{5}{16}$	$7 \frac{1}{8}$	$1 \frac{1}{2}$
L8D8-350	350kcmil – 6AWG	8	$5 \frac{5}{16}$	$9 \frac{9}{16}$	$1 \frac{1}{2}$
L1D2-500	500kcmil – 6AWG	1	$5 \frac{5}{16}$	$1 \frac{1}{4}$	$1 \frac{1}{2}$
L2D4-500	500kcmil – 6AWG	2	$5 \frac{5}{16}$	3	$1 \frac{1}{2}$
L3D4-500	500kcmil – 6AWG	3	$5 \frac{5}{16}$	$3 \frac{7}{8}$	$1 \frac{1}{2}$
L4D6-500	500kcmil – 6AWG	4	$5 \frac{5}{16}$	$5 \frac{1}{4}$	$1 \frac{1}{2}$
L6D10-500	500kcmil – 6AWG	6	$5 \frac{5}{16}$	$7 \frac{29}{32}$	$1 \frac{1}{2}$
L8D12-500	500kcmil – -6AWG	8	$5 \frac{5}{16}$	$10 \frac{19}{32}$	$1 \frac{1}{2}$
L1D2-750	750kcmil – 1/0AWG	1	$6 \frac{3}{16}$	$1 \frac{1}{2}$	$1 \frac{7}{8}$
L2D4-750	750kcmi – 1/0AWG	2	$6 \frac{3}{16}$	$3 \frac{1}{4}$	$1 \frac{7}{8}$
L3D4-750	750kcmil – 1/0AWG	3	$6 \frac{3}{16}$	5	$1 \frac{7}{8}$

Electrical Connectors and Terminations 271

SPADE-TYPE TRANSFORMER AND BUSBAR CONNECTORS *(Cont.)*					
Model Number	Wire Range	Number of Conductors	Dimensions (Inches)		
			Length	Width	Height
L4D8-750	750kcmil – 1/0AWG	4	$6\frac{3}{16}$	$6\frac{5}{8}$	$1\frac{7}{8}$
L6D12-750	750kcmil – 1/0AWG	6	$6\frac{3}{16}$	10	$1\frac{7}{8}$
L8D12-750	750kcmil – 1/0AWG	8	$6\frac{3}{16}$	$13\frac{3}{8}$	$1\frac{7}{8}$

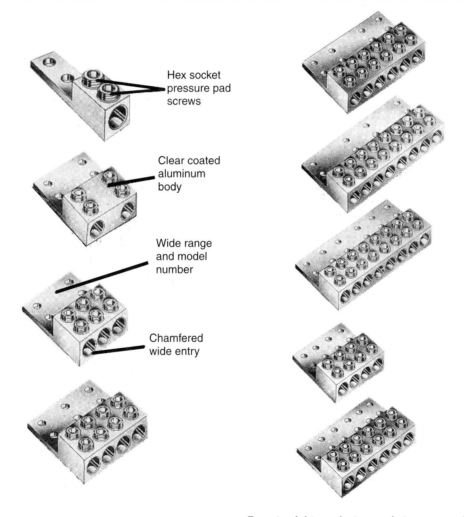

One- to four-conductor spade-type connectors

Four- to eight-conductor spade-type connectors

SPADE-TYPE TRANSFORMER AND BUSBAR CONNECTORS *(Cont.)*

Model Number	Wire Range	Number of Conductors	Dimensions (Inches)		
			Length	Width	Height
L1D2-1000	1000 – 350kcmil	1	6³⁄₁₆	1¾	1⅞
L2D4-1000	1000 – 350kcmil	2	6³⁄₁₆	3½	1⅞
L3D4-1000	1000 – 350kcmil	3	6³⁄₁₆	5⁵⁄₁₆	1⅞
L4D8-1000	1000 – 350kcmil	4	6³⁄₁₆	7⅛	1⅞
L6DI2-l000	1000 – 350kcmil	6	6³⁄₁₆	10¾	1⅞
L8D12-1000	1000 – 350kcmil	8	6³⁄₁₆	14⅜	1⅞

Parallel tap connector with cover

Dual-Rated (Al/Cu) Parallel-Tap Connectors: The connectors are UL listed Al9CU electro tin plated high-strength aluminum alloy and supplied with insulating cover rated at 600 V maximum at 105 degrees C. The cover is attached by a flexible hinge and is self-locking. They are available in the configurations shown in the table on the opposite page.

Parallel tap connector without cover

Dual-Rated Parallel-Tap Connectors Without Cover: Constructed of electro tin plated high-strength aluminum alloy, the connectors are UL 486B listed AL9CU. The bottom of the conductor barrel is serrated to bite into the conductor and form a positive termination. The tap barrel has a machined tongue-and-groove cover plate to facilitate slipping is over the conductor to be tapped. Connectors are available as shown in the table on the opposite page.

Electrical Connectors and Terminations 273

PARALLEL TAP CONNECTORS WITH COVER

Model Number	Wire Range		Strip Length	Dimensions (Inches)		
	Main	Tap		Length	Width	Height
GP-2C	2-12AWG	4-14AWG	5/8	2 7/32	1 29/32	1 7/64
GP-1/OC	1/0-2AWG	1/0-14AWG	3/4	2 5/8	2 5/32	1 1/4
GP-250-0C	250kcmil-1/0AWG	1/0-14AWG	1 1/16	3 7/16	2 7/8	1 5/8
GP-250C	250kcmil-1/0AWG	250kcmil-6AWG	1 1/16	3 7/16	2 7/8	1 5/8
GP-350C	350kcmil-4/0AWG	350kcmil-6AWG	1 1/4	3 3/4	3 3/16	1 3/4
GP-500C	500-350kcmil	500kcmil-2AWG	1 3/8	4 3/8	3 5/16	2 3/32
GP-750C	750-500kcmil	500kcmil-2AWG	1 1/2	4 5/8	3 7/16	2 11/32

PARALLEL TAP CONNECTORS WITHOUT COVER

Model Number	Wire Range		Strip Length	Dimensions		
	Main	Tap		Length	Width	Height
GP-2	2 – 12AWG	4-14AWG	5/8	1 25/64	5/8	7/8
GP-1/0	1/0 – 2AWG	I/0-I4AWG	3/4	1 3/4	3/4	1
GP-250-0	250kcmil – 1/0	1/0-14AWG	1 1/16	2 1/32	1 1/16	1 5/16
GP-250	250kcmil – 1/0	250kcmil-6AWG	1 1/16	2 9/32	1 1/16	1 5/16
GP-350	350kcmil – 4/0	250kcmil-6AWG	1 1/4	2 9/16	1 1/4	1 7/16
GP-500	500 – 350kcmil	500kcmil-2AWG	1 3/8	3 1/8	1 3/8	1 3/4
GP-750	750 – 500kcmil	500kcmil-2AWG	1 1/2	3 3/8	1 1/2	2
GP750-1	750 – 500kcmil	750kcmil-I/0AWG	3 5/32	3 3/8	3 5/32	2

Tee-tap connector with cover

Dual-Rated (Al/Cu) Tee-Tap Connectors With Covers: The connectors are UL 486B listed AL9CU electro tin plated high-strength aluminum alloy. They are UL listed at 600 V, acceptable for use up through 2,000 V, and may be used up through 8,000 V if installed in accordance with Section 310-6 of the NEC. The cover is rated at 600 V maximum at 105 degrees C. The cross of the "T" is machined tongue-and-grooved and designed to slip out of the connector to allow placing over the conductor to be tapped.

TEE-TAP CONNECTORS WITH COVERS

Model Number	Wire Range		Strip Length	Dimensions (Inches)		
	Main	Tap		Length	Width	Height
GT-2C	2-12AWG	4-14AWG	5/8	2 7/32	1 29/32	1 7/64
GT-1/0C	1/0-2AWG	1/0-14AWG	3/4	2 5/8	2 5/32	1 1/4
GT-2500C	250kcmil-1/0	1/0-14AWG	1 1/16	3 7/16	2 7/8	1 5/8
GT-250C	250kcmil-1/0	250kcmil-6AWG	1 1/16	3 7/16	2 7/8	1 5/8
GT-350C	350kcmil-4/0	350kcmil-6AWG	1 1/4	3 3/4	3 3/16	1 3/4
GT-500C	500-350kcmil	500kcmil-2AWG	1 3/8	4 3/8	3 5/16	2 3/32
GT-750C	750-500kcmil	500kcmil-2AWG	1 1/2	4 3/8	3 7/16	2 11/32

Tee-tap connector without cover

Dual-Rated Tee-Tap Connectors Without Covers: These connectors are UL 486B listed AL9CU. The electro tin plated high-strength aluminum alloy provides a corrosion resistant connector with reduced contact resistance and maximum conductivity. Connectors are UL listed at 600 V and are acceptable for use up through 2,000 V and may be used up through 8,000 V if installed in accordance with Section 310-6 of the NEC. Sizes available are:

	TEE-TAP CONNECTORS WITHOUT COVERS					
Model Number	Wire Range		Strip Length	Dimensions (Inches)		
	Main	Tap		Length	Width	Height
GT-2	2-12AWG	4-14AWG	$5/8$	$1^{25}/_{64}$	$5/8$	$7/8$
GT-1/0	1/0-2AWG	1/0-2AWG	$3/4$	$1^{3}/_{4}$	$3/4$	1
GT-250-0	250kcmil-1/0	1/0-14AWG	$1^{1}/_{16}$	$2^{9}/_{32}$	$1^{1}/_{16}$	1
GT-250	250kcmil-1/0	250kcmil-6AWG	$1^{1}/_{16}$	$2^{9}/_{32}$	$1^{1}/_{16}$	$1^{5}/_{16}$
GT-350	350kcmil-4/0	350kcmil-6AWG	$1^{1}/_{4}$	$2^{9}/_{16}$	$1^{1}/_{4}$	$1^{7}/_{16}$
GT-500	500-350kcmil	500kcmil-2AWG	$1^{3}/_{8}$	$3^{1}/_{8}$	$1^{3}/_{8}$	$1^{3}/_{4}$
GT-750	750-500kcmil	500kcmil-2AWG	$1^{1}/_{2}$	$3^{3}/_{8}$	$1^{1}/_{2}$	2

Collar connector

Dual-Rated (Al/Cu) Collar Connectors: The collar connectors CA-110 are UL 486B listed AL9CU, the model CA-60 and CA-10 are listed AL7CU. Collar connectors are designed to be bottom mounted directly to electrical devices or busbars by use of a 10-32 threaded-mounting hole. The CA-10 accommodates wire range 1/0-14AWG and is $19/32$ in long, $17/32$ in wide and $5/8$ in high. Model CA-60 accommodates wire in the range 2-14AWG, dimensions are $9/16$ in × $15/32$ in × $15/32$ in. Model CA-150 will hold wire 2/0-14AWG and is $3/4$ in × $37/64$ in × $5/8$ in.

Grounding Connectors

The NEC states, "The grounding conductor shall be attached to grounding fittings by means of suitable lugs, pressure connectors, clamps or other approved means, except that connectors which depend upon solder shall not be used. Not more than one conductor shall be connected to the grounding electrode by a single clamp or fitting unless the clamp or fitting is a type approved for such use. If conduit is being used as a grounding conductor, it must be anchored securely to one end of the service equipment, using locknut and bushing, and a clamp which can be swiveled for use in conduit should be used."

Heavy-duty ground clamp

Heavy-Duty Ground Clamp: This clamp is designed to maintain proper contact and alignment between ground wire and rod. It is cast of high-strength copper alloy and has brass or plated screws for corrosion resistance. Available sizes are shown in the following table:

DIMENSIONS OF HEAVY-DUTY GROUND CLAMPS		
Model Number	Rod Diameter	Conductor Range
WB12	.500 in	2 – 10
WB58	.625 in	1/0 – 8
WB34	.750 in	1/0 – 8

Standard-duty ground clamp

Standard-Duty Ground Clamp: This clamp is cast of high-strength copper alloy in a lighter version than the heavy-duty clamp and for only a wire range of 2-10AWG. Clamp is available in three models G-4, G-R, and G-6 to fit ground rods .500 in, .625 in and .750 in in diameter.

High-Strength Ground Clamp: This clamp is designed to maintain proper contact and alignment between ground wire and pipe. Mounting screws are either brass or plated for corrosion resistance and is constructed of high-strength copper alloy. Three models are manufactured which are the model J, J2 and J2124 to fit all pipe sizes from $\frac{1}{2}$ to 4 in outside diameter. All clamps will handle conductors from 2-10 AWG.

High-strength ground clamp

Dual-rated ground clamp

Dual-Rated (Al/Cu) Ground Clamp: Constructed of electro tin plated high-strength aluminum, the dual-rated clamp is suitable for use with either aluminum or copper conductors and may be used on $\frac{1}{2}$ in, $\frac{3}{4}$ in or 1 in galvanized or copper water pipe. The clamp will handle wire sizes 1/0-14 AWG and is $2\frac{1}{4}$ in × $\frac{11}{16}$ in. One mounting hole on top of clamp is slotted to permit easy attachment to the pipe. Pipe attachment surface is grooved to assure positive contact.

Two-Bolt and Split-Bolt Connectors

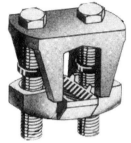

Two-bolt tap connector

Two-Bolt Tap Connectors for Copper-to-Copper Only: The two-bolt tap connector is used to make a tap with both conductors clamped into the connector. The internal serrations and high-strength cast copper alloy construction assure positive terminations. The connectors are available as:

Model Number	CAPACITY OF TWO-BOLT TAP CONNECTORS			
	Maximum Wire Size		Minimum Wire Size	
	Run	Tap	Run	Tap
KB-2/0	2/0	2/0	2	10
KB-4/0	4/0	2/0	1/0	10
KB-350	350kcmil	350kcmil	250kcmil	10
KB-500	500kcmil	500kcmil	400kcmil	10
KB-800	800kcmil	800kcmil	400kcmil	3/0
KB-1000	1000kcmil	1000kcmil	500kcmil	3/0

Dual-rated two-bolt tap connector

Dual-Rated (Al/Cu) Two-Bolt Tap Connectors: The two-bolt dual-rated tap connector is suitable for use with either copper or aluminum due to the tin plated cast copper alloy construction. A spacer is provided to separate the two conductors in the clamp. The connectors are available as specified in the table to follow:

CAPACITY OF DUAL-RATED TWO-BOLT TAP CONNECTORS

Model Number	Maximum Wire Range				
	Copper		C-Weld	Aluminum	
	Solid	Stranded	Solid	AWG	ACSR
KS-2/0	3/0	2/0	3/0	2/0	2/0
KS-4/0	4/0	4/0	4/0	4/0	4/0
KS-350	—	350	—	350	300
KS-500	—	500	—	500	397.5
KS-800	—	800	—	800	715.5
KS-1000	—	1000	—	1000	900

Split-bolt tap connector

Dual-Rated (Al/Cu) Type AS Split-Bolt Tap Connectors: The split-bolt connector has a serrated spacer and bottom plate to retard corrosion and provide a low-contact resistance. The connector is constructed of tin plated aluminum which inhibits oxide formation. The conductors are placed in the four-piece connector and the large nut on the threaded split bolt tightened to secure them. The connector is manufactured as UL 486B listed AL9CU in sizes.

CAPACITY OF DUAL-RATED, TYPE AS SPLIT-BOLT TAP CONNECTORS

Model Number	Wire Range			
	Run		Tap	
	Maximum	Minimum	Maximum	Minimum
AS-6	6 str.	10 solid	6 stranded	10 solid
AS-4	4 str.	8 solid	4 stranded	10 solid

CAPACITY OF DUAL-RATED, TYPE AS SPLIT-BOLT TAP CONNECTORS *(Cont.)*				
Model Number	Wire Range			
	Run		Tap	
	Maximum	Minimum	Maximum	Minimum
AS-2	2 stranded	6 compact	2 stranded	8 stranded
AS-1/0	1/0 stranded	2 compact	1/0 stranded	8 solid
AS-2/0	2/0 stranded	2 compact	2/0 stranded	8 stranded
AS-4/0	4/0 stranded	2 compact	4/0 stranded	6 stranded
AS-350	350kcmil	1/0 compact	350kcmil	4 stranded
AS-500	500kcmil	400 compact	500kcmil	2 compact

Type K split-bolt connector

Type K Split-Bolt Tap Connectors for Copper Conductors: The nut, bolt and body of split-bolt connectors for copper conductors are made of high-strength copper alloy and UL 486A listed. The bottom pad is attached to a nut and guide assembly to simply installation. Connectors are available as specified in the following table:

CAPACITY OF TYPE K SPLIT-BOLT CONNECTORS		
Model Number	Range for Equal Run and Tap	Minimum Tap with Maximum Run
K-8	8 – 10 stranded	14 stranded
K-6	6 solid – 8 stranded	14 stranded
K-4	4 solid – 8 stranded	14 stranded
K-3	2 solid – 6 stranded	14 stranded
K-2	2 – 6 stranded	14 stranded

CAPACITY OF TYPE K SPLIT-BOLT CONNECTORS (Cont.)

Model Number	Range for Equal Run and Tap	Minimum Tap with Maximum Run
K-1/0	1/0 – 4 stranded	14 stranded
K-2/0	2/0 – 2 stranded	14 stranded
K-3/0	3/0 – 1 stranded	14 stranded
K-250	250kcmil – 1 stranded	8 stranded
K-350	350kcmil – 1/0 stranded	1/0 stranded
K-500	500kcmil – 2/0 stranded	1/0 stranded
K-750	750kcmil – 4/0 stranded	4/0 stranded
K-1000	1000 – 350kcmil stranded	4/0 stranded

Mechanical Solderless Lugs

Single-Hole Solderless Lugs for Copper Conductors (Offset): The tang of the lugs are stamped with the wire range and stud size identification. Model numbers CF-25 through CF-90 have slotted-head clamping screws, model CF-125 through CF-225 have hex-head clamping bolts. The floating offset tang and barrel-shaped collar insure a positive contact with the conductor. CF-25 is not C.S.A. certified.

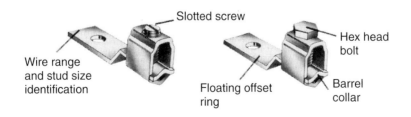

Single-hole offset solderless lugs

Electrical Connectors and Terminations

DIMENSIONS OF SINGLE-HOLE OFFSET SOLDERLESS LUGS					
Model Number	Wire Range	Stud Size	Dimensions (Inches)		
			Length	Width	Height
CF-25	10 – 14 AWG	#8	1	5/16	3/8
CF-35	6 – 14 AWG	3/16 in	1 1/32	3/8	3/4
CF-70	4 – 14 AWG	1/4 in	1 5/16	1/2	13/16
CF-90	2 – 8 AWG	1/4 in	1 15/32	1/2	1
CF-125	1/0 – 2 AWG	1/4 in	1 13/16	5/8	1 11/32
CF-175	3/0 – 2 AWG	3/8 in	2 3/64	3/4	1 9/16
CF-225	4/0 – 2 AWG	5/16 in	2 5/8	1	1 13/16

Straight solderless lugs

Single-Hole Solderless Lugs for Copper Conductors (Straight): The straight connector is UL 486A listed and made of electrolytic copper for use with copper conductors only. The floating straight tang is marked showing wire range and stud size. Model CFS-25 and CFS-90 have slotted screws and CFS-125 and CFS-225 have hex-head bolts. The barrel-shaped collar and arched pressure bar insure positive contact.

DIMENSIONS OF SINGLE-HOLE SOLDERLESS LUGS					
Model Number	Wire Range	Stud Size	Dimensions (Inches)		
			Length	Width	Height
CFS-25	10 – 14 AWG	#8	1	5/16	3/8
CFS-35	6 – 14 AWG	3/16 in	1 9/64	3/8	11/16
CF5-70	4 – 14 AWG	1/4 in	1 1/4	1/2	27/32
CFS-90	2 – 8 AWG	1/4 in	1 15/32	1/2	31/32

DIMENSIONS OF SINGLE-HOLE SOLDERLESS LUGS

Model Number	Wire Range	Stud Size	Dimensions (Inches)		
			Length	Width	Height
CF5-125	1/0 – 2 AWG	1/4 in	1 15/16	5/8	1 1/4
CFS-175	3/0 – 2 AWG	3/8 in	2 1/4	3/4	1 9/16
CF5-225	4/0 – 2 AWG	5/16 in	2 3/8	1	1 21/32

Short-tang solderless lug

Single-Hole Solderless Lugs for Copper Conductors (Short Tang): The short-tang style of solderless lugs is available in three sizes, UL486A listed and constructed of electrolytic copper. The tang is staked to the collar. Models FT-65 and FT-70 use a slotted screw and the FT-90 uses a hex-socket screws.

SPECIFICATIONS FOR SHORT-TANG SOLDERLESS LUGS

Model Number	Wire Range	Stud Size	Dimensions (Inches)		
			Length	Width	Height
FT-65	6-14 AWG	3/16	1	3/8	3/4
FT-70	4-14 AWG	1/4	1 1/8	1/2	13/16
FT-90	1/0-14 AWG	1/4	1 17/16	5/8	1

Above-Ground Transformer Connectors Dual-Rated (Al/Cu): The transformer connectors will connect 4, 6, or 8 conductors and are constructed of clear coated high-strength aluminum alloy. They use hex-slotted pressure pad screws to lock the conductor to the connector. Models available are shown in the following table:

Electrical Connectors and Terminations   283

 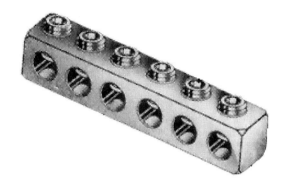

Above-ground transformer connectors

SPECIFICATIONS FOR DUAL-RATED TRANSFORMER CONNECTORS					
Model Number	Wire Range	Number of Conductors	Dimensions (Inches)		
			Length	Width	Height
NA250-4	250kcmil-6AWG	4	$3\frac{5}{8}$	$\frac{7}{8}$	$1\frac{1}{8}$
NA250-6	250kcmil-6AWG	6	$5\frac{3}{8}$	$\frac{7}{8}$	$1\frac{1}{8}$
NA250-8	250kcmil-6AWG	8	$7\frac{1}{8}$	$\frac{7}{8}$	$1\frac{1}{8}$
N4350-4	350kcmil-6AWG	4	$3\frac{27}{32}$	1	$1\frac{3}{8}$
NA350-6	350kcmil-6AWG	6	$5\frac{21}{32}$	1	$1\frac{3}{8}$
NA350-8	350kcmil-6AWG	8	$7\frac{15}{32}$	1	$1\frac{3}{8}$
NA500-4	500kcmil-2AWG	4	$4\frac{27}{32}$	1	$1\frac{5}{8}$
NA500-6	500kcmil-2AWG	6	$7\frac{7}{32}$	1	$1\frac{5}{8}$
NA500-8	500kcmil-2AWG	8	$9\frac{19}{32}$	1	$1\frac{5}{8}$
NA750-4	750kcmil-1/0AWG	4	$6\frac{5}{8}$	2	$2\frac{1}{2}$

SPECIFICATIONS FOR DUAL-RATED TRANSFORMER CONNECTORS (Cont.)

Model Number	Wire Range	Number of Conductors	Dimensions (Inches)		
			Length	Width	Height
NA750-6	750kcmil-1/0AWG	6	10	2	2½
NA750-8	750kcmil-1/0AWG	8	13⅜	2	2½

Noalox Anti-Oxidant

Noalox Anti-Oxidant: Noalox is an anti-oxidant and anti-seizing compound which improves efficiency and service life of aluminum electrical applications. It consists of fine zinc particles suspended in a carrier material. The zinc particles penetrate and "cut" the aluminum oxide, providing additional inter-strand and inter-conductor current paths for improved conductivity and cooler connections. In addition, the carrier material excludes air, and thus prevents further oxidation. Anti-oxidant is for use with all types of pressure wire connectors, including screw-on, tap, service entrance, and split-bolt connectors. It can be used on aluminum conduit joints. It reduces galling and seizing and promotes good ground continuity through the conduit joint.

Compression Lugs

One-Hole Compression Terminals Dual-Rated (Al/Cu): The terminal is UL 486B listed AL9CU and electro tin plated aluminum alloy. The plating resists corrosion, reduces contact resistance and promotes maximum conductivity. Connectors are color coded for size identification. The chamfered barrel is of seamless construction and will not split during installation. Terminals are UL listed at 600 V and are acceptable for use up through 35,000 V if installed in accordance with Section 310-6 of the NEC. Lugs are prefilled and capped with Noalox anti-oxidant. Available lugs are listed in the following table:

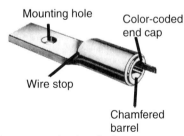

One-hole compression terminal

Electrical Connectors and Terminations

SPECIFICATIONS FOR ONE-HOLE COMPRESSION TERMINALS

Model Number	Wire Range	Strip Length	Plug Color	Stud Size	Dimensions (Inches)		
					Overall Length	Tang Length	Tang Width
CLA-6	6 AWG	3/4 in	Gray	1/4	1 3/4	23/32	21/32
CLA-4	4 AWG	13/16 in	Green	1/4	1 27/32	23/32	13/16
CLA-2	2 AWG	7/8 in	Pink	1/4	1 29/32	23/32	29/32
CLA-1	1 AWG	15/16 in	Gold	5/16	2 1/4	29/32	23/32
CLA-1/0	1/0 AWG	1 1/16 in	Tan	3/8	2 9/16	1 1/16	27/32
CLA-2/0	2/0 AWG	1 1/8 in	Olive	3/8	2 5/8	1 1/16	29/32
CLA-3/0	3/0 AWG	1 1/4 in	Ruby	3/8	2 7/8	1 1/8	1 1/16
CLA-4/0	4/0 AWG	1 3/8 in	White	3/8	3 1/8	1 1/8	1 3/16
CLA-250	250 kcmil	1 7/16 in	Red	1/2	3 17/32	1 3/8	1 9/32
CLA-300	300 kcmil	1 1/2 in	Blue	1/2	3 27/32	1 13/32	1 3/8
CLA-350	350 kcmil	1 5/8 in	Brown	1/2	4 3/32	1 1/2	1 9/16
CLA-400	400 kcmil	1 13/16 in	Green	1/2	4 7/16	1 5/8	1 17/32
CLA-500	500 kcmil	1 15/16 in	Pink	1/2	4 5/8	1 5/8	1 13/16
CLA-600	600 kcmil	2 in	Black	5/8	5	1 29/32	2
CLA-750	750 kcmil	2 1/4 in	Yellow	5/8	5 3/4	2 1/8	2 3/32

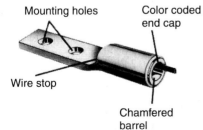

Two-hole compression terminal

Two-Hole Compression Terminal Dual-Rated (Al/Cu): The two holes in the mounting tang are spaced 1 3/4 in on center. The lug is UL 486B listed AL9Cu and electro tin plated to resist corrosion, reduce contact resistance and provide maximum conductivity. The end cap of the barrel is color coded to designate size of the cable. The compression lug is UL listed at 600 V and is acceptable for use up through

35,000 V if installed in accordance with Section 310-6 of the NEC. Sizes are as follows:

SPECIFICATIONS FOR TWO-HOLE, DUAL-RATED, COMPRESSION TERMINALS

Model Number	Wire Range	Strip Length	Plug Color	Stud Size	Dimensions (Inches)		
					Overall Length	Tang Length	Tang Width
2CLA-1/0	1/0AWG	1 1/16	Tan	3/8	3 9/16	2 1/16	27/32
2CLA-2/0	2/0AWG	1 1/8	Olive	1/2	4 13/16	3 1/4	31/32
2CLA-3/0	3/0AWG	1 1/4	Ruby	1/2	5	3 1/4	1 1/16
2CLA-4/0	4/0AWG	1 3/8	White	1/2	5 1/4	3 1/4	1 3/16
2CLA-250	250kcmil	1 7/16	Red	1/2	5 7/16	3 1/4	1 9/32
2CLA-300	300kcmil	1 1/2	Blue	1/2	5 11/16	3 1/4	1 3/8
2CLA-350	350kcmil	1 9/16	Brown	1/2	5 3/4	3 1/4	1 9/16
2CLA-400	400kcmil	1 13/16	Green	1/2	6 1/16	3 1/4	1 17/32
2CLA-500	S00kcmil	2 1/8	Pink	1/2	6 1/2	3 1/4	1 13/32
2CLA-600	60Okcmil	2	Black	1/2	6 11/32	3 1/4	2
2CLA-750	750kcmil	2 1/4	Yellow	1/2	6 7/8	3 1/4	2 3/32

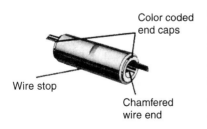

Compression splicer

Compression Splicers Dual-Rated (Al/Cu): The splicers have a wire stop located in the center of the seamless tube to evenly space the compression on each conductor. UL 486B listed AL9CU, the electro tin plated aluminum alloy splicer is prefilled with Noalox anti-oxidant to prevent oxidation. It is UL listed at 600 V and is acceptable for use up through 35,000 V if installed in accordance with Section 310-6 of the NEC. Available compression splicers are described in the following table:

Electrical Connectors and Terminations

SPECIFICATIONS OF COMPRESSION SPLICERS					
Model Number	Wire Range	Strip Length	Cap Color	Sleeve Length	Outside Diameter
SA-6	6 AWG	3/4	Gray	1 5/8	.338
SA-4	4 AWG	1 5/16	Green	1 7/8	.427
SA-2	2 AWG	1 1/8	Pink	2 3/8	.525
SA-1	1 AWG	1 1/8	Gold	2 3/8	.544
SA-1/0	1/0 AWG	1 1/8	Tan	2 11/32	.599
SA-2/0	2/0 AWG	1 13/16	Olive	2 1/2	.674
SA-3/0	3/0 AWG	1 1/4	Ruby	2 5/8	.761
SA-4/0	4/0 AWG	1 1/2	White	3 1/16	.854
SA-250	250kcmil	1 1/2	Red	3 1/8	924
SA-300	300kcmil	1 11/16	Blue	3 1/2	1.010
SA-350	350kcmil	1 15/16	Brown	4 1/16	1.105
SA-400	400kcmil	2 1/4	Green	4 5/8	1.188

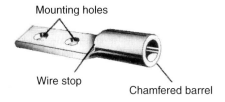

Mounting holes
Wire stop
Chamfered barrel

Two-hole, long-barrel copper compression terminals

Two-Hole Long-Barrel Copper Compression Terminals for Copper Conductors: Copper compression terminals are constructed of bright electro tin plated pure electrolytic copper UL 486A listed for copper conductors. End cap of the chamfered barrel is color coded for easy identification. The lugs are UL listed at 600 V and are acceptable for use up through 35,000 V, if installed in accordance with Section 310-6 of the NEC. Specifications for two-hole, long-barrel copper compression terminals appear in the following table.

SPECIFICATIONS FOR TWO-HOLE, LONG-BARREL COPPER COMPRESSION TERMINALS

Model Number	Wire Size	Stud Size	Strip Length	Die Color	Die Index Code	Dimensions (Inches)		
						Length	Tang Length	Tang Width
CCBL-8	8 AWG	1/4	1 1/8	Red	BB	2 9/16	1 1/4	27/64
CCBL-6	6 AWG	1/4	1 1/8	Blue	CC	2 9/16	1 1/4	1/2
CCBL-4	4 AWG	1/4	1 7/16	Gray	DD	2 13/16	1 1/4	33/64
CCBL-2	2 AWG	5/16	1 7/16	Brown	EE	3 1/4	1 5/8	5/8
CCBL-1	1 AWG	5/16	1 3/4	Green	FF	3 1/2	1 5/8	43/64
CCBL-1/0	1/0AWG	5/16	1 3/4	Pink	GG	3 1/2	1 5/8	23/32
CCBL-2/0	2/0AWG	1/2	1 3/4	Black	HH	5 13/16	3 1/8	13/16
CCBL-3/0	3/0AWG	1/2	2 1/4	Orange	JJ	5 3/4	3 1/8	15/16
CCBL-4/0	4/0kcmil	1/2	2 1/4	Purple	KK	5 3/4	3 1/8	1 1/16
CCBL-250	250kcmil	1/2	2 1/4	Yellow	LL	5 13/16	3 1/8	1 1/8
CCBL-300	300kcmil	1/2	2 1/4	White	NN	5 7/8	3 1/8	1 1/4
CCBL-350	350kcmil	1/2	2 1/4	Red	OO	5 5/16	3 1/8	1 5/16
CCBL-400	400kcmil	1/2	2 1/4	Blue	RR	6	3 1/8	1 13/32
CCBL-500	500kcmil	1/2	2 1/4	Brown	SS	6 1/16	3 1/8	1 9/16
CCBL-600	600kcmil	1/2	2 3/4	Green	TT	6 11/16	3 1/8	1 23/32
CCBL-750	750 kcmil	1/2	3 3/8	Black	VV	7 9/32	3 1/8	1 29/32
CCBL-1000	1000 kcmil	1/2	3 3/8	White	—	7 9/16	3 1/8	2 3/16

One-Hole Long-Barrel Copper Compression Terminals: They are UL 486A listed, constructed of bright electro tin plated pure electrolytic copper. The lugs are for use only on copper conductors. They are color coded to indicate wire and compression die size and UL listed at 600 V but acceptable for use up through 35,000 V if installed in accordance with Section 310-6 of the NEC. Sizes are as follows:

SPECIFICATIONS FOR ONE-HOLE, LONG-BARREL COPPER COMPRESSION TERMINALS								
Model Number	Wire Size	Stud Size	Strip Length	Die Color	Die Index Code	Dimensions (Inches)		
						Length	Tang Length	Tang Width
CCB-8	8 AWG	1/4	1 1/8	Red	BB	1 15/16	11/16	31/64
CCB-6	6 AWG	1/4	1 1/8	Blue	CC	1 15/16	11/16	1/2
CCB-4	4 AWG	1/4	1 7/16	Gray	DD	2 1/4	3/4	33/64
CCB-2	2 AWG	5/16	1 7/16	Brown	EE	2 5/16	13/16	5/8
CCB-1	1 AWG	5/16	1 3/4	Green	FF	2 5/8	13/16	43/64
CCB-1/0	1/0 AWG	5/16	1 3/4	Pink	GG	2 7/8	1 5/8	47/64
CCB-2/0	2/0 AWG	1/2	1 3/4	Black	HH	2 7/8	1 5/8	47/64
CCB-3/0	3/0 AWG	1/2	2 1/4	Orange	JJ	3 1/2	1 5/8	15/16
CCB-4/0	4/0 kcmil	1/2	2 1/4	Purple	KK	3 17/32	1 5/8	1 5/8
CCB-250	250 kcmil	1/2	2 1/4	Yellow	LL	3 9/16	1 5/8	1 1/8
CCB-300	300 kcmil	1/2	2 1/4	White	NN	3 5/8	1 1/8	1 1/4
CCB-350	350 kcmil	1/2	2 1/4	Red	OO	3 3/8	1 1/8	1 5/16
CCB-400	400 kcmil	1/2	2 1/4	Blue	RR	3 7/8	1 1/4	1 13/32
CCB-500	500 kcmil	1/2	2 1/4	Brown	SS	4 7/16	1 1/4	1 9/16
CCB-600	600 kcmil	1/2	2 3/4	Green	TT	5 11/16	1 7/8	1 23/32

Long-barrel copper compression splicer

Long-Barrel Copper Compression Splicers for Copper Conductors: The splicers are UL 486A listed and constructed of bright electro tin plated pure electrolytic copper and suitable for use only on copper conductors. Wire stop in center of barrel aids in proper placement of conductors in the barrel. Models available are as follows:

Model Number	Wir Size	Strip Length	Die Color Code	Die Index Code	Dimensions	
					Overall Length	Outside Diameter
C5L-8	8 AWG	$1^1/_{16}$	Red	BB	2	$8/_{32}$
CSL-6	6 AWG	$1^1/_8$	Blue	CC	$2^3/_8$	$19/_{64}$
CSL-4	4 AWG	$1^1/_8$	Gray	DD	$2^3/_8$	$11/_{32}$
CSL-2	2 AWG	$1^1/_2$	Brown	EE	$2^5/_8$	$27/_{64}$
CSL-1	1 AWG	$1^1/_2$	Green	FF	$2^7/_8$	$15/_{32}$
CSL-1/0	1/0 AWG	$1^9/_{16}$	Pink	GG	$2^7/_8$	$33/_{64}$
CSL-2/0	2/0 AWG	$1^3/_4$	Black	HH	$3^1/_8$	$37/_{64}$
C5L-3/0	3/0 AWG	$1^3/_4$	Orange	JJ	$3^1/_8$	$41/_{64}$
CSL-4/0	4/0 AWG	$1^7/_8$	Purple	KK	$3^3/_8$	$45/_{64}$
CSL-250	250 kcmil	$1^7/_8$	Yellow	LL	$3^3/_8$	$3/_4$
CSL-300	300 kcmil	$2^1/_4$	White	NN	$4^1/_8$	$13/_{16}$
CSL-350	350 kcmil	$2^1/_4$	Red	OO	$4^1/_8$	$7/_8$
CSL-400	400 kcmil	$2^1/_2$	Blue	RR	$4^3/_8$	$31/_{32}$
CSL-500	500 kcmil	$2^1/_2$	Brown	SS	$4^5/_8$	$1^1/_{16}$
CSL-600	600 kcmil	3	Green	TT	$5^3/_8$	$1^3/_{16}$
CSL-750	750 kcmil	$3^3/_8$	Black	VV	$5^7/_8$	$1^{19}/_{64}$
CSL-1000	1000 kcmil	$3^1/_2$	White	—	$6^1/_8$	$33/_{64}$

Two-Hole Short-Barrel Copper Compression Terminal for Copper Conductors: They are constructed of bright electro tin plated pure electrolytic copper and UL 486A listed the terminal is only suitable for use with copper conductors. The lug is color coded to indicate wire and compression die size. Wire stop at end of barrel prevents extension of wire beyond end of barrel.

Short-Barrel Copper Compression Splicers for Copper Conductors: The splicers are UL 486A listed for use with copper conductors and constructed of bright electro tin plated pure electrolytic copper. Indentation at center of length of barrel acts as wire stop to prevent excess length of wire being inserted and help locate correct compression point. The splicer is UL listed at 600 V and is acceptable for use up through 35,000 V if installed in accordance with Section 310-6 of the NEC. Specifications follow:

					Dimensions (Inches)	
Model Number	Wire Size	Strip Length	Die Color Code	Die Index Code	Overall Length	Outside Diameter
CCS-8	8 AWG	7/8	Red	BB	1 5/8	1/4
CCS-6	6 AWG	1	Blue	CC	1 3/4	19/64
CCS-4	4 AWG	1	Gray	DD	1 3/4	11/32
CCS-2	2 AWG	1 1/8	Brown	EE	1 7/8	27/64
CC5-1	1 AWG	1 1/8	Green	FF	1 7/8	15/32
CCS-1/0	1/0 AWG	1 3/16	Pink	GG	1 7/8	33/64
CCS-2/0	2/0 AWG	1 3/16	Black	HH	2	37/64
CCS-3/0	3/0 AWG	1 1/4	Orange	JJ	2 1/8	41/64
CCS-4/0	4/0 AWG	1 1/4	Purple	KK	2 1/8	45/64
CCS-250	250 kcmil	1 3/8	Yellow	LL	2 1/4	3/4
CCS-300	300 kcmil	1 3/8 m	White	NN	2 1/4	13/16
CCS-350	350 kcmil	1 7/16	Red	OO	2 3/8	7/8
CCS-400	400 kcmil	1 1/2	Blue	RR	2 1/2	31/32
CCS-500	500 kcmil	1 3/4	Brown	SS	2 7/8	1 5/8
CCS-600	600 kcmil	1 13/16	Green	TT	3	1 13/16

SHORT-BARREL COPPER COMPRESSION SPLICERS

SHORT-BARREL COPPER COMPRESSION SPLICERS

Model Number	Wire Size	Strip Length	Die Color Code	Die Index Code	Dimensions (Inches)	
					Overall Length	Outside Diameter
CCS-750	750 kcmil	2	Black	VV	3 3/8	1 19/64
CCS-1000	1000 kcmil	2 1/4	White	—	3 7/8	1 33/64

One-Hole Short-Barrel Copper Compression Terminals for Copper Conductors: They are built of bright electro tin plated pure electrolytic copper and are UL 486A listed for use only on copper conductors. Wire stop at inside end of barrel prevents over insertion of wire. Color coding indicates wire size and correct compression dye for secure connection. Listed UL at 600 V, the terminal may be used up through 35,000 V if installed in accordance with Section 310-6 of the NEC.

Chapter 23

Introduction to Wiring Devices

The term "wiring device" may be defined as a part of the electrical system used to stop, start, transmit or alter the form of transmission of electrical energy but not to use it. There are some electrically-operated devices that use small amounts of electricity but the amount of energy used to operate them is incidental, considering the energy that is transmitted.

The National Electrical Manufacturers Association (NEMA) has established rating classifications to define the applications for which a device is environmentally suited.

NEMA-3

NEMA-3 devices are suitable for use indoors or outdoors to protect against windblown water and dust. They are raintight, sleet-resistant and dusttight.

NEMA-4

NEMA-4 devices are suitable for use indoors or outdoors to protect against water applied during cleaning operations, using pressurized water streams. They are watertight and dusttight.

NEMA-4X

NEMA-X devices are suitable for use indoors or outdoors to protect against water applied during cleaning operations, using pressurized water streams, or when conditions result in severe condensation. They are sleet-resistant, watertight, dusttight and corrosion-resistant.

NEMA-7 (A, B, C, or D)

NEMA-7 devices are suitable for use in hazardous locations, for Class I air-brake explosion-proof applications.

NEMA-9 (E, F, or G)

NEMA-9 devices are suitable for use in hazardous locations, Class II dust- and ignition-proof applications.

NEMA-12

NEMA-12 devices are suitable for use indoors. They are designed to protect against lint, dust, dirt, fibers and light splashing or dripping of non-corrosive fluids.

The application conditions vary somewhat on NEMA-7 and NEMA 9 devices, in that most of the previously classified devices have been designed primarily to protect life and property from electrocution or electrical fires. The case with NEMA-7 and NEMA-9 is that they are designed to be installed and operated in hazardous atmospheres and protect an explosive environment from the possibility of an explosion due to electric arc, sparks or overheating. Existing environments and their characteristics are classified as Class I, II, or III.

Class I

Class I environments are those in which gases or vapors are, or could be, present in the air in levels sufficient to produce and explosion or fire. Class I is further divided into divisions.

Class I, Division 1

Class I, Division 1 environments are those in which the hazardous conditions exist continuously, intermittently, or periodically during normal operating conditions. The term normal "operating" is meant to include periodic maintenance, leakage and malfunction of equipment which could cause hazardous concentrations of explosive vapors to be released.

Class I, Division 2

Class I, Division 2 environments are those in which hazardous fluids or gases are processed, used or stored, but confined in closed systems or containers. Only in the case of accidental rupture of containers or major malfunction of the handling equipment could hazardous conditions exist. Environments which would normally be considered as Class I, Division 1 may in some instances be considered by the local authorities having jurisdiction under the National Electrical Code as Class I, Division 2, if adequate positive mechanical ventilation is present provided, of course, that adequate safeguards such as standby ventilation systems, are used to back up the main ventilation system, should it fail.

Class II

Class II environments are those in which hazardous dust is or could be present in the air in levels sufficient to produce and explosion or fire. Class II is further divided into divisions.

Class II, Division 1

Class II, Division 1 environments are those in which the hazardous conditions exist continuously, intermittently or periodically during normal operating conditions. The term "normal operating" is meant to include periodic maintenance, leakage and malfunction of equipment which could cause hazardous concentrations of explosive or electrically-conductive dust to be released.

Class II, Division 2

Class II, Division 2 environments are those in which hazardous dust is not normally present in the air, but where accumulations of dust may be present in sufficient quantities to interfere with heat dissipation from electrical apparatus, thereby causing a hazard or where electrical arcs are normally associated with such devices as magnetic starters or switches could cause fire.

Class III

Class III environments are those in which ignitable fibers or materials are present in quantities sufficient to present a hazard. Class II is further divided into divisions.

Class III, Division 1

Class III, Division 1 environments are those in which ignitable fibers or materials are manufactured, handled or stored. Such environments usually exist in the process or handling related to cotton gins and mills, flax plants, rayon and textile plants.

Class III, Division 2

Class III, Division 2 environments are those in which ignitable fibers or materials are handled or stored but not manufactured.

Temperature Characteristics

Some gases have an extremely low combustion point and because of this, devices and enclosures used in hazardous environments must have operating temperatures below this flash or combustion point. Accordingly, electrical apparatus designed for use in hazardous areas are rated and identified as to their operating temperatures under full load. These rating vary from the T-1 rating, which is 45 degrees C to the T-6 rating with an 85 degrees C operating temperature based upon a 40 degree C ambient temperature. Chapter 5 of the National Electrical Code should be consulted for detailed information concerning hazardous locations and temperature characteristics.

Chapter 24

Plugs and Receptacles

A receptacle is a device usually installed in an outlet box and used to provide a point of attachment for a portable cord, thus allowing portable equipment to be connected to a power source. The plug is a device installed on the portable cord to enable connection between the cord and receptacle. Receptacles are equipped with a number of slots which receive the male blades of the plug. The number of slots, blades and their configuration depends upon the number of phases used, the available voltage, and the amperage.

There are two main categories into which receptacles and plugs are divided:

- Nonlocking
- Locking

The nonlocking configuration uses straight blades and allows the plug to be inserted straight into the receptacle. The locking configuration is commonly referred to as "twist lock." Twist-lock receptacles are connected by inserting the plug into the receptacle and then rotating the plug to complete the connection.

The twist-lock receptacle is considered the safer of the two because the electrical connection is not completed until the plug is rotated in the receptacle. This eliminates the possibility of electrical shock which is present with the straight-blade configuration if the plug has not been inserted completely into the receptacle; that is, should fingers come in contact with the bare energized blades of the straight-blade type, the person will receive a shock. In addition, the twist-lock plug cannot be removed from the receptacle without first rotating the plug. Consequently, the possibility of an accidental disconnection is eliminated when using twist-lock receptacles.

The National Electrical Manufacturing Association (NEMA) has established certain requirements and configurations for receptacles and plugs. For example, the NEMA configuration for a residential nonlocking 120-V, 15 A receptacle is 5-15-R.

- 5 = NEMA configuration
- 15 = rating in amperes
- R = receptacle

A plug to match this receptacle would be designated 5-15-P.

NEMA Configurations for Nonlocking Plugs and Receptacles

Voltage	15 A	20 A	30 A	50 A	60 A
TWO-POLE, TWO-WIRE RECEPTACLES/PLUGS					
125 V	1-15R-P	—	—	—	—
250 V	2-15R-P	2-20R-P	2-30R-P	—	—
TWO-POLE, THREE-WIRE GROUNDING-TYPE RECEPTACLES/PLUGS					
125 V	5-15R-P	5-20R-P	5-30R-P	5-50R-P	—
250 V	6-15R-P	6-20R-P	6-30R-P	6-50R-P	—
277 V	7-15R-P	7-20R-P	7-30R-P	7-50R-P	—
THREE-POLE, THREE-WIRE RECEPTACLES/PLUGS					
125/250 V	14-15R-P	14-20R-P	14-30R-P	14-50R-P	14-60R-P
250 V, 3Ø	15-15R-P	15-20R-P	15-30R-P	15-50R-P	15-60R-P

FOUR-POLE, FOUR-WIRE RECEPTACLES/PLUGS

Voltage	15 A	20 A	30 A	50 A	60 A
120/208 V, 3Ø	18-15R-P	18-20R-P	18-30R-P	18-50R-P	18-60R-P

NEMA Configurations for Locking Plugs and Receptacles

TWO-POLE, TWO-WIRE LOCKING RECEPTACLES/PLUGS

Voltage	15 A	20 A	30 A	50 A	60 A
125 V	L1-15R-P	—	—	—	—
250 V	—	L2-20R-P	—	—	—

TWO-POLE, THREE-WIRE, GROUNDING-TYPE LOCKING RECEPTACLES/PLUGS

Voltage	15 A	20 A	30 A	50 A	60 A
125 V	L5-15R-P	L5-20R-P	L5-30R-P	—	—
250 V	L6-15R-P	L6-20R-P	L6-30R-P	—	—
277 V	L7-15R-P	L7-20R-P	L7-30R-P	—	—
480 V	—	L8-20R-P	L8-30R-P	—	—
600 V	—	L9-20R-P	L9-30R-P	—	—

THREE-POLE, THREE-WIRE LOCKING RECEPTACLES/PLUGS

Voltage	15 A	20 A	30 A	50 A	60 A
125/250 V	—	L10-20R-P	L10-30R-P	—	—
250 V, 3Ø	L11-15R-P	L11-20R-P	L11-30R-P	—	—
480 V, 3Ø	—	L12-20R-P	L12-30R-P	—	—
600 V, 3Ø	—	—	L13-30R-P	—	—

THREE-POLE, FOUR-WIRE, GROUNDING-TYPE LOCKING RECEPTACLES/PLUGS

Voltage	15 A	20 A	30 A	50 A	60 A
125/250 V	—	L14-20R-P	L14-20R-P	—	—

THREE-POLE, FOUR-WIRE, GROUNDING-TYPE LOCKING RECEPTACLES/PLUGS (Cont.)					
Voltage	15 A	20 A	30 A	50 A	60 A
250 V, 3Ø	—	L10-20R-P	L10-30R-P	—	—
480 V, 3Ø	—	L11-20R-P	L11-30R-P	—	—
680 V, 3Ø	—	L22-20R-P	L12-30R-P	—	—
FOUR-POLE, FOUR-WIRE, GROUNDING-TYPE LOCKING RECEPTACLES/PLUGS					
120/208 V, 3Ø	—	L18-20R-P	L18-30R-P	—	—
277/480 V, 3Ø	—	L19-20R-P	L19-30R-P	—	—
347/600 V, 3Ø	—	L20-20R-P	L20-30R-P	—	—
FOUR-POLE, FIVE-WIRE, GROUNDING-TYPE LOCKING RECEPTACLES/PLUGS					
120/208 V, 3Ø	—	L21-20R-P	L21-30R-P	—	—
277/480 V, 3Ø	—	L22-20R-P	L22-30R-P	—	—
347/600 V, 3Ø	—	L23-20R-P	L23-30R-P	—	—

Hospital Grade Devices

Underwriters Laboratories (UL) has established strict standards and tests to apply to "Hospital Grade" wiring devices.

Receptacles

A standard plug attached to a test weight is fully inserted in the receptacle. The weight is dropped from a minimum height of 12 in, causing severe and abrupt removal. The test is repeated 8 times. The abrupt removals are made in all directions parallel to the plug face. After abuse,

Plugs and Receptacles

		15 ampere		20 ampere		30 ampere	
		Receptacle	Plug cap	Receptacle	Plug cap	Receptacle	Plug cap
2 - pole 2 - wire	1 125 V	1-15R	1-15P				
	2 250 V		2-15P	2-20R	2-20P	2-30R	2-30P
2 - pole 3 - wire grounding	5 125 V	5-15R	5-15P	5-20R	5-20P	5-30R	5-30P
	6 250 V	6-15R	6-15P	6-20R	6-30P	6-30R	6-30P
3 - pole 3 - wire	7 277 V	7-15R	7-15P	7-20R	7-30P	7-30R	7-30P
	10 125/ 250 V			10-20R	10-20P	10-30R	10-30P
	11 3φ Δ 250 V	11-15R	11-15P	11-20R	11-20P	11-30R	11-30P
3 - pole 4 - wire grounding	14 125/ 250 V	14-15R	14-15P	14-20R	14-20P	14-30R	14-30P
	15 3φ Δ 250 V	15-15R	15-15P	15-20R	15-20P	15-30R	15-30P
4 - pole 4 - wire	18 3φ Y 120/ 208 V	18-15R	18-15P	18-20R	18-20P	18-30R	18-30P

NEMA configurations for general-purpose nonlocking receptacles and plug caps

the ground contact shall retain a ground pin weighing four ounces inserted in the receptacle, positioned face down.

There also shall be no breakage of the receptacle that interferes with the receptacle function or the integrity of the enclosure as a result of the abrupt removals of the plug in any direction.

Ground Pin Retention: Using a hardened steel pin, .204 in diameter, the grounding contact shall be abused by 20 insertions and withdrawals. After abuse, a .184 in diameter pin weighing 4 oz is to be inserted and retained by the receptacle without any displacement.

Assembly Security: The receptacle, while supported by the screw-mounted yoke, shall have a force of 100 lbs applied to the base through the power slots. "There shall be no indication of any cracking of these bodies nor permanent deformation of the yoke."

Receptacles must also pass tests for terminal strength, impact, grounding contact temperature, resistance, fault current, mold stress relief, as well as requirements of UL 498.

Plugs and Connectors: The cord clamp alone shall limit the displacement of conductors to $1/32$ in or less following these tests:

1. Static cord pull test consisting of a gradually-applied straight pull of 30 lbs force for a period of one minute.

2. Rotary cord pull test consisting of a 10-lb steady pull for 2 hours.

NEMA configurations for general-purpose nonlocking receptacles and plug caps

3. Abrupt removal testing. There shall be no discontinuity in the cord insulation (cuts, rips or tears) as a result of this test.

Crush Test: Wired plugs shall be placed between horizontal steel plates and a crushing force applied, increasing gradually, to a value of 500 lbs. "There shall be no breakage, deformation, or other effect that may interfere with the function of the device."

Plugs must also pass test for impact, mechanical drop, mold stress relief, and UL 498. Connectors must also pass tests for power blade retention, ground pin retention, impact and UL 498.

Isolated Grounding Circuits

The use of delicate electronic instruments and devices has grown tremendously over the past decade. Consequently, it has become necessary to change established methods of grounding. This change was made to isolate the transient signals that can be created by other noise-producing sources which may disrupt the operation of sensitive electronic equipment.

The conventional receptacle is grounded to the building grounding system at the receptacle. When mounted in the box, grounding contacts are connected to the box, fittings, conduit and all other building ground system components. This method can act as a large antenna for electromagnetic interference (EMI) and radio frequency interference (RFI).

With the isolated ground receptacle, the grounding contacts are bonded directly to the service entrance grounding system. An insulating barrier isolates the grounding contacts from the mounting strap. The insulated grounding conductor is bonded to the service entrance grounding system per NEC Article 250-74, Exception No. 4.

Devices with isolated ground are identified in this book with the prefix IG in the catalog number.

Corrosion Resistant Devices

Corrosive environments are common in many commercial, industrial and marine applications creating a demand for wiring devices which will resist the destructive effects of corrosion. Special design features enable these devices to better withstand the destructive atmospheres of plating and chemical facilities as well as the salt atmospheres in marina and shipboard applications. Devices with these design features are also well suited for areas exposed to moisture like food processing facilities, dairies and laundries.

This added protection from adverse conditions involving acids, alkalies, grease, oil and solvents, as well as chemicals, salt, and excessive moisture is obtained through the use of carefully selected and tested materials and/or plating.

Depending upon the intended end use, materials such as chrome plated brass, fiberglass, stainless steel and various plastics such as neoprene, Lexan or melamine may be used to provide the protection required.

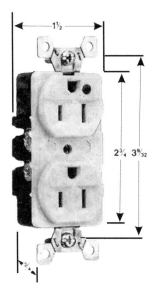

Hospital grade duplex receptacle

Hospital Grade Duplex Receptacles – 2-Pole, 3-Wire Grounding, 15 A, 125V: The Arrow-Hart #8200 has heavy-duty, virtually indestructible nylon face and rigid, glass reinforced back body. Wide frame construction shrouds terminations and adds strength. One piece nickel plated brass mounting strap/grounding system provides reliable low resistance grounding. A nickel-plated, wraparound strap interlocks the cover at four points.

An automatic grounding system allows easy installation. Receptacle is UL listed to Federal Specification WC596F, CSA certified, and available in brown or ivory colors.

Extra Heavy-Duty Duplex Receptacle – Back and Side Wired: The Arrow-Hart # 5262 is constructed of a heavy-duty, virtually indestructible nylon face and rigid glass reinforced nylon back body. Wide frame design shrouds terminations and adds strength. A one-piece heavy-gauge brass mounting strap/grounding system provides reliable low-resistance grounding. The shallow design allows more room in the box for easier wiring. A wraparound strap interlocks with cover at four points. All terminal, ground and mounting screws are combination Phillips/straight-slot head. Heavy-duty, high-grade brass-alloy, triple-wipe contacts. There are eight back and side conductor slots (holes) that accept up to No. 10 AWG conductors. This receptacle type is UL listed to Federal Specification WC596-F and CSA certified. Available colors include brown, ivory, yellow and gray.

Tamper-Resistant Duplex Receptacle – Back and Side Wired: The Arrow-Hart #TR82 receptacle has a unique locking shutter mechanism which will not open if an object is inserted into only one blade opening. Two or three contact plugs release the shutter, permitting full insertion of the plug. These receptacles satisfy NEC Section 517-18C; that is, requirements for hospital pediatric care areas. The receptacle cover is constructed of heavy-duty nylon for durability. The shutter mechanism is high-strength thermoplastic. Other specifications are the same as the receptacles listed above.

Heavy-Duty Back- and Side-Wired Duplex Receptacle: The Arrow-Hart #5252 receptacle has a heavy-duty nylon face and a rigid, glass-reinforced nylon back. The one-piece, heavy-gauge, brass grounding shunt provides reliable low-resistance grounding. The shallow design provides more room in the box for easier wiring. It also has a plated steel wraparound strap that interlocks with the cover at four points for assembly integrity. The terminal, grounding, and mounting screws are combination Phillips/straight-slot heads. Contacts are heavy-duty, high-grade brass alloy of the triple-wipe type. Besides the conventional colors, black, gray, white, or red are available on special order.

Heavy-Duty, Side-Wired Duplex Receptacle: The Arrow-Hart #5242 has the same general specifications as other hospital-grade receptacles except it is designed for side wiring, using the integral terminal screws.

Isolated Grounding Duplex Receptacle: The Arrow-Hart #IG5262 duplex receptacle has a total of eight wire terminations — four on the side and four on the back — for feed-through wiring of No. 14 – 10 AWG solid or stranded copper conductors. The isolated grounding terminals are insulated from the mounting strap to prevent transmission of transient currents, often associated with the normal grounding circuit in a building; that is, conduit, metallic boxes, raceways or fittings. The automatic grounding system of this receptacle type eliminates the need for a bonding jumper as per NEC Section 250-74. This receptacle type is UL verified to Federal Specification WC595-F and is available in orange, gray, or ivory. It is also available in hospital grade #IG5262HG.

Combination Rating, NEMA 5-15R and 6-15R Receptacle: The Arrow-Hart #5292 receptacle will accept one 125 V, 15 A circuit and one 250 V, 15 A circuit. This receptacle type is the binding-screw type for back and side wiring. It is also available in the isolated-grounding type #5292IG with an orange face.

Tamper-Resistant, Side-Wired Receptacle: The Arrow-Hart #6352 is designed to accept 2- and 3-wire grounding plugs rated at 125 V, 15 A. Power tools and other portable equipment with grounding plugs are safely grounded for maximum protection. It shunts current harmlessly around objects, such as screwdrivers, nails,etc. when inserted in slots by children and others. Current flows to paired slots only when both blades of a suitable plug are in position. The moment that the plug blades are withdrawn, internal connections with the supply line are broken, preventing energizing of foreign objects. This receptacle type has a break-off fin for two-circuit use and has silver-plated contacts, locked-in mounting strap, and a safety ground clip provided for fast installation. This receptacle type is available in the standard brown and ivory colors, and on special order, gray, black, and white.

Single Receptacle, Back- and Side-Wired: The Arrow-Hart #8210 series of receptacles are of hospital grade and are made of heavy-duty nylon bodies and faces with an automatic grounding clip. The wraparound mounting strap is locked into the base. This receptacle type meets NEMA WD1-1983 heavy-duty requirements and is UL verified to Federal specificaitons WC596-F. Contacts and mounting straps are corrosion resistant nickel plated.

Single back- and side-wired receptacle

This type of receptacle is also available as an isolated ground model #IG8210. Brown and ivory colors are standard, with orange the color of the isolated ground model.

The ivory standard model is available with a lighted face; this type is designated #8210L.

Single Receptacle, Side-Wired: The Arrow-Hart # 5251 is designed for side wiring only, constructed with heavy-duty nylon body and face, has an automatic grounding clip and wraparound mounting strap locked into place. It meets NEMA W01-1983 heavy duty requirements and is UL verified to Federal Specification WC596-F. It is also available as model IG5251 isolated ground model and with standard or short mounting straps. Standard model colors are brown or ivory, whereas the isolated ground model color is orange. Standard mounting strap holes are $3^9/_{32}$ in on center and the short strap mounting holes are $1^{15}/_{16}$ in on center. Extra-short mounting strap model #5258 is available with mounting holes $1^3/_4$ in on center and model 10118 extra short mounting strap model with .250 × .032 spade terminals.

Single side-wired receptacle

Arrow-Hart #6266HG, #6265, #6265HG Dead-Front Plugs: These 125 V, 15 A nylon/Lexan plugs are back wired and of dead-front construction. They provide for fast wiring and comply with NEC dead-front requirements. The nylon body is designed for high abuse areas, suitable for 105 degrees C/221 degrees F. Individual wire pockets contain stranded wire and dual range cord grip

Dead-front plug

Dead-front angle plug

secures cord without crushing. The are UL verified to Federal Specification WC596-F, meets NEMA WD1-1983 requirements. All models are gray in color. The #6265 plugs are angle style.

Arrow-Hart 125 V, 15 A #5266N, 5266NHG, 5266NCR, Dead-Front Plugs: Lexan body is designed for long life in high abuse areas, and suitable for 105 degrees C/221 degrees F continuous use. The automatic cord grip clamps cord when assembly screws are tightened. The back-wired construction provides for fast-wiring and complies with "dead-front" NEC requirements on plugs. Tapered cover design eliminates snagging on objects. One-piece grounding contacts provide superior continuity while individual wire pockets contain stranded wire. A neoprene gasket seals cord hole opening and prevents dirt, moisture, and metal chips from entering terminal opening. Meets NEMA WL1-1983 requirements. UL verified to Federal Specification WC596 F. The #5266N is available in gray, white, yellow and black. The 5266NHG (hospital grade) is only available in gray. The # 5266NCR (corrosion resistant) is available in yellow. Plugs are also manufactured in an isolated ground model both in the standard grade and the hospital grade as #IG5266N and IG5266-HG.

Arrow-Hart # 5965V, 5965VCR2, IG59659, 115V Dead Front: The twist-on cover eliminates assembly screws. Nylon backwired dead front construction meets NEMA WD1-1983 requirements and is UL listed, CSA certified. It has easy-to-wire clamp-type terminals, with all terminal screws located on same side for fast wiring. Nylon body is suitable for 65 degrees C / 140 degrees F continuous use. The # 5965V is available in gray and yellow, while the #5965VCR (corrosion resistant) is available in yellow. The isolated ground model # IG5965V is available only in orange. The hospital grade plug #8115V is made in gray or white.

Safety-Grip connector

Arrow-Hart 125 V, 15 A #6269, 6269HG, 6269HGC Connectors: The nylon/Lexon body is designed for high abuse areas suitable for 105 degrees C /221 degrees F continuous use. The dual-range cord grip secures cord without crushing. Neoprene cord hole seal prevents dirt, moisture and metal chips from entering terminal area. Meets NEMA WD1-1983 requirements and UL verified to Federal Specification WC569-F. The #6269 and #6269HG (hospital grade) are available in gray and # 6269HGC (hospital grade clear) has a clear cover.

NEMA 6-15R 250 V duplex receptacle

Arrow-Hart #8400 Hospital Grade, Back- and Side-Wired, 2-Pole, 3-Wire Grounding 15 A, 250 V Duplex Receptacle: This receptacle type is constructed of a glass-reinforced nylon back body with a heavy-duty nylon face. The wide frame construction shrouds terminations and adds strength to the receptacle body. The one-piece nickel-plated brass mounting strap offers reliable low-resistance grounding. The strap further interlocks with the cover at four points for assembly integrity. It is available in brown and ivory.

Arrow-Hart #5662 Back- and Side-Wired, 2-Pole, 3-Wire Grounding 15 A, 250 V Duplex Receptacle: This receptacle type is constructed of a glass-reinforced nylon back body with a heavy-duty nylon face. The wide frame construction shrouds terminations and adds strength to the receptacle body. The one-piece nickel-plated brass mounting strap offers reliable low-resistance grounding. The strap further interlocks with the cover at four points for assembly integrity. All terminal, ground, and mounting screws are combination Phillips/straight-slot heads. They have heavy-duty, high-grade brass alloy triple-wipe contacts. Automatic grounding system aids in easy installataion. UL listed to Federal specification WC596-F; also CSA certified. Available in brown and ivory. The Model 3IG5662 (isolated ground) is only available in orange. Compact designs (#5662C and #5662CI) are available in brown and ivory.

Arrow-Hart #5652 and 5652I Back- and Side-Wired, 15 A, 250 V Heavy-Duty Duplex Receptacle: This receptacle type is constructed with a heavy-duty nylon face and a rigid, glass-reinforced nylon back body. The one-piece, heavy-gauge brass grounding shunt is of shallow design for easy wiring. The strap further interlocks with the cover at four points for assembly integrity. All terminal, ground, and mounting screws are combination Phillips/straight-slot heads. It has heavy-duty, high-grade brass alloy triple-wipe contacts. Automatic grounding system aids in easy installataion. UL listed to Federal specification WC596-F; also CSA certified. It is available in brown and ivory. The Model 3IG5662 (isolated ground) is only available in orange. Compact designs (#5662C and #5662CI) are available in brown and ivory.

Plugs and Receptacles

NEMA 5-15P inlet

NEMA 5-15R outlet

NEMA 6-15R, 250-V, 15-A duplex receptacle

Arrow-Hart #5278C and #5279C Flanged Inlet and Outlet Devices: The NEMA 5-15R 2-pole grounding receptacle and the NEMA 5-15P plug are constructed of gray nylon/Lexan suitable for 105-degree C continuous temperatures. The mounting holes on this receptacle type are interchangeable with other designs. The back-wired, clamp-type terminals are easy to wire.

Arrow-Hart #5278 and #5289 Metal-Shell Flanged Inlet and Outlet Devices: The metal case and flange provide rough-duty protection and locking capability. The #5378 inlet is back-wired and the #5289 outlet is also equipped with a cord grip suitable for cord diameters from .296 in to .625 in. This outlet type is also available in a side-wired style, designated #5246.

Metal-shell flanged inlet (top) and outlet (bottom)

Arrow-Hart #8400 Hospital Grade Duplex Receptacle, 240V – 15A, NEMA 6-15R: This receptacle type is constructed of a heavy-duty nylon face and glass-reinforced nylon back body. The wide frame construction shrouds terminations and adds strength. One-piece, nickel-plated brass mounting strap provides low-resistance grounding. The nickel-plated, wraparound strap interlocks with the cover at four points for assembly integrity. This receptacle type also has heavy-duty, nickel-plated, brass-alloy, triple-wipe contacts and an automatic grounding system. It is UL listed to Federal Specifications WC496F and CSA certified. The standard colors are brown and ivory with optional colors of black, gray, white, or red on special order. This receptacle may be either back- or side-wired.

Arrow-Hart #8410 Hospital-Grade Single Receptacle: This is a 2-pole, 3-wire, grounding 15 A, 250 V heavy-duty receptacle. It has a nylon body and face which makes it virtually indestructible. The automatic grounding clip and wraparound mounting strap locks into the base. It meets NEMA WD1-1983 heavy-duty requirements and is UL verified to Federal Specification WC596-F. The #8410 series has corrosion-resistant nickel-plated contacts and mounting strap. The standard version is available in brown, ivory, black, gray, white or red. It is available in orange as #IG8410 with isolated ground, with a lighted face in ivory as #8410IL and in short mounting strap style as #5684. The device is back- and side-wired.

Arrow-Hart #5651 Side-Wired Single Receptacle: This heavy-duty receptacle is 2-pole, 3-wire grounding 15 A, 250 V. It has a nylon body and face which makes it almost indestructible. The automatic grounding terminal and wraparound mounting strap locks into the base. It meets NEMA WD1-1983 heavy-duty requirements. It is UL verified to Federal Specification WC596F. Brown and ivory are the available colors. The isolated ground version is available as #IG5651 in orange.

Arrow-Hart #6666, #6669 "Safety Grip" Dead-Front Plugs and Connectors: These devices are 2-pole, 3-wire grounding, 15 A, 250 V. Their nylon/Lexan back-wired dead-front construction is designed for high-abuse areas and suitable for use up to 105 degrees C/221 degrees F. A neoprene gasket seals the cord hole opening. A dual range cord grip design secures the cord without crushing. The one-piece contacts provide superior grounding continuity. The individual wire pockets contain stranded wire and meets NEMA WD1-1983 requirements. They are UL verified to Federal Specification WC596F. The plug also complies with NEC requirements. The connectors and plugs are available in gray. The plug is also available as angle style #6665HG hospital grade and as #6665HGC hospital grade with a clear cover.

Arrow-Hart #5666N, #5669N and #IG5666N "Autogrip" Dead-Front Plugs and Connectors: These devices are 2-pole, 3-wire grounding 15 A, 250 V. The Lexan body is designed for long life in high-abuse areas and is suitable for continuous use up to 221 degrees F. An automatic cord grip clamps the cord when assembly screws are tightened. The back-wired construction provides for fast wiring and complies with "dead front" NEC requirements on plugs. The tapered cover design eliminates snagging on objects. One-piece grounding contacts provide superior continuity. Individual wire pockets contain stranded wire. A neoprene gasket seals the cord hole opening and prevents dirt, moisture, and metal chips from entering the terminal opening. The standard plug and connectors are available in gray and the isolated ground version (#IG5666N) comes in orange. The devices meet NEMA WD1-1983 requirements and are UL verified to Federal Specification WC596-F.

Arrow-Hart #5666V "Quickeze" Dead-Front Plugs: These devices are 2-pole, 3-wire grounding, 15 A, 250 V, nylon body, back-wired dead-front construction. The twist-on cover eliminates assembly screws. The easy wire terminals with all terminal screws are located on the same side for fast wiring. They will accommodate 2.18 in minimum to 2.36 in maximum cord diameter. The nylon body is suitable for 140 degrees F continuous use. It is UL listed CSA certified and meets NEMA WD1-1983 heavy-duty requirement. The only color available is gray.

Arrow-Hart #5666Y, #5669Y "Quickgrip" Dead-Front Plug and Connector: These devices are 2-pole, 3-wire grounding 15 A, 250 V. The one-piece design and quick drive screws provide for fast wiring. The highly visible yellow nylon body offers excellent chemical and impact resistance. It is UL listed and CSA certified.

Arrow-Hart #5678C, #5679C, #5656 Flanged Inlet and Outlet Devices: They are 2-pole, 3-wire grounding 15 A, 250 V. The gray nylon/Lexan body is suitable for 221 degrees F continuous use. The mounting holes are interchangeable with other designs. The terminals are back-wired clamp type. The #5656 outlet has a metal shell. Mounting holes are $3/32$ in × $1/4$ in on $2 3/32$-in diameter circular flange.

Arrow-Hart #5302 Heavy-Duty, Back- and Side-Wired Duplex Receptacle and #5766, #5768, plugs, 277 V, 15 A: The NEMA 7-15P heavy-duty receptacle has a totally-enclosed molded cover and base with a safety ground clip for fast installation. Standard colors are brown and ivory.

The related plugs are constructed of black rubber with cord grips that will handle cords from .325 in to .625 in diameter.

Arrow-Hart #8300 and #5362 Duplex Receptacles: This hospital grade 2-pole, 3-wire grounding receptacle type is rated at 125 V, 20 A. Otherwise, the specifications are the same as the 125-V, 15-A version. Standard colors are brown and ivory. Compact sizes are also available that are designated #6362HG and 5362HGI.

NEMA 5-20R hospital grade duplex receptacle

Arrow-Hart #5362 Series, Extra Heavy-Duty 2-Pole, 3-Wire Grounding Receptacle, 120 V, 20 A: This receptacle type has a glass reinforced nylon back body and a heavy-duty nylon face. The wide-frame construction shrouds terminations and adds strength. The one-piece heavy-gauge brass mounting strap provides for a reliable low-resistance ground. The wrap-around strap also interlocks with the cover at four points. All terminal, ground and mounting screws are combination Phillips/straight-slot heads. Heavy-duty, high-grade brass-alloy triple-wipe contacts are used throughout. The back- and side-wiring contacts accept conductor sizes up to No. 10 AWG. This receptacle type is UL listed to Federal Specification WC-596-F. This model is vailable in both brown or ivory colors. A corrosion-resistant style is available in yellow or gray and a compact model is also available (#5362C). In addition, both the standard and compact designs are available in black, gray, white or red on special order.

Arrow-Hart # TR83 Tamper Resistant, Duplex Receptacles, Hospital Grade, 2-Pole, 3-Wire Grounding 125 V, 20 A: A Unique locking shutter mechanism will not open if an object is inserted into only one blade opening. Two or three contact plugs release the shutter permitting full insertion of the plug. These receptacles satisfy the NEC 517-18C requirements for hospital pediatric care areas. Receptacles have heavy-duty nylon covers for durability and also a high strength thermoplastic shutter mechanism. They are back- and side-wired for flexible installation. UL listed to Federal Specification WC 556-F. Available in brown or white.

Arrow-Hart #5352 Heavy-Duty Duplex Receptacle: This 2-pole, 3-wire grounding receptacle type is rated for 125 V, 20A and has a heavy-duty nylon face and a rigid, glass-reinforced nylon back body. The one-piece, heavy-gauge brass grounding shunt and the plated steel wraparound strap interlocks with cover at four points. Terminal, grounding and mounting screws are combination Phillips/straight-slot heads. It has heavy-duty, high-grade brass alloy triple-wipe contacts. The back- and side-wiring terminals accept conductors up to #10 AWG. Listed to Federal Specification WC596-F. Available in brown or ivory.

Arrow-Hart #5342 Heavy-Duty, Side-Wired Duplex Receptacle: 2-Pole, 3-Wire Grounding type rated at 125 V, 20 A. Same specifications as above except it has a shallow design for easy wiring. UL listed Federal Specification 596-5.

Arrow-Hart #IG5362HG Isolated Ground Hosital Grade Duplex Receptacle: This 2-pole, 3-wire grounding receptacle type is rated at 125 V, 20 A and has an automatic grounding system provided for fast installation. The back- and side-wiring accepts #14 to 10 AWG. They have a heavy-duty nylon face and reinforced nylon back body. Available in orange or ivory.

Arrow-Hart #5492 Combination Rating NEMA 5-20R and 6-20 Duplex Receptacle: Construced of heavy-duty nylon for abusive areas with an automatic grounding system for fast installation. Back- and side-wired construction. Available in brown, ivory, gray, white or red.

Arrow-Hart #8310, #IG8310, and #8310M Single Receptacle: This 2-pole, 3-wire grounding receptacle type is rated at 120 V, 20 A and has a heavy-duty strong nylon body and face. Automatic grounding clip. Wraparound mounting strap locked into base. Meets NEMA WD1-1983 heavy-duty requirements UL listed to Federal Specificatio WC596-F. Corrosion resistant nickel plated contacts and mounting straps on the 8310 series. The IG represents isolated ground model. The 8310M model has a short mounting strap. Standard colors are brown or ivory; the isolated ground model is orange or ivory. Back- and side-wired.

Arrow-Hart #5362 Extra-Heavy Duty Duplex Receptacles: These receptacles are 2-pole, 3-wire grounding, 20 A, 125 V. They have a rigid, glass reinforced nylon back body and heavy-duty virtually indestructible nylon face. The wide frame construction shrouds terminations and adds strength. They have a one-piece heavy-gauge brass mounting strap/grounding system for reliable low resistance grounding. The wraparound strap interlocks with the cover at four points. All terminal, ground and mounting screws are combination Phillips/straight-slot head. The receptacles have heavy-duty, high-grade brass alloy triple wipe contacts. The 8-hole back and side wiring accepts up to No. 10 AWG. They are UL listed to Federal Spec WC-596-F. They are available in brown or ivory standard and a corrosion resistant style is available in yellow and gray. A compact model is available as #53262C in brown and ivory. Both the standard and compact design are also available in black, gray, white or red on order.

Arrow-Hart #TR83 Duplex Receptacles: These receptacles are tamper resistant, hospital grade, 2-pole, 3-wire grounding 20 A, 125 V. They have a unique locking shutter mechanism that will not open if an object is inserted into only one blade opening. The two or three contact plugs release the shutter permitting full insertion of the plug. These receptacles satisfy the NEC 517-18(c) requirements for hospital pediatric care areas. The receptacles have a heavy-duty nylon cover for durability and high-strength thermoplastic shutter mechanism. The 8-hole back and side wiring is for flexible installation. It is UL listed to Federal Specification WC-596-F and available in brown or ivory.

Arrow-Hart #5352 Duplex Heavy-Duty Receptacles: These are back- and side-wired, 20A, 125 V, 2-pole, 3-wire grounding receptacles. The nylon face is practically indestructible because of the heavy-duty construction. They also have a rigid, glass reinforced nylon back body and a one-piece, heavy-gauge brass grounding shunt. A plated steel wraparound strap interlocks with the cover at four points. The terminal, grounding and mounting screws are combination Phillips/straight-slot head. They have heavy duty, high-grade brass alloy triple wipe contacts. The 8-hole back and side wiring accepts up to No. 10 AWG. They are available in brown or ivory and listed to Federal Spec. WC-596-F.

Arrow-Hart #5342 Heavy-Duty, Side-Wired Duplex Receptacle: These receptacles are 2-pole, 3-wire grounding, 20 A, 125 V with a heavy, rough-duty nylon face and rigid, glass reinforced nylon back body. They have a one-piece, heavy-gauge brass grounding shunt. The shallow design permits easy wiring. The plated steel wraparound strap interlocks with the cover at four points. The terminal, grounding and mounting screws are a combination Phillips/straight-slot head. They have heavy-duty, high-grade brass alloy triple wipe contacts. An automatic grounding system allows easy installation. They are available in brown or ivory and listed to Federal Spec. WC-596-F.

Arrow-Hart #IG5362, IG5362HG Isolated Ground Hospital Grade Duplex Receptacles: They are 2-pole, 3-wire grounding, 20 A, 125 V with an automatic grounding system for fast installation. The 8-hole back and side wiring accepts No. 14 to No. 10 AWG. They have a heavy-duty nylon face and reinforced nylon back body. They are available in orange or ivory.

Arrow-Hart #5492 Combination Rating NEMA 5-20R & 6-20 Duplex Receptacles: One circuit is 20 A, 125 V and another circuit is 20 A, 250 V. The heavy-duty nylon face is good for abusive areas. The automatic grounding system permits fast installation. They are heavy-duty, back and side wired. They are available in brown, ivory, gray, white or red.

Arrow-Hart #8310, IG8310, 8310L and 8310M Single Receptacles: These receptacles are 20 A, 125 V, 2-pole, 3-wire grounding with a heavy-duty strong nylon body and face. They have an automatic grounding clip and a wraparound mounting strap locked into the base. They meet NEMA WD1-1983 heavy-duty requirements and are UL listed to Federal Spec WC-596-F. They have corrosion-resistant nickel plated contacts and mounting straps on the 8310 series. The IG in the number represents an isolated ground model which is available in orange or ivory. The 8310M model has a short mounting strap and comes in brown or ivory. They are back and side wired.

Arrow-Hart #5351, IG5351 Side-Wired, Single Receptacles: They are 20 A, 125 V, 2-pole, 3-wire grounding, with a heavy-duty virtually indestructible nylon body and face. They come with an automatic grounding terminal and a wraparound mounting strap locked into the base. They meet NEMA WD1-1983 heavy-duty requirements and are UL verified to Federal Spec. 596-F. Standard colors are brown or ivory; isolated ground model is available in orange or ivory. A short strap style is made in either the standard or isolated ground model.

Arrow-Hart #5358 Side-Wired, Extra-Short Mounting Strap Single Receptacles: These receptacles are 20 A, 125 V, 2-pole, 3-wire grounding and available only in brown. Mounting strap holes are $1\frac{3}{4}$-in center to center spacing with 8-32 screws.

Arrow-Hart #6766 & 6766HG "Safety Grip" Dead Front Plugs: These plugs are 20 A, 125 V, 2-pole, 3-wire grounding with nylon/Lexan back wired dead-front construction. A neoprene gasket seals the cord hole opening. The dual range cord grip design will handle cord .250 in to .655 in diameter. They meet NEMA WD1-1983 requirements and are UL verified to Federal Spec. WC-596-F. They are available in gray in both the standard and hospital grade models.

Arrow-Hart #6765HGC & 6765HG "Safety Grip" Angle Dead-Front Plugs: These plugs are 20 A, 125 V, 2-pole, 3-wire grounding with a nylon/Lexan back wire, dead-front construction. A neoprene gasket seals the cord hole opening. A dual range cord grip design will handle a cord .300 in to .655 in diameter. They meet NEMA WD1-1983 requirements and are UL verified to Federal Spec. WC-596-F. The #6765 HGC is hospital grade with clear cover and #6765 HG is hospital grade and gray in color.

Arrow-Hart #5366N, 5366NHG, 5366NHGC, 5366NCR, IG5366N, 5366NY Auto Grip, Dead-Front Plugs: The plugs are 2-pole, 3-wire grounding, 20 A, 125 V with neoprene gaskets that seals a cord hole opening. They also have thermoplastic dead-front construction and an automatic cord grip that clamps the cord during assembly. They meet NEMA WD1-1983 requirements and are UL verified to Federal Spec. WC-596-F. #5366N is gray and will accommodate a cord .250 in to .656 in diameter. #5366NHG is hospital grade, gray, and will handle a cord .300 in to .656 in diameter. #5366NHGC is hospital grade, clear cover and will take cord .330 in to .656 in diameter. #5366NY is yellow and will take a cord .250 in to .656 in diameter. #IG5366N is orange, has an isolated ground and handles a cord .250 in to .656 in diameter.

Arrow-Hart #5364V "Quickeze" Dead-Front Plugs: These plugs are 2-pole, 3-wire grounding, 20 A, 125 V with nylon back-wired, dead-front construction. The twist-on cover eliminates assembly screws. They have easy-wire clamp-type terminals that are located on same side for fast wiring. They are UL listed and CSA certified. The nylon is suitable for 65 degrees C (140 degrees F) continuous use. They meet NEMA WD1-1983 requirements. They are available in gray in the standard model, yellow in #5464VCR corrosion resistant model and #8364V hospital grade in gray.

Arrow-Hart #5364Y "Quickgrip" Dead-Front Plugs: These are 2-pole, 3-wire grounding, 20 A, 125 V and one-piece design with quick drive screws for fast wiring. The high-visibility yellow nylon is excellent for chemical and impact resistance. They are UL listed and CSA certified and will accept cord .230 in to .720 in diameter.

Arrow-Hart #6769, 6769HG "Safety Grip" Connectors: These connectors are 2-pole, 3-wire grounding, 20A, 125 V with nylon/Lexan construction suitable for 105 degree C/221 degree F continuous use. A neoprene gasket seals the cord hole opening. They have a dual range cord grip design. They meet NEMA WD1-1983 requirements and are UL listed. The standard #6769 is gray and will handle a cord .250 in to .656 in diameter. #6769HG hospital grade connector is gray and will handle a cord .300 in to .656 in diameter.

Arrow-Hart #5369N "Autogrip" Connectors: These connectors are 2-pole, 3-wire grounding, 20A, 125 V. They have a neoprene gasket that seals the cord hole opening. The thermoplastic construction is suitable for 105 degrees C/221 degrees F continuous use. An automatic cord grip clamps the cord during assembly. They meet NEMA WD1-1983 requirements and are UL listed. The standard plug is available in gray, #5369NHG (hospital grade) is available in gray or with a clear cover and #5369NCR (corrosion resistant model) is available in yellow.

Arrow-Hart #5369Y "Quickgrip" Connectors: These are 2-pole, 3-wire grounding 20 A, 125 V with one-piece design and quick drive screws for fast wiring. The high visibility yellow nylon is excellent for chemical and impact resistance They are UL listed and CSA certified.

Arrow-Hart #5778, 5779, 5787 Flanged Inlets and Outlets: These devices are 2-pole, 3-wire grounding 20 A, 125 V gray nylon/Lexan suitable for 105 degrees C/221 degrees F continuous use. The mounting holes are interchangeable with other designs and have back wired clamp-type terminals. #5778C is a flanged inlet constructed of gray nylon/Lexan, #5779C is a matching outlet. #5778 is a metal flanged inlet and #5787 is the matching outlet. The #5779 is a metal flanged inlet with reversed shell.

Arrow-Hart #5392 Heavy-Duty Duplex Receptacles: These devices are 2-pole, 3-wire grounding, 20 A, 125 V with CSA configuration. They have a heavy-duty back and side wired with a virtually indestructible nylon face and a rigid, glass reinforced nylon back body. They have a one-piece, heavy-gauge brass grounding shunt for reliable low resistance grounding. The plated steel wraparound strap interlocks with the cover at four points for assembly integrity. The terminal, grounding and mounting screws are a combination Phillips/straight-slot head. They have heavy-duty, high-grade brass alloy triple wipe contacts. The 8-hole back and side wiring accepts up to No. 10 AWG. They are CSA certified and UL listed. Colors are brown and ivory or orange in the #IG5392.

Other 125/250 V, 20 A, 2-pole grounding-type receptacles, connectors, and plugs include the following:

- Arrow-Hart #5361CSA single receptacle, 2-pole, 3-wire grounding type
- Arrow-Hart #5361CSA single receptacle, 2-pole, 3-wire grounding type
- Arrow-Hart #5369NCSA "safety grip" connector
- Arrow-Hart #8500 hospital grade duplex receptacle

- Arrow-Hart #5462 and #5462C extra heavy-duty duplex receptacle
- Arrow-Hart #5452 heavy-duty receptacle
- Arrow-Hart #8510 and IG8510 hospital grade single receptacle
- Arrow-Hart #5451 single receptacle
- Arrow-Hart #5878C flanged inlet and outlet
- Arrow-Hart #6866 and 6869 "safety grip" plug and connector
- Arrow-Hart #5466N and 5469N "auto grip" dead-front plug and connector
- Arrow-Hart #5464Y and 5469Y "quickgrip" dead-front plug and connector
- Arrow-Hart #5464V "quickeze" dead-front plug
- Arrow-Hart #6865HG angle dead-front plug

Arrow-Hart #7621 277 Vac, 20 A Single Receptacle: The #7621 NEMA 7-20 single receptacle is provided with a safety ground clip and fits a standard single wall plate. It is available in either black or ivory. The corresponding 37642 dead-front plug is constructed of nylon/Lexan and is back-wired in the dead-front style. A neoprene gasket seals the cord opening. The plug is UL verified to Federal Specification WC596-F and accepts cords from .375 in to 1 in diameter. The plugs are provided only in gray color.

Arrow-Hart #5716N, #IG5716N, and 5716P Receptacles: These NEMA 5-30R receptacles are rated at 125 V, 30 A and have a high-strength, glass-reinforced nylon body that resists breakage in high abuse areas. These receptacles are UL verified to Federal Specification WC596-F and the terminals are approved for either copper or aluminum conductors.

NEMA 5-30R single receptacle

Other plugs, connectors, and flanged inlets rated at 125 V, 30 A include the following:

- Arrow-Hart #5717N plug
- Arrow-Hart #5717AN plug
- Arrow-Hart #6716N connector
- Arrow-Hart #5717NFI flanged inlet

Additional Receptacles Rated At 125 V

- Arrow-Hart #5711N, IG5711N, 5711P, single receptacles, 2-pole, 3-wire grounding, 50 A, 125 V
- Arrow-Hart #5712N, 5712AN, 6711N, 5712NFI plugs, connector, flanged inlet, 2-pole, 3-wire grounding, 50 A, 125 V
- Arrow-Hart #7465N, 7428N, 7479N, 2-pole, 2-wire, 15 A, 125 V plugs
- Arrow-Hart #7464, 7427 connectors, 2-pole, 2-wire, 15 A, 125 V
- Arrow-Hart #7466, 7467, 7468 flanged inlets & outlet, 2-pole, 2-wire, 15 A, 125 V
- Arrow-Hart #7540 duplex, #7537 single receptacle, 2-pole, 2-wire, 15 A, 125 V
- Arrow-Hart #7546, 7548 plugs, 2-pole, 2-wire 15 A, 125 V
- Arrow-Hart #7506 connector, 2-pole, 2-wire, 15 A, 125 V
- Arrow-Hart #7523, 7526 flanged inlet & outlet, 2-pole, 2-wire, 15 A, 125 V
- Arrow-Hart #7594, 7570 plugs, 2-pole, 3-wire grounding, 15 A, 125 V
- Arrow-Hart #7593, 7571 connectors, 2-pole, 3-wire grounding, 15 A, 125 V
- Arrow-Hart #4700, IG4700, 4700CR, 4701, 4702, 4703 duplex receptacles, 2-pole, 3-wire grounding, 15 A, 125 V

- Arrow-Hart #5792, IG5792, 5792CR, 5793 combination duplex receptacles, 2-pole, 3-wire grounding, 15 A, 125 V
- Arrow-Hart #4710, IG4710, 4711, 4712, 4713, IG4712, IG4713 single receptacles, 2-pole, 3-wire grounding, 15 A, 125 V
- Arrow-Hart #4721, 4721AN, 4731, 4731AN safety grip plugs & connector, 2-pole, 3-wire grounding, 15 A, 125 V
- Arrow-Hart #4721N, IG4721N, 4721NCR, 4731N, 4731NCR, autogrip plugs and connectors, 2-pole, 3-wire grounding, 15 A, 125 V
- Arrow-Hart #4716C, 4715C flanged inlet & outlet, 2-pole, 3-wire grounding, 15 A, 125 V
- Arrow-Hart #6200, IG6200, 6200QC receptacles, nylon 2-pole, 3-wire grounding 20 A, 125 V
- Arrow-Hart #6202, 6202CR, IG6202, 6204, 6204CR plugs and connectors, 2-pole, 3-wire grounding, 20 A, 125 V
- Arrow-Hart #6205, 6206 flanged inlet & outlet, 2-pole, 3-wire grounding, 20 A, 125 V
- Arrow-Hart #23030 receptacle, #23035N plug, #23032N connector, #23007 cover, 2-pole, 3-wire grounding 20 A, 125 V
- Arrow-Hart #23056, 23054 armored plug and connector, 2-pole, 3-wire grounding, 20 A, 125 V
- Arrow-Hart #6140 single flush receptacle, 2-pole, 3-wire grounding, 30 A, 120 V, 400 Hertz
- Arrow-Hart#6142 plug, #6144 connector, 2-pole, 3-wire grounding 30 A, 120 V, 400 Hertz
- Arrow-Hart #6330, 6330CR, IG6330, 6330QC receptacle, nylon 2-pole, 3-wire grounding, 30 A, 125 V
- Arrow-Hart #6332, 6332CR, IG6332 plug, #6334, 6334CR connector, 2-pole, 3-wire grounding 30 A, 125 V

- Arrow-Hart #6335, 6336 flanged inlets & outlet, 2-pole, 3-wire grounding, 30A, 125 V
- Arrow-Hart #63CR61N plug, #63CR6ON connector, 2-pole, 3-wire grounding, 50 A, 125 V corrosion resistant
- Arrow-Hart #63CR70 receptacle 2-pole, 3-wire grounding, corrosion resistant, 50 A, 125 V
- Arrow-Hart #63CR72 hull inlet, 2-pole, 3-wire grounding, 50 A, 125 V
- Arrow-Hart #6130 single flush receptacle, 3-pole, 4-wire grounding, 30 A, 3-phase, 120 V, 400 Hertz
- Arrow-Hart #6132 plug, 6134 connector, 3-pole, 4-wire grounding, 30 A, 3-phase, 120 V, 400 Hertz

Additional Receptacles Rated At 250 V

- Arrow-Hart #5700N receptacle, 2-pole, 3-wire grounding, 30 A, 250 V
- Arrow-Hart #5701N, 5701AN, 6700N, 5701NFI plugs, connector flanaged inlet, 30 A, 250 V, 2-pole, 3-wire grounding
- Arrow-Hart #5709N, IG5709N, 5709P single receptacles, 2-pole, 3-wire grounding, 50 A, 250 V
- Arrow-Hart #5710N, 5710AN, 6709N, 5710NFI plugs, connector, flanged inlet 50 A, 250 V, 2-pole, 3-wire grounding
- Arrow-Hart #5705N, 5705AN, 6796 N, 5705 NFI plugs, connector, flanged inlet, 2-pole, 3-wire grounding, 50 A, 250 Vac
- Arrow Hart #8430N, single receptacle, #8432N, 8432N plugs, 3-pole, 4-wire grounding, 30 A, 250 V, 3-phase
- Arrow-Hart #8450N receptacle, 8452N, 8452AN plugs, 50 A, 250 V, 3-phase, 3-pole, 4-wire grounding
- Arrow-Hart #8460N receptacle, #8462N, 8462AN plugs, 3-pole, 4-wire grounding, 60A, 250 V, 3-phase

- Arrow-Hart #7210B single receptacle, 2-pole, 2-wire, 20 A, 250 V
- Arrow-Hart #7102N, 9102N plugs and #7101N connector, 2-pole, 2-wire, 20 A, 250 V
- Arrow-Hart #9808 flanged inlet, #8809 flanged outlet, 2-pole, 2-wire, 20 A, 250 V
- Arrow-Hart #6580, IG6580, 6581 duplex receptacles, 2-pole, 3-wire grounding 15 A, 250 V
- Arrow-Hart #6565, 6565AN, 6566 safety grip plugs & connector, 2-pole, 3-wire grounding, 15 A, 250 V
- Arrow-Hart #6565N, 6565NY, 6566N, 6566NY, IG6565N, autogrip plugs & connector, 2-pole, 3-wire grounding, 15 A, 250 V
- Arrow-Hart #6586C, 6585C flanged inlet & outlet, 2-pole, 3-wire grounding, 15 A, 250 V
- Arrow-Hart #6210, IG 6210, 6210AC nylon receptacle, 2-pole, 3-wire grounding, 20 A, 250 V
- Arrow-Hart #6212, 6214, IG6412 plugs and connector, 2-pole, 3-wire grounding, 20 A, 250 V
- Arrow-Hart #6215, 6216 flanged inlets & outlet, 2-pole, 3-wire grounding, 20 A, 250 V
- Arrow-Hart #6340, IG6340, 6340QC receptacle, nylon 2-pole, 3-wire grounding, 30 A, 250 V
- Arrow-Hart #6342 plug, 6344 conector, 2-pole, 3-wire grounding, 30 A, 250 V
- Arrow-Hart #6345, 6346 flanged inlets & outlet, 2-pole, 3-wire grounding, 30 A, 250 V
- Arrow-Hart #CS8269 receptacle, pressure-type terminals, 2-pole, 3-wire grounding, 50 A, 250 V
- Arrow-Hart #CS8265N plug, #CS8264 connector, nylon shell, 2-pole, 3-wire grounding, 50 A, 250 V
- Arrow-Hart #CS8275, CS8277 flanged inlets, 2-pole, 3-wire grounding, 250 V

- Arrow-Hart #25505, 25605 receptacle & stainless steel plate assembly, 2-pole, 3-wire grounding, 50 A, 60 A, 250 Vac
- Arrow-Hart #25503, 25603 receptacle 7 cast aluminum plate assembly, 2-pole, 3-wire grounding, 50 A, 60 A, 250 Vac
- Arrow-Hart #6250 receptacle, 3-pole, 3-wire, 20 A, 3-phase, 250 V
- Arrow-Hart #6252 plug, #6254 connector, 3-pole, 3-wire, 20 A, 3-phase, 250 V
- Arrow-Hart #6255 flanged inlet, #6256 flanged outlet, 3-pole, 3-wire, 20 A, 3-phase, 250 V
- Arrow-Hart #6390 receptacle, nylon, 3-pole, 3-wire, 30 A, 3-phase, 250 V
- Arrow-Hart #6392 plug, #6394 connector, 3-pole, 3-wire, 30 A, 3-phase, 250 V
- Arrow-Hart #6395 flanged inlet, #6396 flanged outlet, 3-pole, 3-wire, 30 A, 3-phase, 250 V
- Arrow-hart #6420, IG6420 single receptacle, nylon, 3-pole, 4-wire grounding, 20 A, 3-phase, 250 V
- Arrow-Hart #6422 plug, #6424 connector, 3-pole, 4-wire grounding, 20 A, 3-phase, 250 V
- Arrow-Hart #6520, IG6520 receptacles, nylon, 3-pole, 4-wire grounding, 30 A, 3-phase, 250 V
- Arrow-Hart #6522 plug, #6524 connector, 3-pole, 4-wire gounding, 30 A, 3-phase, 250 V
- Arrow-Hart #6525 flanged inlet, #6526 flanged outlet, 3-pole, 4-wire grounding, 30 A, 3-phase, 250 V
- Arrow-Hart #5781 receptacle, #5783 plug, 3-pole, 4-wire grounding, 20 A, 3-phase, 250 V

Additional 277-V Receptacles

- Arrow-Hart #5759N single receptacle, 2-pole, 3-wire grounding, 30 A, 277 Vac

- Arrow-Hart #5703N, 5703AN, 6795N, 5703NFI plugs, connector, flanged inlet, 2-pole, 3-wire grounding, 30 A, 277 Vac
- Arrow-Hart #5796N single receptacle, 2-pole, 3-wire grounding, 50 A, 277 Vac
- Arrow-Hart #4750, IG4750, 4751, 4752, 4753 duplex receptacle, 2-pole, 3-wire grounding 15 A, 277 Vac
- Arrow-Hart #4760, IG4760, 5761, 4763 single receptacles, 2-pole, 3-wire grounding, 15 A, 277 Vac
- Arrow-Hart #4771, 4771AN, 4772 safety grip plugs & connector, 2-pole, 3-wire grounding, 15 A, 277 Vac
- Arrow-Hart #4771N, 4772N autogrip plug & connector, 2-pole, 3-wire grounding, 15 A, 277 Vac
- Arrow-Hart 4786C, 4785C flanged inlet & outlet, 2-pole, 3-wire grounding, 15 A, 277 Vac
- Arrow-Hart #6220, IG6220 nylon receptacles, 2-pole, 3-wire grounding, 20 A, 277 Vac
- Arrow-Hart #6225, 6226 flanged inlet and outlet 2-pole, 3-wire grounding, 20 A, 277 Vac
- Arrow-Hart #6360, IG6360 receptacle, nylon, 2-pole, 3-wire grounding, 30 A, 277 Vac
- Arrow-Hart #6362 plug, #6364 connector, 2-pole, 3-wire grounding, 30 A, 277 Vac
- Arrow-Hart #6365, 6366 flanged inlets & outlet, 2-pole, 3-wire grounding, 30 A, 277 Vac

Receptacles Rated At 480 V

- Arrow-Hart #6232 plug, #6234 connector, 2-pole, 3-wire grounding, 20 A, 480 Vac
- Arrow-Hart #6370, IG6370 receptacles, nylon, 2-pole, 3-wire grounding, 30 A, 480 Vac
- Arrow-Hart #6372 plug, #6374 connector, 2-pole, 3-wire grounding 30 A, 480 Vac

- Arrow-Hart #6375, 6376 flanged inlet & outlet, 2-pole, 3-wire grounding, 30 A, 480 Vac
- Arrow-Hart #6940 single receptacle, 3-pole, 20 A, 3-phase, 480 V
- Arrow-Hart #6942 plug, 6944 connector, 3-pole, 3-wire, 20 A, 3-phase, 480 V
- Arrow-Hart #6945, 6946 flanged inlet and outlet, 3-pole, 3-wire, 20 A, 3-phase, 480 V
- Arrow-Hart #6960 receptacle, nylon, 3-pole, 3-wire, 30 A, 3-phase, 480 V
- Arrow-Hart #6962 plug, #6964 connector, 3-pole, 3-wire, 30 A, 3-phase, 480 V
- Arrow-Hart #6965 flanged inlet, #6966 flanged outlet, 3-pole, 3-wire, 30 A, 3-phase, 480 V
- Arrow-Hart #6430, IG6430 receptacle, nylon, 3-pole, 4-wire grounding 20 A, 3-phase, 480 V
- Arrow-Hart #6432 plug, #6434 connector, 3-pole, 4-wire grounding, 20 A, 3-phase, 480 V
- Arrow-Hart #6435 flanged inlet, #6436 flanged outlet, 3-pole, 4-wire grounding, 20 A, 3-phase, 480 V
- Arrow-Hart #6530, IG6530 receptacles, nylon, 3-pole, 4-wire grounding, 30 A, 3-phase, 480 V
- Arrow-Hart #6532 plug, #6534 connector, 3-pole, 4-wire grounding, 30 A, 3-phase, 480 V
- Arrow-Hart #6535 flanged inlet, #6536 flanged outlet, 3-pole, 4-wire grounding, 30 A, 3-phase, 480 V
- Arrow-Hart #20443 receptacle, 3-pole, 4-wire grounding, 30 A, 480 Vac
- Arrow-Hart #20445N plug, #20444N connector, 3-pole, 4-wire grounding, 30 A, 490 Vac
- Arrow-Hart #CS8169 receptacle, 3-pole, 4-wire grounding, 50 A, 3-phase, 480 Vac
- Arrow-Hart #CS8165N, #CS8164 connector, 3-pole, 4-wire grounding, 50 A, 3-phase, 480 Vac

- Arrow-Hart #CS8175, CS8177 flanged inlets, 3-pole, 4-wire grounding, 50 A, 3-phase, 480 Vac

Receptacles Rated at 600 V

- Arrow-Hart #6920 receptacle, nylon, 2-pole, 3-wire grounding, 20 A, 600 Vac
- Arrow-Hart #6922 plug, 6924 connector, 2-pole, 3-wire grounding, 20 A, 600 Vac
- Arrow-Hart #6925, 6926 flanged inlet and outlet, 2-pole, 3-wire grounding, 20 A, 600 Vac
- Arrow-Hart #6930 receptacle, nylon, 2-pole, 3-wire grounding, 30 A, 600 Vac
- Arrow-Hart #6932 plug, #6934 connector, 2-pole, 3-wire grounding, 30 A, 600 Vac
- Arrow-Hart #6935, #6936 flanged inlet and outlet, 2-pole, 3-wire grounding, 30 A, 600 Vac
- Arrow-Hart #6980 single receptacle, nylon, 3-pole, 3-wire, 30 A, 600 V, 3-phase
- Arrow-Hart #6982 plug, #6984 connector, 3-pole, 3-wire, 30 A, 600 V, 3-phase
- Arrow-Hart #6985 flanged inlet, #6986 flanged outlet, 3-pole, 3-wire, 30 A, 600 V, 3-phase
- Arrow-Hart #6540 receptacle, nylon, 3-pole, 4-wire grounding, 30 A, 600 V, 3-phase
- Arrow-Hart #6542 plug, #6544 connector, 3-pole, 4-wire grounding, 30 A, 600 V, 3-phase
- Arrow-Hart #6545 flanged inlet, #6546 flanged outlet, 3-pole, 4-wire grounding, 30 A, 600 V, 3-phase
- Arrow-Hart #26420, #26421 receptacles, 3-pole, 4-wire equipment grounding, 60 A, 600 Vac
- Arrow-Hart #26415, #26426 plugs, 3-pole, 4-wire equipment grounding, 60 A, 600 Vac
- Arrow-Hart #26414, 26427 connectors, 3-pole, 4-wire equipment grounding, 60 A, 600 Vac

- Arrow-Hart #26414A flanged outlet, #26415A inlet, 3-pole, 4-wire equipment grounding, 60 A, 600 Vac
- Arrow-Hart 26520, 26521 heavy-duty industrial-type receptacle, 4-pole, 5-wire equipment grounding, 60 A, 600 Vac
- Arrow-Hart #26515, #26526 heavy-duty industrial-type plugs, 4-pole, 5-wire equipment grounding, 60 A, 600 Vac
- Arrow-Hart #26514 #26527 connectors, heavy-duty industrial-type, 4-pole, 5-wire equipment grounding, 60 A, 600 Vac
- Arrow-Hart #26514A flanged outlet, #26515A flanged inlet, 4-pole, 5-wire equipment grounding, 60 A, 600 Vac

Combination Voltage Receptacles

- Arrow-Hart #6051 single receptacle 3-pole, 3-wire, 10 A, 250 V; 15 A, 125 V
- Arrow-Hart #4475, 4472 plug and connector 10 A, 250 V; 15 A, 125 V, 3-pole, 3-wire
- Arrow-Hart #9140I single receptacle and #9151N plug, 20 A, 125/250 V, 3-pole, 3-wire
- Arrow-Hart #9344N, 9344P, 9344, 9350 single receptacles, 3-pole, 3-wire, 30 A, 125/250 V
- Arrow-Hart #9337N, 9352AN, 9341N, 9337NFI plugs, connectors, flanged inlet, 3-pole, 3-wire, 30 A, 125/250 V
- Arrow-Hart #7985N, 7985P, 7950, 7986 single receptacles, 3-pole, 3-wire, 50 A, 125/250 V
- Arrow-Hart #4524N, 7952AN, 4526N, 4526NFI plugs, connector, flanged inlet, 3-pole, 3-wire, 50 A, 125/250 V
- Arrow-Hart #5759 receptacle, #5757 plug, 3-pole, 4-wire grounding, 20 A, 125/250 V

- Arrow-Hart #5744N, IG5744N, 5744P, 5784 receptacles, 3-pole, 4-wire grounding, 30 A, 125/250 V
- Arrow-Hart #5746N, 5732AN plugs, 3-pole, 4-wire grounding, 30 A, 125/250 V
- Arrow-Hart #5754N, IG5754N, 5754P, 5785 receptacles, #5745N, 5752AN plugs, 3-pole, 4-wire grounding, 50 A, 125/250 V
- Arrow-Hart #9460N, IG9460N receptacles, #9462N, 9462AN plugs, 3-pole, 4-wire grounding, 60 A, 125/250 V
- Arrow-Hart #7250 receptacle, #7251N plug, 4-pole, 4-wire, 20 A, 120/208 V, 3-phase
- Arrow-Hart #8330N receptacle, #8332N, 8332AN plugs, 4-pole, 4-wire, 30 A, 120/208 V, 3-phase Y
- Arrow-Hart #8350N receptacle, #8352N, 8352AN plugs, 4-pole, 4-wire, 50 A, 120/208 V, 3-phase Y
- Arrow-Hart #5515N receptacle, #5517N, 4516AN plugs, 4-pole, 4-wire, 60 A, 120/208 V, 3-phase Y
- Arrow-Hart #3771 receptacle, 2-pole, 3-wire equipment grounding, 50 A, 250 Vac and 600 Vac
- Arrow-Hart #3763N, plug and #3762N connector, 2-pole, 3-wire equipment grounding, 50 A, 250 Vdc and 600 Vac
- Arrow-Hart #3777, 3767 flanged inlets, 2-pole, 3-wire equipment grounding, 50 A, 250 Vdc and 600 Vac
- Arrow-Hart #7485, 7432 plugs, 3-pole, 3-wire, 15 A, 125/250 V
- Arrow-Hart #7484, 7433 connectors, 3-pole, 3-wire, 15 A, 125/250 V
- Arrow-Hart #7486, 7487, 7486N, 7487N flanged inlets and outlets, 3-pole, 3-wire, 15 A, 125/250 V
- Arrow-Hart #7580 duplex recpetacle, #7582 single receptacle, 3-pole, 3-wire, 10 A, 250 V; 15 A, 125 V

- Arrow-Hart #4767, 4767AN plugs, #4755 connector, safety grip, 3-pole, 3-wire, 10 A, 250 V; 15 A, 125 V
- Arrow-Hart #7567N plug, #7565 connector, autogrip style, 3-pole, 3-wire, 10 A, 250 V; 15 A, 125 V
- Arrow-Hart #7556C, 7557C flanged inlet and outlet, 3-pole, 3-wire, 10 A, 250 V; 15 A, 125 V
- Arrow-Hart #7310B, 7317B single receptacles, 3-pole, 3-wire, 20 A, 125/250 V
- Arrow-Hart #9965C, plug, #7314C connector, 3-pole, 3-wire, 20 A, 125/250 V
- Arrow-Hart #7327N flanged inlet, #7328N flanged outlet, 3-pole, 3-wire, 20 A, 125/250 V
- Arrow-Hart #6240 single receptacle, 3-pole, 3-wire, 20 A, 125/250V
- Arrow-Hart #6242 plug, #6244 connector, 3-pole, 3-wire, 20 A, 125/250 V
- Arrow-Hart #6245, #6246 flanged inlet and outlet, 3-pole, 3-wire, 20 A, 125/250 V
- Arrow-Hart #23000, 23000G receptacle, 3-pole, 3-wire, 20 A, 125 V; 10 A, 250 Vdc, 10 A, 480 Vac
- Arrow-Hart #23005N plug, #23002N connector, safety-grip style, 3-pole, 3-wire, 20 A, 125 V; 10 A, 250 Vdc, 10 A, 480 Vac
- Arrow-Hart #6380 single receptacle, nylon, 3-pole, 3-wire, 30 A, 125/250 V
- Arrow-Hart #6382 plug, #6384 connector, 3-pole, 3-wire, 30 A, 125/250 V
- Arrow-Hart #6385 flanged inlet, #6386 flanged outlet, 3-pole, 3-wire, 30 A, 125/250 V
- Arrow-Hart #3330 single receptacle, Lexan, 3-pole, 3-wire, 30 A, 125/250 V
- Arrow-Hart #3331N plug, #3333N connector, 3-pole, 3-wire, 30 A, 125/250 V

- Arrow-Hart #3337N flanged inlet, #3336 flanged outlet, 3-pole, 3-wire, 30 A, 125/250 V
- Arrow-Hart #6400, 6400QC, IG6400 single receptacle, nylon, 3-pole, 4-wire grounding, 20 A, 125/250 V
- Arrow-Hart #6402 plug, #6404 connector, 3-pole, 4-wire grounding, 20 A, 125/250 V
- Arrow-Hart #6405 flanged inlet, #6406 flanged outlet, 3-pole, 4-wire grounding, 20 A, 125/250 V
- Arrow-Hart #6510, 6510QC, IG6510 receptacles, nylon, 3-pole, 4-wire grounding, 30 A, 125/250 V
- Arrow-Hart #6512 plug, #6514 connector, 3-pole, 4-wire grounding, 30 A, 125/250 V
- Arrow-Hart #6515 flanged inlet and outlet, 3-pole, 4-wire grounding, 30 A, 125/250 V
- Arrow-Hart #63CR69 corrosion resistant receptacle, 3-pole, 4-wire grounding, 50 A, 125/250 V
- Arrow-Hart #63CR65N corrosion resistant plug, #6CR64N connector, 3-pole, 4-wire grounding, 50 A, 125/250 V
- Arrow-Hart #63CR74 hull inlet, #60CR75 adapter, 3-pole, 4-wire grounding, 50 A, 125/250 V
- Arrow-Hart #CS6369 receptacle, 3-pole, 4-wire grounding, 50 A, 125/250 V
- Arrow-Hart #CS6365N plug, #CS6364 connector, 3-pole, 4-wire grounding, 50 A, 125/250 V
- Arrow-Hart #CS6375 flanged inlet, #6376 flanged outlet, 3-pole, 4-wire grounding, 50 A, 125/250 V
- Arrow-Hart #3769 receptacle, 3-pole, 4-wire equipment grounding, 50 A, 250 Vdc and 600 Vac
- Arrow-Hart #3765N plug, #3764N connector, 3-pole, 4-wire equipment grounding, 50 A, 250 Vdc and 600 Vac

- Arrow-Hart #3775, #3768 flanged inlets, 3-pole, 4-wire equipment grounding, 50 A, 250 Vdc and 600 Vac
- Arrow-Hart #7379 receptacle, 3-pole, 4-wire equipment grounding, 50 A, 250 Vdc and 600 Vac (for replacement use only)
- Arrow-Hart #7765N plug, #7764N connector, 3-pole, 4-wire equipment grounding, 50 A, 250 Vdc and 600 Vac (for replacement use only)
- Arrow-Hart #7958, #7968 flanged inlets, 3-pole, 4-wire equipment grounding, 50 A, 250 Vdc and 600 Vac
- Arrow-Hart #7410B, #7417B receptacles, 4-pole, 4-wire, 20 A, 120/208 V, 3-phase Y
- Arrow-Hart #7411C plug, #7413C connecor, 4-pole, 4-wire, 20 A, 120/208 V, 3-phase Y
- Arrow-Hart #7408N flanged inlet, #7409N flanged outlet, 4-pole, 4-wire, 20 A, 120/208 V, 3-phase Y
- Arrow-Hart #6440 receptacle, 4-pole, 4-wire, 20 A, 120/208 V, 3-phase Y
- Arrow-Hart #6442 plug, #6444 connector, 4-pole, 4-wire, 20 A, 120/208 V, 3-phase Y
- Arrow-Hart #6445 flanged inlet, #6446 flanged outlet, 4-pole, 4-wire, 20 A, 120/208 V, 3-phase Y
- Arrow-Hart #6450 receptacle, nylon, 4-pole, 4-wire, 20 A, 277/480 V, 3-phase Y
- Arrow-Hart #6452 plug, #6454 connector, 4-pole, 4-wire, 20 A, 277/480V, 3-phase Y
- Arrow-Hart #6455 flanged inlet, #6456 flanged outlet, 4-pole, 4-wire, 20 A, 277/480 V, 3-phase Y
- Arrow-Hart #6460 receptacle, nylon, 4-pole, 4-wire, 20 A, 347/600 V, 3-phase Y
- Arrow-Hart #6462 plug, #6464 connector, 4-pole, 4-wire, 20 A, 347/600 V, 3-phase Y

- Arrow-Hart #6465 flanged inlet, #6466 flanged outlet, 4-pole, 4-wire, 20 A, 347/600 V, 3-phase Y
- Arrow-Hart #3430 receptacles, nylon, 4-pole, 4-wire, 30 A, 120/208 V, 3-phase Y
- Arrow-Hart #3431N plug, #3433 connector, 4-pole, 4-wire, 30 A, 120/208 V, 3-phase Y
- Arrow-Hart #3434 flanged inlet, #3436 flanged outlet, 4-pole, 4-wire, 30 A, 120/208 V, 3-phase Y
- Arrow Hart #6570 receptacle, nylon, 4-pole, 4-wire, 30 A, 120/208 V, 3-phase Y
- Arrow-Hart #6578 plug, #6574 connector, 4-pole, 4-wire, 30 A, 120/208 V, 3-phase Y
- Arrow-Hart #6575 flanged inlet, #6576 flanged outlet, 4-pole, 4-wire, 30 A, 120/208 V, 3-phase Y
- Arrow-Hart #6590 receptacle, nylon, 4-pole, 4-wire, 30 A, 277/480 V, 3-phase Y
- Arrow-Hart #6592 plug, #6594 connector, 4-pole, 4-wire, 30 A, 277/480 V, 3-phase Y
- Arrow-Hart #6595 flanged inlet, #6596 flanged outlet, 4-pole, 4-wire, 30 A, 277/480 V, 3-phase Y
- Arrow-Hart #6600 receptacle, nylon, 4-pole, 4-wire, 30 A, 347/600 V, 3-phase Y
- Arrow-Hart #6602 plug, #6604 connector, 4-pole, 4-wire, 30 A, 347/600 V, 3-phase Y
- Arrow-Hart #6605 flanged inlet, #6606 flanged outlet, 4-pole, 4-wire, 30 A, 347/600 V, 3-phase Y
- Arrow-Hart #6470, #IG6470 receptacle, nylon, 4-pole, 5-wire grounding, 20 A, 120/208 V, 3-phase Y
- Arrow-Hart #6472 plug, #6474 connector, 4-pole, 5-wire grounding, 20 A, 120/208 V, 3-phase Y
- Arrow-Hart #6475 flanged inlet, #6476 flanged outlet, 4-pole, 5-wire grounding, 20 A, 120/208 V, 3-phase Y

- Arrow-Hart #6110 single flush receptacle, 4-pole, 5-wire grounding, 20 A, 120/208 V, 3-phase Y, 400 Hertz
- Arrow-Hart #6112 plug, #6114 connector, 4-pole, 5-wire grounding, 20 A, 120/208 V, 3-phase Y, 400 Hertz
- Arrow-Hart #6116 flanged inlet, #6118 flanged outlet, 4-pole, 5-wire grounding, 20 A, 120/208 V, 3-phase Y, 400 Hertz
- Arrow-Hart #6490, IG6490 receptacles, nylon, 4-pole, 5-wire grounding, 20 A, 277/208 V, 3-phase Y
- Arrow-Hart #6492 plug, #6494 connector, 4-pole, 5-wire grounding, 20 A, 277/480 V, 3-phase Y
- Arrow-Hart #6495 flanged inlet, #6496 flanged outlet, 4-pole, 5-wire grounding, 20 A, 277/208 V, 3-phase Y
- Arrow-Hart #6500, IG6500 receptacle, nylon, 4-pole, 5-wire grounding, 20 A, 347/600 V, 3-phase Y
- Arrow-Hart #6502 plug, #6504 connector, 4-pole, 5-wire grounding, 20 A, 347/600 V, 3-phase Y
- Arrow-Hart #6505 flanged inlet, #6506 flanged outlet, 4-pole, 5-wire grounding, 20 A, 347/600 V, 3-phase Y
- Arrow-Hart 36610, IG6610 receptacles, nylon, 4-pole, 5-wire grounding, 30 A, 120/208 V, 3-phase
- Arrow-Hart 36612 plug, #6614 connector, 4-pole, 5-wire grounding, 30 A, 120/208 V, 3-phase
- Arrow-Hart #6615 flanged inlet, #6616 flanged outlet, 4-pole, 5-wire grounding, 30 A, 120/208 V, 3-phase Y
- Arrow-Hart #6620, IG6620 receptacle, nylon, 4-pole, 5-wire grounding, 30 A, 277/480 V, 3-phase Y
- Arrow-Hart #6622 plug, #6624 connector, 4-pole, 5-wire grounding, 30 A, 277/480 V, 3-phase Y
- Arrow-Hart #6625 flanged inlet, #6626 flanged outlet, 4-pole, 5-wire grounding, 30 A, 277/480 V, 3-phase Y

- Arrow-Hart #6120 single flush receptacle, 4-pole, 5-wire grounding, 30 A, 120/208 V, 3-phase, 400 Hertz
- Arrow-Hart #6122 plug, #6124 connector, 4-pole, 5-wire grounding, 30 A, 120/208 V, 3-phase, 400 Hertz
- Arrow-Hart #6635 flanged inlet, #6636 flanged outlet, 4-pole, 5-wire grounding, 30 A, 347/600 V, 3-phase Y
- Arrow-Hart 36630, IG6630 receptacles, nylon, 4-pole, 5-wire grounding, 30 A, 347/600 V, 3-phase Y
- Arrow-Hart #6632 plug, #6634 connector, 4-pole, 5-wire grounding, 30 A, 347/600 V, 3-phase Y
- Arrow-Hart #20403 receptacle, 3-pole, 4-wire equipment grounding, 30 A, 600 Vac; 20 A, 250 Vdc
- Arrow-Hart #20451N plug, #20414N connector, 3-pole, 4-wire equipment grounding, 30 A, 600 Vac, 20 A, 250 Vdc
- Arrow-Hart #21420 receptacle, 3-pole, 4-wire equipment grounding, 30 A, 600 Vac; 20 A, 250 Vdc
- Arrow-Hart #21447 weather-tight flanged inlet, 3-pole, 4-wire equipment grounding, 30 A, 600 Vac; 20 A, 250 Vdc
- Arrow-Hart #20418 nonweather-tight inlet, 3-pole, 4-wire equipment grounding, 30 A, 600 Vac; 20 A, 250 Vdc
- Arrow-Hart #25403, #25250 receptacles, 4-wire, 5-wire equipment grounding, 30 A, 600 Vac; 20 A, 250 Vdc
- Arrow-Hart #25415N plug, #25414N connector, 4-pole, 5-wire equipment grounding, 30 A, 600 Vac; 20 A, 250 Vdc
- Arrow-Hart #25416 and #25418 flanged inlet, #25406 flanged outlet, 4-pole, 5-wire equipment grounding, 30 A, 600 Vac; 20 A, 250 Vdc

Receptacle Characteristics

Receptacles have various symbols and information inscribed on them that help to determine their proper use and ratings. For example, the illustration below shows a standard duplex receptacle that contains the following printed inscriptions:

- The testing laboratory label
- The CSA (Canadian Standards Association) label
- Type of conductor for which the terminals are designed
- Current and voltage ratings, listed by maximum amperage, maximum voltage, and current restrictions

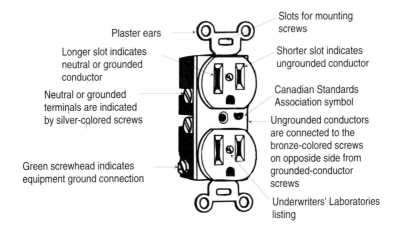

Characteristics of a typical duplex receptacle

The testing laboratory label is an indication that the device has undergone extensive testing by a nationally recognized testing lab and has met with the minimum safety requirements. The label does not indicate any type of quality rating. The receptacle in the above illustration is marked with the "UL" label which indicates that the device type was tested by Underwriters' Laboratories, Inc. of Northbrook, IL. ETL Testing Laboratories, Inc. of Cortland, NY is another nationally recognized testing laboratory. They provide a labeling, listing and follow-up service for the safety testing of electrical products to nationally recognized safety standards or specifically designated requirements of jurisdictional authorities.

The CSA (Canadian Standards Association) label is an indication that the material or device has undergone a similar testing procedure by the Canadian Standards Association and is acceptable for use in Canada.

Current and voltage ratings are listed by maximum amperage, maximum voltage and current restriction. On the device under discussion, the maximum current rating is 15 amperes at 125 volts — the latter of which is the maximum voltage allowed on a device so marked.

Conductor markings are also usually found on duplex receptacles. Receptacles with quick-connect wire clips will be marked "Use #12 or #14 solid wire only." If the inscription "CO/ALR" is marked on the receptacle, either copper, aluminum, or copper-clad aluminum wire may be used. The letters "ALR" stand for "aluminum revised." Receptacles marked with the inscription "CU/AL" should be used for copper only, although they were originally intended for use with aluminum also. However, such devices frequently failed when connected to 15- or 20-ampere circuits. Consequently, devices marked with "CU/AL" are no longer acceptable for use with aluminum conductors.

The remaining markings on duplex receptacles may include the manufacturer's name or logo, "Wire Release" inscribed under the wire-release slots, and the letters "GR." beneath or beside of the green grounding screw.

The screw terminals on receptacles are color-coded. For example, the terminal with the green screw head is the equipment ground connection and is connected to the U-shaped slots on the receptacle. The silver-colored terminal screws are for connecting the grounded or neutral conductors and are associated with the longer of the two vertical slots on the receptacle. The brass-colored terminal screws are for connecting the ungrounded or "hot" conductors and are associated with the shorter vertical slots on the receptacle.

Note: The long vertical slot accepts the grounded or neutral conductor while the shorter vertical slot accepts the ungrounded or hot conductor.

Mounting Receptacles

Although no actual NEC requirements exist on mounting heights and positioning receptacles, there are certain installation methods that have become "standard" in the electrical industry. The illustration on the opposite page shows mounting heights of duplex receptacles used on conventional residential and small commercial installations. However, do not take these dimensions as gospel; they are frequently varied to suit the

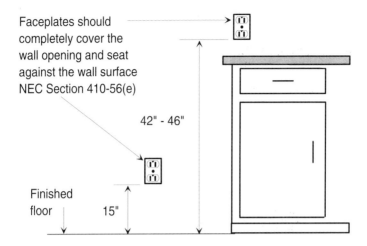

Recommended mounting heights of duplex receptacles under various situations

building structure. For example, ceramic tile might be placed above a kitchen or bathroom countertop. If the recommended dimensions puts the receptacle part of the way out of the tile, say, half in and half out, the mounting height should be adjusted to either place the receptacle completely in the tile or completely out of the tile as shown in the illustration below.

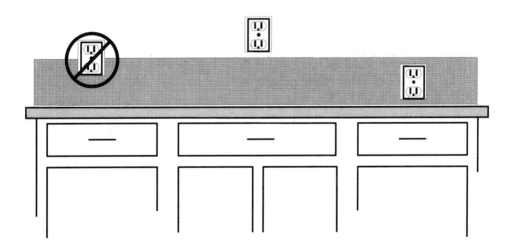

Adjust mounting heights so receptacles will either be completely in or completely our the tile

Refer again to the recommended mounting heights and note that the mounting heights are given to the bottom of the outlet box. Many dimensions on electrical drawings are given to the center of the outlet box or receptacle. However, during the actual installation, workers installing the outlet boxes can mount them more accurately (and in less time) by lining up the bottom of the box with a chalk mark rather than trying to "eyeball" this mark to the center of the box.

A decade or so ago, most electricians mounted receptacle outlets 12 inches from the finished floor to the center of the outlet box. However, a recent survey taken of over 500 homeowners shows that they prefer a mounting height of 15 inches from the finished floor to the bottom of the outlet box. It is easier to plug and unplug the cord-and-plug assemblies at this height — especially among senior citizens and those homeowners who are confined to wheelchairs. However, always check the working drawings, written specifications, and details of construction for measurements that may affect the mounting height of a particular receptacle outlet.

There is always the possibility of a metal receptacle cover coming loose and falling downward onto the blades of an attachment plug cap that may be loosely plugged into the receptacle. By the same token, a hairpin, fingernail file, metal fly-swatter handle, or any other metal object may be knocked off a table and fall downward onto the the plug blades. Any of these objects could cause a short-circuit if the falling metal object fell on both the "hot" and grounded neutral blades of the plug at the same time. For these reasons, it is recommended that the equipment grounding slot in receptacles be placed at the top. In this position, any falling metal object would fall onto the grounding blade which would more than likely prevent a short-circuit. See the illustration on the opposite page.

When duplex receptacles are mounted in a horizontal position, the grounded neutral slots should be on top for the same reasons as discussed previously.

NEC Section 370-20 requires all outlet boxes installed in walls or ceiling of concrete, tile, or other noncombustible material such as plaster or drywall to be installed in such a matter that the front edge of the box or fitting will not set back of the finished surface more than $\frac{1}{4}''$. Where walls and ceilings are constructed of wood or other combustible material, outlet boxes and fittings must be flush with the finished surface of the wall.

Wall surfaces such as drywall, plaster, etc. that contain wide gaps or are broken, jagged, or otherwise damaged must be repaired so there will be no gaps or open spaces greater than $\frac{1}{8}''$ between the outlet box and wall material. These repairs should be made prior to installing the faceplate.

Such repairs are best made with a noncombustible caulking or spackling compound.

Locating Receptacles

Several NEC Sections specify specific requirements for locating receptacles in all types of installations. A summary of these requirements is presented herein.

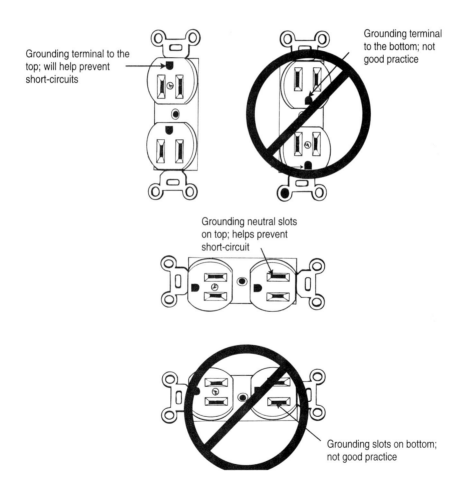

Recommended mounting positions for duplex receptacles

340 Handbook of Electrical Construction Tools and Materials

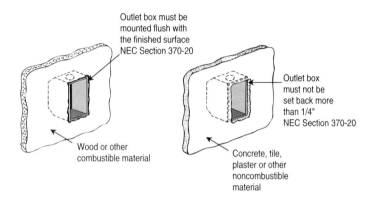

Summary of NEC requirements for mounting outlet boxes in walls and ceilings

Residential Occupancies

NEC Section 210-52 should be referred to when laying out outlets for residential and some commerical installations. This section details the general provisions along with small-appliance circuit requirements, laundry requirements, unfinished basements, attached garages, and other areas of the home.

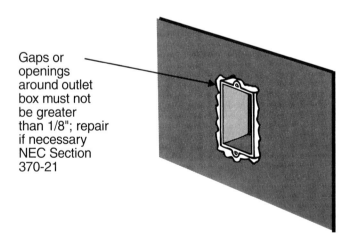

Gaps or openings around outlet boxes must be repaired

In general, every dwelling — regardless of its size — must have receptacles located in each habitable area so that no point along the floor line in any wall space (2 feet wide or wider) is more than 6 feet in that space. The purpose of this requirement is to prevent the need for extension cords longer than 6 feet and to minimize the use of cords across doorways, fireplaces, and similar openings.

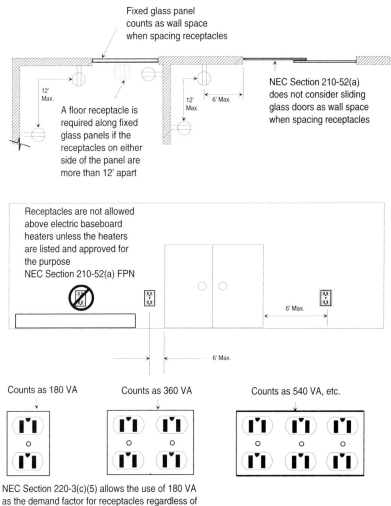

Summary of NEC requirements for locating receptacles

342   Handbook of Electrical Construction Tools and Materials

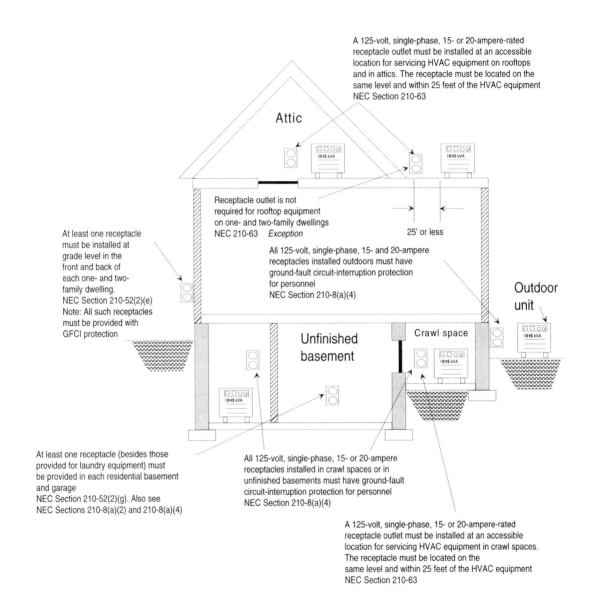

More NEC requirements concerning placement of receptacles

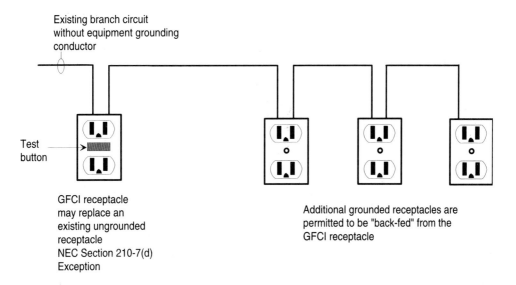

A GFCI may replace an ungrounded receptacle

In addition, a minimum of two 20-ampere small appliance branch circuits are required to serve all receptacle outlets, including refrigeration equipment, in the kitchen, pantry, breakfast room, dining room, or similar area of the dwelling unit. Such circuits, whether two or more are used, must have no other outlets connected to them.

At least one receptacle is required in each laundry area, on the outside of the building at the front and back, in each basement, in each attached and detached garage, in each hallway 10 feet or more in length, and at an accessible location for servicing any HVAC equipment.

When upgrading existing electrical systems, the NEC permits the use of a GFCI receptacle in place of a grounded receptacle. With such an arrangement, additional grounded receptacles may be connected on the downstream side of the GFCI.

Other receptacles and related circuits are provided as needed according to the load to be served. For example, receptacles are normally provided in residential occupancies for electric ranges, clothes dryers, and similar appliances. Most operate on 120/240-volt branch circuits using 30- to 60-ampere receptacles.

Commercial Applications

Receptacle requirements for commercial installations follow the same general requirements set forth for residential occupancies, with some exceptions. For example, guest rooms in hotels, motels, and similar occupancies must have receptacle outlets installed in accordance with Section 210-52. However, some leaway is given commercial installations. NEC Section 210-60 permits receptacle outlets to be located conveniently for permanent furniture layout.

The only other "must" requirement for commercial installations deals with the placement of receptacle outlets in show windows. NEC Section 210-62 requires at least one receptacle for each 12 linear feet of show window area measured horizontally at its maximum width. To calculate the number of receptacles required at the top of any show window, measure the total linear feet, and then divide this figure by 12 and any remainder or "fraction thereof" requires an additional receptacle. For example, the show window shown below is 18 feet in length. Consequently, the number of receptacles required may be calculated as follows:

$$\frac{18\ feet}{12\ feet} = 1.5\ receptacles$$

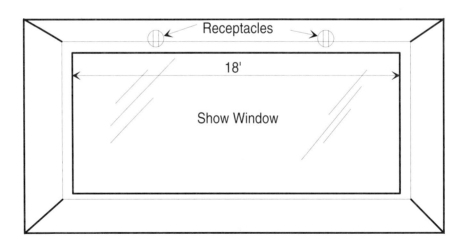

NEC Section 310-62 deals with commercial show windows

Since "1.5" is more than one, to comply with NEC Section 210-62, "or major fraction thereof," two receptacles are required in this area. Had the calculation resulted in a figure of, say, 1.01, the local inspection authorities would probably require only one receptacle in the show widow.

Of course, GFCIs are required on all 15- and 20-ampere receptacles installed in commercial bathrooms or toilets, in commercial garages, receptacles installed outdoors where there is direct grade-level access (below 6 feet 6 inches), crawl spaces, boathouses, and all receptacles installed on roofs.

Other receptacles and related circuits are provided as needed according to the load to be served.

Chapter 25

Switches

The purpose of a switch is to make and break an electrical circuit, safely and conveniently. In doing so, a switch may be used to manually control lighting, motors, fans, and other various items connected to an electrical circuit. Switches may also be activated by light, heat, chemicals, motion, and electrical energy for automatic operation. NEC Article 380 covers the installation and use of switches.

Although there is some disagreement concerning the actual definitions of the various switches that might fall under the category of *wiring devices*, the most generally accepted ones are as follows:

Bypass Isolation Switch: This is a manually operated device used in conjunction with a transfer switch to provide a means of directly connecting load conductors to a power source, and of disconnecting the transfer switch.

General-Use Switch: A switch intended for use in general distribution and branch circuits. It is rated in amperes, and it is capable of interrupting its rated current at its rated voltage.

General-Use Snap Switch: A form of general-use switch so constructed that it can be installed in flush device boxes or on outlet box covers, or otherwise used in conjunction with wiring systems recognized by the NEC.

Isolating Switch: A switch intended for isolating an electric circuit from the source of power. It has no interrupting rating, and it is intended to be operated only after the circuit has been opened by some other means.

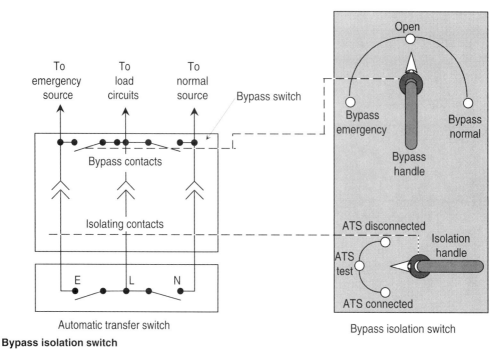

Bypass isolation switch

Motor-Circuit Switch: A switch, rated in horsepower, capable of interrupting the maximum operating overload current of a motor of the same horsepower rating as the switch at its rated voltage.

Transfer Switch: A transfer switch is a device for transferring one or more load conductor connections from one power source to another. This type of switch may be either automatic or nonautomatic.

Common Terms

Although basic switch terms are covered to some extent in earlier chapters, a brief review of these terms is warranted here. In general, the major terms used to identify the characteristics of switches are:

- Pole or poles
- Throw

The term *pole* refers to the number of conductors that the switch will control in the circuit. For example, a single-pole switch breaks the connection on only one conductor in the circuit. A double-pole switch breaks the connection to two conductors, and so forth.

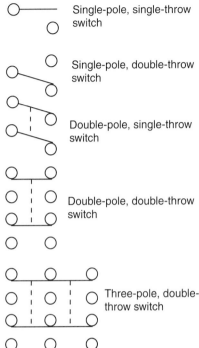

Common switch configurations

The term *throw* refers to the number of internal operations that a switch can perform. For example, a single-pole, single-throw switch will "make" one conductor when thrown in one direction — the "ON" direction — and "break" the circuit when thrown in the opposite direction; that is, the "OFF" position. The commonly used ON/OFF toggle switch is an SPST switch (single-pole, single-throw). A two-pole, single-throw switch opens or closes two conductors at the same time. Both conductors are either open or closed; that is, in the ON or OFF position. A two-pole, double-throw switch is used to direct a two-wire circuit through one of two different paths. One application of a two-pole, double-throw switch is in an electrical transfer switch where certain circuits may be energized from either the main electric service, or from an emergency standby generator. The double-throw switch "makes" the circuit from one or the other and prevents the circuits from being energized from both sources at once.

Switch Identification

Switches vary in grade, capacity, and purpose. It is very important that proper types of switches are selected for the given application. For example, most single-pole toggle switches used for the control of lighting are restricted to ac use only. This same switch is not suitable for use on, say, a 32-Vdc emergency lighting circuit. A switch rated for ac only will not extinguish a dc arc quickly enough. Not only is this a dangerous practice (causing arcing and heating of the device), the switch contacts would probably burn up after only a few operations of the handle, if not the first time.

The illustration on the next page shows a typical single-pole toggle switch — the type most often used to control ac lighting in all installations. Note the identifying marks.

- The testing laboratory label
- The CSA (Canadian Standards Association) label

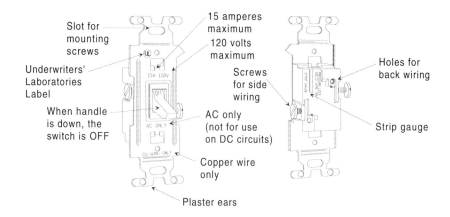

Typical identifying marks on a single-pole switch

- Type of conductor for which the terminals are designed
- Current and voltage ratings, listed by maximum amperage, maximum voltage, and current restrictions

The testing laboratory label is an indication that the device has undergone extensive testing by a nationally recognized testing lab and has met with the minimum safety requirements. The label does not indicate any type of quality rating. The switch shown is marked with the "UL" label which indicates that the device type was tested by Underwriters' Laboratories, Inc. of Northbrook, IL. ETL Testing Laboratories, Inc. of Cortland, NY is another nationally recognized testing laboratory. They provide a labeling, listing and follow-up service for the safety testing of electrical products to nationally recognized safety standards or specifically designated requirements of jurisdictional authorities.

The CSA (Canadian Standards Association) label is an indication that the material or device has undergone a similar testing procedure by the Canadian Standards Association and is acceptable They are similar to those on the duplex receptacle discussed previously. The main difference is the "T" rating which means that the switch is rated for switching lamps with tungsten filaments (incandescent lamps).

Screw terminals are color-coded on conventional toggle switches. Switches are typically constructed with a ground screw attached to the metallic strap of the switch. The ground screw is usually a green-colored hex-head screw. This screw is for connecting the equipment-grounding conductor to the switch. On three-way switches, the common or pivot terminal usually has a black or bronze screw head.

The switch shown is the type normally used for residential construction. Heavier-duty switches are usually the type used on commercial wiring — some of which are rated for use on 277-V circuits with current-carrying ratings up to 30 A. Therefore, it is important to check the rating of each switch before it is installed.

The exact type and grade of switch to be used on a specific installation is often dictated by the project drawings or written specifications. Sometimes wall switches are specified by manufacturer and catalog number; other times they are specified by type, grade, voltage, current rating, and the like, leaving the contractor or electrician to select the manufacturer. The naming of a certain brand of switch for a particular project, does not necessarily mean that this brand must be used. A typical paragraph from an electrical specification (concerning the substitution of materials) may read as follows:

The naming of a certain brand or make or manufacturer in the specifications is to establish a quality standard for the article desired. The contractor is not restricted to the use of the specific brand of the manufacturer named unless so indicated in the specifications. However, where a substitution is requested, a substitution will be permitted only with the written approval of the engineer. No substitute material or equipment shall be ordered, fabricated, shipped, or processed in any manner prior to the approval of the architect-engineer. The contractor shall assume all responsibility for additional expenses as required in any way to meet changes from the original material or equipment specified. If notice of substitution is not furnished to the architect-engineer within ten days after the contract is awarded, the equipment and materials named in the specifications are to be used.

Electrical specifications dealing with wall switches are covered in at least two sections of the specifications:

- 16100 Basic Materials and Methods
- 16500 Lighting

Brief excerpts from these two sections follow:

SECTION 16B - BASIC MATERIALS AND WORKMANSHIP

1. Portions of the sections of the Documents designated by the letters "A", "B" & "C" and "DIVISION ONE - GENERAL REQUIREMENTS" apply to this Division.

2. Consult Index to be certain that set of Documents and Specifications is complete. Report omissions or discrepancies to the architect.

   c. SWITCH OUTLET BOXES: Wall switches shall be mounted approximately 54 inches above the finished floor (AFF) unless otherwise noted. When the switch is mounted in a masonry wall the bottom of the outlet box shall be in line with the bottom of a masonry unit. Where more than two switches are located, the switches shall be mounted in a gang outlet box with gang cover. Dimmer switches shall be individually mounted unless otherwise noted. Switches with pilot lights, switches with overload motor protection and other special switches that will not conveniently fit under gang wall plates may be individually mounted.

13. EQUIPMENT AND INSTALLATION WORKMANSHIP:

   a. All equipment and material shall be new and shall bear the manufacturer's name and trade name. The equipment and material shall be essentially the standard product of a manufacturer regularly engaged in the production of the required type of equipment and shall be the manufacturer's latest approved design.

   b. The Electrical Contractor shall receive and properly store the equipment and material pertaining to the electrical work. The equipment shall be tightly covered and protected against dirt, water, chemical or mechanical injury and theft. The manufacturer's directions shall be followed completely in the delivery, storage, protection and installation of all equipment and materials.

c. The Electrical Contractor shall provide and install all items necessary for the complete installation of the equipment as recommended or as required by the manufacturer of the equipment or required by code without additional cost to the Owner, regardless whether the items are shown on the plans or covered in the Specifications.

d. It shall be the responsibility of the Electrical Contractor to clean the electrical equipment, make necessary adjustments and place the equipment into operation before turning equipment over to the Owner. Any paint that was scratched during construction shall be "touched-up" with factory color paint to the satisfaction of the Architect. Any items that were damaged during construction shall be replaced.

6. **WIRING DEVICES:**

   a. GENERAL: The wiring devices specified below with ARROW HART numbers may also be the equivalent wiring device as manufactured by BRYANT ELECTRIC, HARVEY HUBBELL or PASS & SEYMOUR. All other items shall be as specified.

   b. WALL SWITCHES: Where more than one flush wall switch is indicated in the same location, the switches shall be mounted in gangs under a common plate.

(1) Single Pole	AH#1991
(2) Three-Way	AH#1993
(3) Four-Way	AH#1994
(4) Switch with pilot light	AH#2999-R
(5) Motor Switch - Surface	AH#6808
(6) Motor Switch - Flush	AH#6808-F

   c. WALL PLATE: Stainless steel wall plates with satin finish minimum .030 inches shall be provided for all outlets and switches.

In general, the preceding electrical specifications give the grade of materials to be used on the project and the manner in which the electrical system must be installed. Most specification writers use an abbreviated language; although it is relativley difficult for beginners to understand, experience makes possible a proper interpretation with little difficulty. However, electricians involved with any project should make certain that everything is clear. If it is not, contact the architectural or engineering firm and clarify the problem prior to installing the work, not after a system has been completed.

NEC Requirements For Switches

The NEC requirements for installing light switches are many and are scattered in various locations in the NEC book. For example, wall-switch controlled lighting outlets are required in each habitable room of all residential occupancies. Wall-switch controlled lighting is also required in each bathroom, hallways, stairways, attached garages, and at outdoor entrances. A wall-switch controlled receptacle may be used in place of the lighting outlet in habitable rooms other than the kitchen and bathrooms. Providing a wall switch for room lighting is intended to prevent an occupant's groping in the dark for table lamps or pull chains. In stairways with six or more steps, the stairway lighting must be controlled at two locations — at the top and also the bottom of the stairway. This is accomplished by using two 3-way switches and connected as discussed later in this chapter.

Lighting outlets are also required in attics, crawl spaces, utility rooms, and basements when these spaces are used for storage or contain equipment such as HVAC equipment. Again, if the basement or attic stairs have more than 6 steps, a 3-way switch is required at each landing.

At least one wall switch-controlled lighting outlet is required in each guest room in hotels, motels, or similar locations. Note that a wall switch-controlled receptacle is permitted in lieu of the lighting outlet.

At least one wall switch-controlled lighting outlet must be installed at or near equipment requiring servicing such as HVAC equipment. The wall switch must be located at the point of entry to the attic or underfloor space.

In many commercial installations, circuit breakers in panelboards are permitted to control main-area lighting where the areas are constantly illuminated during operating hours. Consequently, wall switches are not required in these areas. However, wall switches are normally installed at outdoor entrances, entrances to store rooms, small offices, toilets, and similar locations.

Lighting Control

Many lighting-control devices have been developed since Edison's first lamp. They have been designed to make the best use of the lighting equipment provided by the lighting industry. These include:

- Automatic timing devices for outdoor lighting
- Dimmers for residential lighting
- The common single-pole, 3-way, and 4-way switches used in nearly every home in the nation.

Switches for lighting control fall into the following basic categories:

- Snap-action switches
- Mercury switches
- Quiet switches

Snap-Action Switches: A single-pole snap-action switch consists of a device containing two stationary current-carrying elements, a moving current-carrying element, a toggle handle, a spring, and a housing. When the contacts are open, the circuit is "broken" and no current flows. When the moving element is closed, by manually flipping the toggle handle, the contacts complete the circuit and the lamp will be energized.

Mercury Switch: Mercury switches consist of a sealed capsule containing mercury. Inside the capsule are contacting surfaces "A" and "B," which may be part of the wall of the capsule. The switch is operated by means of a handle which moves the capsule.

The capsule is tilted so that the mercury "C" has collected at one end of the capsule. Here, it bridges two contact points, "A" and "B," to complete the circuit and light the lamp. However, if the capsule is tilted the opposite way, the circuit between contacts "A" and "B" will

When the contacts are open, the circuit is "broken" and no current flows

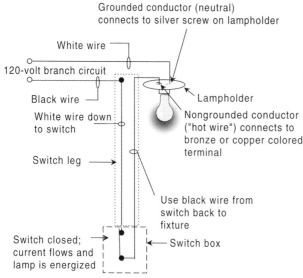

When the moving element is closed, by manually flipping the toggle handle, the contacts complete the circuit and the lamp will be energized

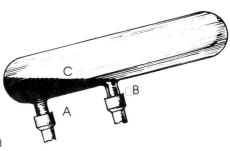

Interior view of a mercury capsule used in a mercury switch

not be completed, and the lamp will be de-energized (be turned off). Mercury switches offer the ultimate in silent operation and are recommended where the "clicking" of a light switch may be annoying.

Quiet Switch: The quiet switch is a compromise between the snap-action switch and the mercury switch. Its operation is much quieter than the snap-action switch, yet it is not as expensive as the mercury switch.

The quiet switch consists of a stationary contact and a moving contact that are close together when the switch is open. Only a short, gentle movement is applied to open and close the switch, producing very little noise. This type of switch may be used only on alternating current, since the arc will not extinguish on direct current.

The quiet switch is the most commonly used switch for modern lighting practice. These switches are common for loads from 10 to 20 A, in single-pole, three-way, four-way, etc., configurations.

Many other types of switches are available for lighting control. One type of switch used mainly in residential occupancies is the door-actuated type which is generally installed in the door jamb of a closet to control a light inside the closet. When the door is open, the light comes on; when the door is closed, the light goes out. Refrigerator and oven lights are usually controlled by door switches.

The Despard switch is another special switch. Due to its small size, up to three may be mounted in a standard single-gang switch box. Weatherproof switches are made for outdoor use. Combination switch-indicator light assemblies are also available for use where the light cannot be seen

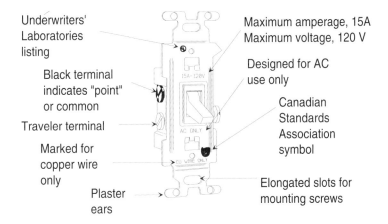

Characteristics of a typical three-way switch

from the switch location, such as an attic or garage. Switches are also made with small neon lamps in the handle that light when the switch is off. These low-current-consuming lamps make the switches easy to find in the dark.

Three-Way Switches

Three-way switches are used to control one or more lamps from two different locations, such as at the top and bottom of stairways, in a room that has two entrances, etc. A typical three-way switch is shown above. Note that there are no ON/OFF markings on the handle. Furthermore, a three-way switch has three terminals. The single terminal at one end of the switch is called the common (sometimes hinge point). This terminal is easily identified because it is darker than the two other terminals. The feeder ("hot" wire) or switch leg is always connected to the common dark or black terminal. The two remaining terminals are called traveler terminals. These terminals are used to connect three-way switches together.

The connections of three-way switches begin on the next page. By means of the two three-way switches, it is possible to control the lamp from two locations. By tracing the circuit, it may be seen how these three-way switches operate.

A 120-V circuit emerges from the left side of the drawing. The white or neutral wire connects directly to the neutral terminal of the lamp. The "hot" wire carries current, in the direction of the arrows, to the common terminal of the three-way switch on the left. Since the handle is in the up position, the current continues to the top traveler terminal and is carried by this

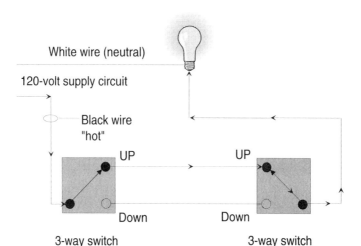

Completed circuit by means of 2 three-way switches with both switch handles in the UP position

traveler to the other three-way switch. Note that the handle is also in the up position on this switch; this picks up the current flow and carries it to the common point, which, in turn, continues on to the ungrounded terminal of the lamp to make a complete circuit. The lamp is energized.

Moving the handle to a different position on either three-way switch will break the circuit, which in turn, deenergizes the lamp. For example, if a person leaves the room at the point of the three-way switch on the left. The switch handle is flipped down, giving a condition as shown below. Note that the current flow is now directed to the bottom traveler terminal, but

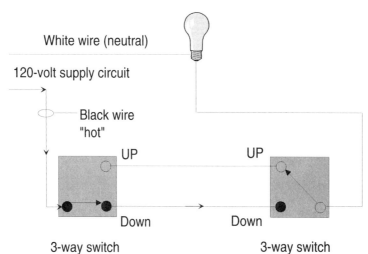

Any change of either switch handle from the previous diagram will deenergize the circuit

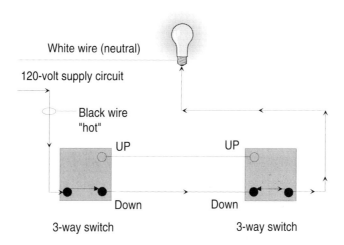

Both handles in the down position once again energizes the circuit

since the handle of the three-way switch on the right is still in the up position, no current will flow to the lamp.

Another person enters the room at the location of the three-way switch on the right. The handle is flipped downward which gives the condition as shown above. This change provides a complete circuit to the lamp which causes it to be energized. In this example, current flow is on the bottom traveler. Again, changing the position of the switch handle (pivot point) on either three-way switch will deenergize the lamp.

In actual practice, the exact wiring of the two three-way switches to control the operation of a lamp will be slightly different than the routing shown in these three diagrams. There are several ways that two three-way switches may be connected. One solution is shown in the illustration at the top of the next page. Here, 14/2 w/ground NM cable (Romex) is fed to the three-way switch on the left. The black or "hot" conductor is connected to the common terminal on the switch, while the white or neutral conductor is spliced to the white conductor of 14/3 w/ground NM cable leaving the switch. This 3-wire cable is necessary to carry the two travelers plus the neutral to the three-way switch on the right. At this point, the black and red wires connect to the two traveler terminals, respectively. The white or neutral wire is again spliced — this time to the white wire of another 14/2 w/ground NM cable. The neutral wire is never connected to the switch itself. The black wire of the 14/2 w/ground NM cable connects to the common terminal on the three-way switch. This cable — carrying the

360　Handbook of Electrical Construction Tools and Materials

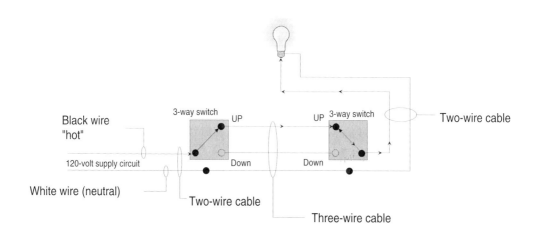

Practical wiring method for 2 three-way switches controlling a lamp or series of lamps

"hot" and neutral conductors are routed to the lighting fixture outlet for connection to the fixture. The illustration below shows how this wiring arrangement will appear on an electrical floor-plan drawing.

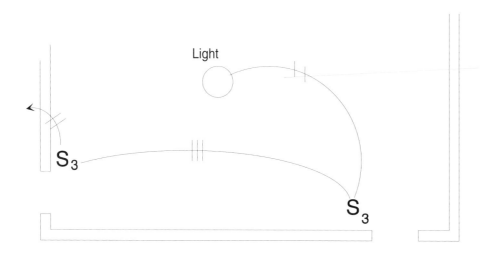

Floor-plan layout of three-way switches controlling a lighting outlet

Another solution is to feed the lighting-fixture outlet with 2-wire cable. Run another 2-wire cable — carrying the "hot" and neutral conductors — to one of the three-way switches. A three-wire cable is pulled between the two three-way switches, and then another 2-wire cable is routed from the other 3-way switch to the lighting-fixture outlet.

Some electricians use a short-cut method by eliminating one of the 2-wire cables in the preceding method. Rather, a 2-wire cable is run from the lighting-fixture outlet to one three-way switch. Three-wire cable is pulled between the two three-way switches — two of the wires for travelers and the third for the common-point return. This method should not be used with a metallic conduit system.

Four-Way Switches

Two three-way switches may be used in conjunction with any number of four-way switches to control a lamp, or a series of lamps, from any number of positions. When connected correctly, the actuation of any one of these switches will change the operating condition of the lamp(s); that is, either turn the lamp on or off.

The illustration below shows how a four-way switch may be used in combination with two three-way switches to control a device from three locations. In this example, note that the "hot" wire is connected to the common terminal on the three-way switch on the left. Current then travels

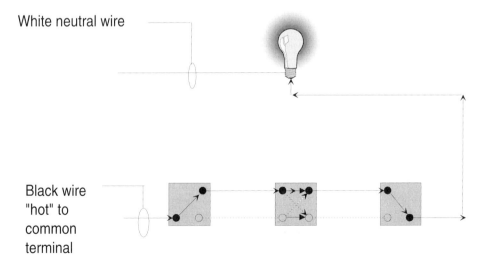

Two three-way switches and one four-way switch connected to control a lighting outlet from three locations

to the top traveler terminal and continues on the top traveler conductor to the four-way switch. Since the handle is up on the four-way switch, current flows through the top terminals of the switch and onto the traveler conductor going to the other three-way switch. Again, the switch is in the up position. Therefore, current is carried from the top traveler terminal to the common terminal and on to the lighting fixture to energize it. Under this condition, if the position of any one of the three switch handles are changed, the circuit will be broken and no current will flow to the lamp.

Any number of four-way switches may be used in combination with two three-way switches, but two three-way switches are always necessary for the correct operation of one or more four-way switches.

Photoelectric Switches

The chief application of the photoswitch is to control outdoor lighting, especially the "dusk-to-dawn" lights found in suburban areas.

This interesting switch has an endless number of possible uses and is a great tool for electricians dealing with outdoor lighting situations.

Relays

Next to switches, relays play the most important part in the control of light. An electric relay is a device whereby an electric current causes the opening or closing of one or more pairs of contacts. These contacts are usually capable of controlling much more power than is necessary to operate the relay itself. This is one of the main advantages of relays.

One popular use of the relay in residential lighting systems is that of remote-control lighting. In this type of system, all relays are designed to operate on a 24-V circuit and are used to control 120-V lighting circuits. They are rated at 20 A which is sufficient to control the full load of a normal lighting branch circuit, if desired.

Remote-control switching makes it possible to install a switch wherever it is convenient and practical to do so or wherever there is an obvious need for having a switch — no matter how remote it is from the lamp or lamps it is to control. This method enables lighting designs to achieve new advances in lighting control convenience at a reasonable cost. Remote-control switching is also ideal for rewiring existing homes with finished walls and ceilings.

One relay is required for each fixture or for each group of fixtures that are controlled together. Switch locations for remote-control follow the same rules as for conventional direct switching. However, since it is easy to add switches to control a given relay, no opportunities should be overlooked for adding a switch to improve the convenience of control.

Remote-controlled lighting also has the advantage of using selector switches at certain locations. For example, selector switches located in the master bedroom or in the kitchen of a home enable the owner to control every lighting fixture on the property from this location. The selector switch may turn on and off an outside or other light which customarily would be left on until bedtime and which might otherwise be forgotten.

Dimmers

Dimming a lighting system provides control of the quantity of illumination. It may be done to create various atmospheres and moods or to blend certain lights with others for various lighting effects.

For example, in homes with separate dining rooms, a chandelier mounted directly above the dining table and controlled by a dimmer/switch becomes the centerpiece of the room while providing general illumination. The dimmer adds versatility since it can set the mood of the activity — low brilliance (candlelight effect) for formal dining or bright for an evening of cards. When chandeliers with exposed lamps are used, the dimmer is essential to avoid a garish and uncomfortable atmosphere. The size of the chandelier is also very important; it should be sized in proportion to the size of the dining area.

Chapter 26

Contactors and Relays

The general classification of contactors covers a type of electromagnetic apparatus designed to handle relatively high currents. The conventional contactor is identical in appearance, construction and current carrying ability to the equivalent NEMA size magnetic motor starter. The magnet assembly and coil, contacts, holding circuit interlock and other structural features are the same.

The significant difference is that the contactor does not provide overload protection. Contactors, therefore, are used to switch high current, nonmotor loads, or are used in motor circuits if overload protection is separately provided. A typical application of the latter is in a reversing motor starter.

A relay is an electromagnetic device whose contacts are used in control circuits of magnetic starters, contactors, solenoids, timers and other relays. Relays are generally used to amplify the contact capability or multiply the switching functions of a pilot device.

A tremendous variety of relays are used for lighting, motor, and HVAC control circuits, but the magnetic relay is still probably the most common, although solid-state control devices are rapidly finding their way into all types of applications.

Besides controlling electrical circuits, some types of relays are also used for circuit and equipment protection. For example, thermal overload relays are intended to protect motors, controllers, and branch-circuit conductors against excessive heating due to prolonged motor overcurrents up to and including locked rotor currents. These relays sense motor current by converting this current to heat in a resistance element. The heat generated is used to open a normally closed contact in series with a starter coil causing the motor to be disconnected from the line.

There is also a great variety of electrical protective relays installed in modern power systems, including:

- Overcurrent — time-delay and instantaneous
- Over/under voltage
- Directional over-current
- Percentage differential

MAGNETIC CONTACTORS

The typical magnetic contactor contains a coil of wire wound around an iron core or a laminated iron core and an armature. The coil acts as an electromagnet, and when the proper electric current is applied, it attracts, or picks up, the armature. Since the movable contacts are attached via an iron bar to the armature, when the armature moves upwards, the movable contacts make contact with a set of stationary contacts to close and complete the circuit.

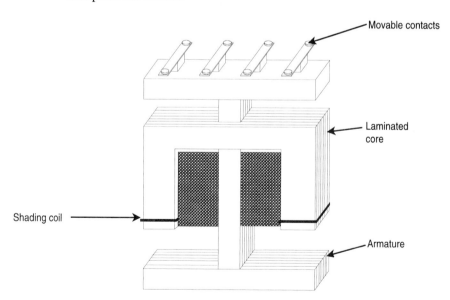

Operating principles of a magnetic contactor

When power is being applied to the coil, the device is said to be energized. The motion of the armature moves contact points together or apart, according to their arrangement. When the power is disconnected from the coil, a spring returns the armature to its original position. When no power is being applied, the device is in its de-energized, or normal, condition. Contacts that are together when the relay is de-energized are said to be normally closed (N.C.). Contacts that are apart when the relay is de-energized are said to be normally open (N.O.).

Notice that a shading coil has been added to the iron core in this arrangement. The shading coil is used with ac relays to prevent contact chatter and hum. The shading coil operates in the same way it does in the shaded-pole motor covered in an earlier module. In general, the shading coil is used to provide a continuous magnetic flow to the armature when the voltage of the ac waveform is zero.

The relay operates on the solenoid principle; that is, it converts electrical energy to linear motion as shown below. In this example, a coil of wire is wound around an iron core. When current flows through the coil, a magnetic field is developed in the iron core. The magnetic field of the iron core attracts the movable arm, known as the armature. In its present position, the movable contact makes connection with a stationary contact. This contact set is normally closed. When the armature is attracted to the iron core, the movable contact breaks connection with one stationary

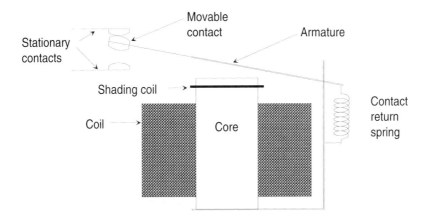

Contactor operating on the solenoid principle

contact and makes connection to another. This relay has both a normally open and normally closed set of contacts. The movable contact would be the common and the stationary contacts would be labeled normally open and normally closed.

Lighting Contactors

Filament type lamps (tungsten, infrared, quartz) have inrush currents of approximately 15-17 times the normal operating currents. Standard motor control contactors must be derated if used to control this type of load, to prevent welding of the contacts on the high initial current.

Lighting contactors differ from standard contactors in that the contact tip material is a silver tungsten carbide which resists welding on high initial currents. A holding circuit interlock is not normally provided, since this type of contactor is frequently controlled by a 2-wire pilot device such as a time clock or photo-electric relay.

Unlike standard contactors, lighting contactors are not horsepower rated or categorized by NEMA size, but are designated by ampere ratings (20, 30, 60, 100, 200, 300 amperes). It should be noted that lighting contactors are specialized in their application, and should not be used on motor loads.

A basic circuit using a magnetic relay and serving as a control for a high amperage lamp is shown below. The purpose is to control the lamp from two different locations. The relay coil is indicated by the standard symbol

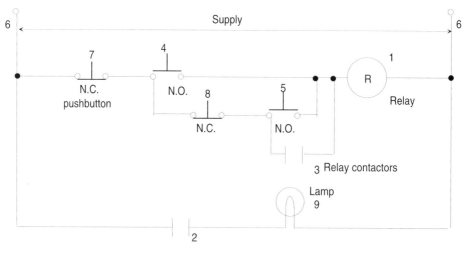

Pushbutton control circuit with magnetic relay

at 1 and the relay contactors are indicated at 2 and 3 in the diagram by two parallel lines. The contactors are drawn separated from the coil 1 to show their function. Actually, they are next to the coil 1, in the relay itself.

When either of the normally open pushbuttons (4 and 5) is momentarily pressed down, it closes the circuit from the supply-line terminals 6 through the coil 1. The normally closed pushbuttons (7 and 8) are in the path, but since they are not actuated, the current is not interrupted. The now energized coil 1 acts on the contactors 2 and 3 and closes their contacts. The contacts of the contactor 2 have closed a path for current from the supply line through the lamp 9, and the lamp is lighted. The contacts of the contactor 3 keep the relay coil 1 energized, since the button 4 or 5 was pressed only momentarily and was then released. Contactor 3 is known as holding contacts.

The lamp will remain lighted until either of the normally closed pushbuttons (7 and 8) is momentarily pressed. Pressing either of these buttons opens the circuit to the coil 1. The coil will drop out, or become de-energized, and will open the contacts of the contactors 2 and 3. The circuit through the lamp 9 is now open and the lamp is turned off. The lamp remains off until one of the pushbuttons (4 and 5) is pressed again. The pushbuttons 4 and 7 (start and stop) are placed close together at one location where the remote control of the lamp 9 is desired, and pushbuttons 5 and 8 (another set of start/stop buttons) in another convenient location. From either one of these locations the lamp can be turned on and off.

Another application of a magnetic relay is its use in an emergency-lighting circuit. The lamps in such a circuit should be lighted instantly after the failure of the main-power supply. A simple emergency-lighting system controlling one lamp is shown in the illustration at the top of the next page. The main power is supplied through terminals 1 to the relay coil 2. The relay has a normally open contactor 3 and a normally closed contactor 4. As long as the main power is available, it energizes the relay coil, which in turn keeps the contacts of the contactor 3 closed. The power is therefore supplied to the transformer 5, whose primary is connected in series with contactor 3. The secondary of the transformer keeps the storage battery 6 charged through the rectifier 7. Since the coil 2 is energized, the contactor 4 is open and the emergency lamp 8 is not lighted.

In the event of main-power failure or low voltage in the line, the relay will drop out, and the contactor 3 will open and contactor 4 close. The lamp will be connected to the battery and give light as long as the relay coil 2 remains de-energized. Such relay and circuit arrangements are frequently used in emergency battery packs found in many commercial, industrial, and public buildings of all types.

370　Handbook of Electrical Construction Tools and Materials

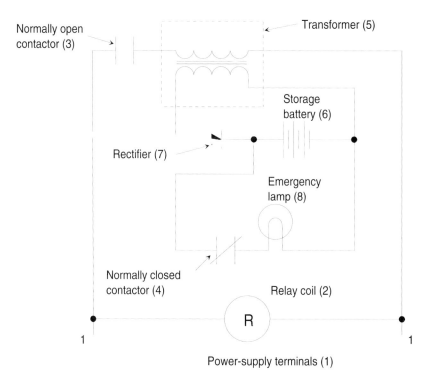

Emergency-lighting circuit using a relay for operation

The following example demonstrates how a relay and one single-pole, 15-A toggle switch can be used to control 60 A of lighting.

The illustration on the opposite page shows a diagram of a relay controlled by a single-pole switch. The relay operates six sets of contacts, each of which is connected to a lighting circuit with lamps totaling 1200 watts (10 A). When the single-pole switch is turned to the ON position (contacts closed), the relay coil is energized and moves the armature, which in turn closes all of the six contacts simultaneously. When these contacts are closed, this completes the circuit to all of the lamps. Thus, all lamps will come on at the same time.

Six single-pole switches could have been used in place of the relay, but simultaneous switching of the lamps would then not be possible.

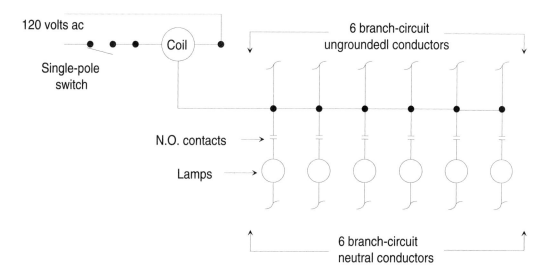

Several lighting circuits controlled by one relay

Mechanically Held Contacts

In a conventional contactor, current flow through the coil creates a magnetic pull to seal in the armature and maintain the contacts in a switched position; that is, N.O. contacts will be held closed while N.C. contacts will be held open. Because the contactor action is dependent on the current flow through the coil, the contactor is described as electrically held. As soon as the coil is de-energized, the contacts will revert to their initial position.

Mechanically held versions of contactors and relays are also available and used in some applications. The action is accomplished through use of two coils and a latching mechanism. Energizing one coil (latch coil) through a momentary signal causes the contacts to switch, and a mechanical latch holds the contacts in this position, even though the initiating signal is removed, and the coil is de-energized. To restore the contacts to their initial position, a second coil (unlatch coil) is momentarily energized.

Mechanically held contactors and relays are used where the slight hum of an electrically held device would be objectionable, as in auditoriums, hospitals, churches, and the like.

Relays

A control relay is an electromagnetic device, similar in operating characteristics to a contactor. The contactor, however, is generally employed to switch power circuits, or relatively high current loads.

Relays, with few exceptions, are used in control circuits, and consequently their lower ratings (15 A maximum at 600 V) reflect the reduced current levels at which they operate.

Contactors generally have from one to five poles. Although normally open and normally closed contacts are frequently encountered, the great majority of applications use the normally open contact configuration, and there is little, if any, conversion of contact operation on the job.

As compared to contactors, it is not uncommon to find relays used in applications requiring 10 or 12 poles per device, with various combinations of normally open and normally closed contacts. In addition, some relays have convertible contacts, permitting changes to be made in the field from N.O. to N.C. operation, or vice versa, without requiring modification kits or additional components.

Relays are commonly used in complex controllers to provide the logic or "brains" to set up and initiate the proper sequencing and control of a number of interrelated operations.

Relays differ in voltage ratings (150, 300, 600 V), number of contacts, contact convertability, physical size, and in attachments to provide accessory functions such as mechanical latching and timing.

In selecting a relay for a particular application, one of the first steps should be a determination of the control voltage at which the relay will operate. Once the voltage is known, the relays that have the necessary contact rating can be further reviewed, and a selection made, on the basis of the number of contacts and other characteristics needed.

Timers and Timing Relays

A pneumatic timer or timing relay is similar to a control relay, except that certain of its contacts are designed to operate at a preset time interval after the coil is energized or deenergized. A delay on energization is also referred to as "on delay." A time delay on de-energization is also called "off delay."

A timed function is useful in applications such as the lubricating system of a large industrial machine, in which a small oil pump must deliver lubricant to the bearings of the main motor for a set period of time before the main motor starts.

In pneumatic timers, the timing is accomplished by the transfer of air through a restricted orifice. The amount of restriction is controlled by an adjustable needle valve, permitting changes to be made in the timing period.

Timing relays are often motor driven (clock type mechanism); solid state RC timing circuits will also be frequently encountered.

Solid-State Relays

The solid-state relay is a device that has become increasingly popular for switching applications — the same as described for magnetic relays. The solid-state relay has no moving parts, it is resistant to shock and vibration, and is sealed against dirt and moisture. The greatest advantage of the solid-state relay, however, is the fact that the control input voltage is isolated from the load circuit which is controlled by the relay.

Solid-state relays can be used to control either a dc load or an ac load. If the relay is designed to control a dc load, a power transistor is used to connect the load to the line.

The control voltage for most solid-state relays ranges from about 3 to 32 V and can be either dc or ac. If a triac is used as the control device, load voltage ratings of 120 to 240 Vac are common and current ratings can range from 5 to 25 A. Many solid-state relays have a feature known as zero-switching. Zero switching means that if the relay is programmed to turn off when the ac voltage is in the middle of a cycle, it will continue to conduct until the ac voltage drops to a zero level and then turns off. For example, assume the ac voltage is at its positive peak value when the gate instructs the triac to turn off. The triac will continue to conduct until the ac voltage drops to a zero level before it actually turns off. Zero switching can be a great advantage when used with some inductive loads such as transformers. The core material of a transformer can be left saturated on one end of the flux swing if power is removed from the primary winding when the ac voltage is at its positive or negative peak. This can cause inrush currents of up to 600 percent of the normal operating current when power is restored to the primary.

Solid-state relays are available in different styles and ratings. Some are designed to be used as time-delay relays. One of the most common uses for a solid-state relay is the I/O track of a programmable controller—a controller than can be changed by reprogramming rather than requiring an extensive rewiring of the control system. In application, as an input, the relay switches a large (120V) signal to a low (5Vdc) level signal.

Low-Voltage Remote-Control Switching

In applications where lighting must be controlled from several points, or where there is a complexity of lighting or power circuits, or where flexibility is desirable in certain systems, low-voltage remote-controlled relay systems have been applied. Basically, these systems use special low-voltage components, operated from a transformer, to switch relays which in turn control the standard line voltage circuits. Because the control wiring does not carry the line load directly, small lightweight cable can be used. It can be installed wherever and however convenient — placed behind moldings, stapled to woodwork, buried in shallow plaster channels, or installed in holes bored in wall studs.

Chapter 27

Motor Controls

Electric motors provide one of the principal sources for driving all types of equipment and machinery. Every motor in use, however, must be controlled, if only to start and stop it, before it becomes of any value.

Motor controllers cover a wide range of types and sizes, from a simple toggle switch to a complex system with such components as relays, timers, and switches. The common function, however, is the same in any case, that is, to control some operation of an electric motor. A motor controller will include some or all of the following functions:

- Starting and stopping
- Overload protection
- Overcurrent protection
- Reversing
- Changing speed
- Jogging
- Plugging
- Sequence control
- Pilot light indication

The controller can also provide the control for auxiliary equipment such as brakes, clutches, solenoids, heaters, and signals, and may be used to control a single motor or a group of motors.

The term *motor starter* is often used and means practically the same thing as a *controller*. Strictly, a motor starter is the simplest form of controller and is capable of starting and stopping the motor and providing it with overload protection.

Types of Motor Controllers

A large variety of motor controllers are available that will handle almost every conceivable application. However, all of them can be grouped in the following categories.

Plug-and-Receptacle

The National Electrical Code (NEC) defines a controller as any switch or device normally used to start and stop a motor by making and breaking the motor circuit current. The simplest form of controller allowed by the NEC is an attachment plug and receptacle. See the illustration on the opposite page. However, such an arrangement is limited to portable motors rated at $1/3$ horsepower (hp) or less.

Referring again to the illustration, note that drawing (A) is a pictorial view of a portable drill motor with a cord-and-plug assembly attached. If this motor is portable and less than $1/3$ horsepower (hp), then the plug and receptacle may act as the motor's controller as permitted in NEC Section 430-81(c).

Drawing (B) is the same circuit, but this time depicted in the form of a wiring diagram. Note that symbols have been used to represent the various circuit items rather than actually drawing the items in life form; yet they are arranged on the basis of their physical relationship to each other. This simplifies the drawing — both from a drafter's point of view and also those who must interpret the drawing.

Another form of drawing for this same circuit is shown in (C). This type of drawing has become known as *ladder diagram*; it's a schematic representation of the electrical circuit in question — the same as the drawing in (B). However, ladder diagrams are drawn in an "H" format, with the energized power conductors represented by vertical lines and the individual circuits represented by horizontal lines. Rather than physically representing the circuit items as in drawing (B), a ladder diagram arranges the conductors and electrical components according to their electrical function

Motor Controls

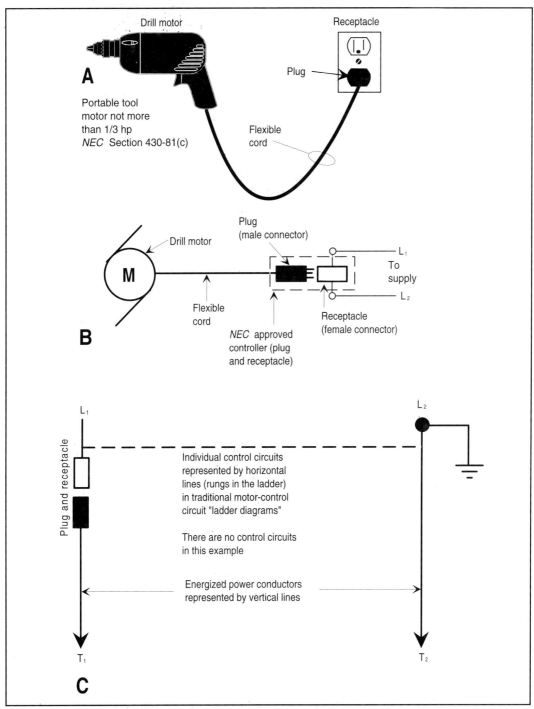

Various types of drawings for a plug-and-receptacle motor controller

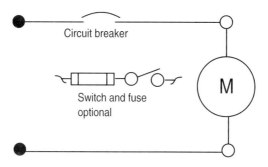

Branch-circuit protective device serving as the motor controller

in the circuits; that is, schematically. Therefore, ladder diagrams merely represent the current paths (shown as the rungs of a ladder) to each of the controlled or energized output devices.

Where stationary motors rated at $\frac{1}{8}$ hp or less and which are normally left running (clock motors, fly fans, and the like), and are so constructed that they cannot be damaged by overload or failure to start, the branch-circuit protective device may serve as the the controller. Consequently, the branch-circuit breaker or fusible disconnect serves as both branch-circuit overcurrent protection and motor controller. Such a circuit appears above.

Manual Starters

A manual starter is a motor controller whose contact mechanism is operated by a mechanical linkage from a toggle handle or pushbutton, which is in turn operated by hand. A thermal unit and direct-acting overload mechanism provide motor running overload protection. Basically, a manual starter is an ON-OFF switch with overload relays.

Manual starters are used mostly on small machine tools, fans and blowers, pumps, compressors, and conveyors. They have the lowest cost of all motor starters, have a simple mechanism, and provide quiet operation with no ac magnet hum. The contacts, however, remain closed and the lever stays in the ON position in the event of a power failure, causing the motor to automatically restart when the power returns. Therefore, low-voltage protection and low-voltage release are not possible with these manually operated starters. However, this action is an advantage when the starter is applied to motors that run continuously.

Fractional Horsepower Manual Starters

Fractional-horsepower manual starters are designed to control and provide overload protection for motors of 1 hp or less on 120- or 240-V single-phase circuits. They are available in single- and two-pole versions and are operated by a toggle handle on the front. When a serious overload occurs, the thermal unit trips to open the starter contacts, disconnecting the motor from the line. The contacts cannot be reclosed until the overload relay has been reset by moving the handle to the full OFF position, after allowing about 2 minutes for the thermal unit to cool. The open-type starter will fit into a standard outlet box and can be used with a standard flush plate. The compact construction of this type of device makes it possible to mount it directly on the driven machinery and in various other places where the available space is small.

Note in the diagrams that the single-pole FHP starter has only one contact to trip and disconnect the motor from the line; the grounded or neutral conductor is not opened when the handle is in the OFF position. This single-pole starter also has one overload relay connected in series with the ungrounded conductor.

The two-pole FHP starter has two contacts to open both phases when connected to a 240-V circuit. When the toggle handle is in the off position, no current flows to the motor. However, only one overload relay is needed, since one will shut down the motor if the relay detects an overload and opens.

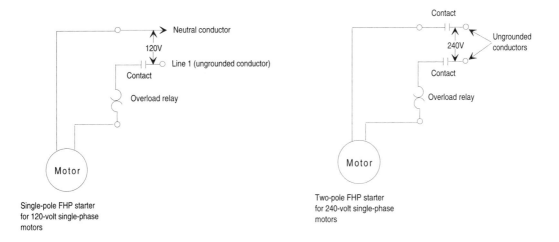

Single-pole FHP starter for 120-volt single-phase motors

Two-pole FHP starter for 240-volt single-phase motors

Wiring diagrams for fractional-horsepower manual motor starters

Manual Motor-Starting Switches

Manual motor starting switches provide ON-OFF control of single- or three-phase ac motors where overload protection is not required or is separately provided. Two- or three-pole switches are available with ratings up to 10 hp, 600 V, three phase. The continuous current rating is 30 A at 250 V maximum and 20 A at 600 V maximum. The toggle operation of the manual switch is similar to the fractional-horsepower starter, and typical applications of the switch include pumps, fans, conveyors, and other electrical machinery that have separate motor protection. They are particularly suited to switch nonmotor loads, such as resistance heaters.

Integral Horsepower Manual Starters

The integral horsepower manual starter is available in two- and three-pole versions to control single-phase motors up to 5 hp and polyphase motors up to 10 hp, respectively.

Two-pole starters have one overload relay and three-pole starters usually have three overload relays. When an overload relay trips, the starter mechanism unlatches, opening the contacts to stop the motor. The contacts cannot be reclosed until the starter mechanism has been reset by pressing the STOP button or moving the handle to the RESET position, after allowing time for the thermal unit to cool.

Integral horsepower manual starters with low-voltage protection prevent automatic start-up of motors after a power loss. This is accomplished with a continuous-duty solenoid, which is energized whenever the line-side voltage is present. If the line voltage is lost or disconnected, the solenoid de-energizes, opening the starter contacts. The contacts will not automatically close when the voltage is restored to the line. To close the contacts, the device must be manually reset. This manual starter will not function unless the line terminals are energized. This is a safety feature that can protect personnel or equipment from damage

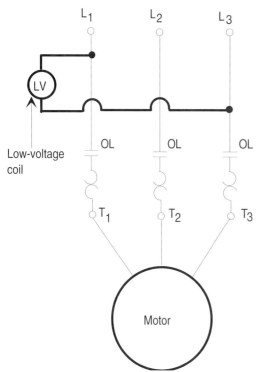

Integral hp manual starter with low-voltage protection

and is used on such equipment as conveyors, grinders, metal-working machines, mixers, woodworking, etc.

Magnetic Controllers

Magnetic motor controllers use electromagnetic energy for closing switches. The electromagnet consists of a coil of wire placed on an iron core. When current flows through the coil, the iron of the magnet becomes magnetized and attracts the iron bar, called the armature. An interruption of the current flow through the coil of wire causes the armature to drop out due to the presence of an air gap in the magnetic circuit.

Line-voltage magnetic motor starters are electromechanical devices that provide a safe, convenient, and economic means for starting and stopping motors, and they have the advantage of being controlled remotely. The great bulk of motor controllers are of this type. Therefore, the operating principles and applications of magnet motor controllers should be fully understood.

In the construction of a magnetic controller, the armature is mechanically connected to a set of contacts so that, when the armature moves to its closed position, the contacts also close. When the coil has been energized and the armature has moved to the closed position, the controller is said to be picked up and the armature is seated or sealed in. Some of the magnet and armature assemblies in current use are as follows:

- Clapper type: In this type, the armature is hinged. As it pivots to seal in, the movable contacts close against the stationary contacts.

- Vertical action: The action is a straight line motion with the armature and contacts being guided so that they move in a vertical plane.

- Horizontal action: Both armature and contacts move in a straight line through a horizontal plane.

- Bell crank: A bell crank lever transforms the vertical action of the armature into a horizontal contact motion. The shock of armature pickup is not transmitted to the contacts, resulting in minimum contact bounce and longer contact life.

The magnetic circuit of a controller consists of the magnet assembly, the coil, and the armature. It is so named from a comparison with an electrical

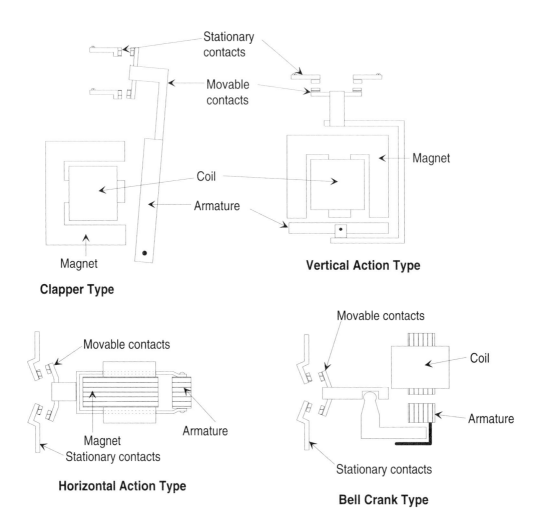

Several types of magnetic-armature assemblies

circuit. The coil and the current flowing in it causes magnetic flux to be set up through the iron in a similar manner to a voltage causing current to flow through a system of conductors. The changing magnetic flux produced by alternating currents results in a temperature rise in the magnetic circuit. The heating effect is reduced by laminating the magnet assembly and armature by placing a coil of many turns of wire around a soft iron core.

The magnetic flux set up by the energized coil tends to be concentrated; therefore, the magnetic field effect is strengthened. Since the iron core is the path of least resistance to the flow of the magnetic lines of force, magnetic attraction will concentrate according to the shape of the magnet.

The magnetic assembly is the stationary part of the magnetic circuit. The coil is supported by and surrounds part of the magnet assembly in order to induce magnetic flux into the magnetic circuit.

The armature is the moving part of the magnetic circuit. When it has been attracted into its sealed-in position, it completes the magnetic circuit. To provide maximum pull and to help ensure quietness, the faces of the armature and the magnetic assembly are ground to a very close tolerance.

When a controller's armature has sealed in, it is held closely against the magnet assembly. However, a small gap is always deliberately left in the iron circuit. When the coil becomes de-energized, some magnetic flux (residual magnetism) always remains, and if it were not for the gap in the iron circuit, the residual magnetism might be sufficient to hold the armature in the sealed-in position.

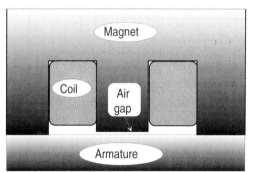

A small air gap is always deliberately left in the iron-core circuit

The shaded-pole principle is used to provide a time delay in the decay of flux in dc coils, but it is used more frequently to prevent a chatter and wear in the moving parts of ac magnets. A shading coil is a single turn of conducting material mounted in the face of the magnet assembly or armature. The alternating main magnetic flux induces currents in the shading coil, and these currents set up auxiliary magnetic flux that is out of phase from the pull due to the main flux, and this keeps the armature sealed-in when the main flux falls to zero (which occurs 120 times per second with 60-cycle ac). Without the shading coil, the armature would tend to open each time the main flux goes through zero. Excessive noise, wear on magnet faces, and heat would result. A magnet assembly and armature showing shading coils is shown in the illustration at the top of the next page.

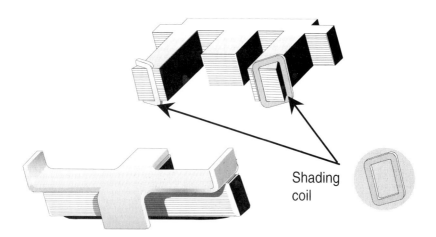

Magnet assembly and armature along with shading coils

Magnetic Coils

The magnetic coil used in motor controllers has many turns of insulated copper wire wound on a spool. Most coils are protected by an epoxy molding which makes them very resistant to mechanical damage.

When the controller is in the open position there is a large air gap (not to be confused with the built-in gap discussed previously) in the magnet circuit; this is when the armature is at its furthest distance from the magnet. The impedance of the coil is relatively low, due to the air gap, so that when the coil is energized, it draws a fairly high current. As the armature moves closer to the magnet assembly, the air gap is progressively reduced, and with it, the coil current, until the armature has sealed in. The final current is referred to as the sealed current. The inrush current is approximately 6 to 10 times the sealed current. The ratio varies with individual designs. After the controller has been energized for some time, the coil will become hot. This will cause the coil current to fall to approximately 80 percent of its value when cold.

AC magnetic coils should never be connected in series. If one device were to seal in ahead of the other, the increased circuit impedance will reduce the coil current so that the "slow" device will not pick up or, having picked up, will not seal. Consequently, ac coils are always connected in parallel.

Magnet coil data is usually given in volt-amperes (va). For example, given a magnetic starter whose coils are rated at 600 va inrush and 60 va sealed, the inrush current of a 120-V coil is 600/120 or 5 A. The same starter with a 480-V coil will only draw 600/480 or 1.25 A inrush and 60/480 or .125 A sealed.

Pick-up Voltage: The minimum voltage which will cause the armature to start to move is called the pick-up voltage.

Sealed-in Voltage: The seal-in voltage is the minimum control voltage required to cause the armature to seat against the pole faces of the magnet. On devices using a vertical action magnet and armature, the seal-in voltage is higher than the pick-up voltage to provide additional magnetic pull to insure good contact pressure.

Control devices using the bell-crank armature and magnet arrangement are unique in that they have different force characteristics. Devices using this operating principle are designed to have a lower seal-in voltage than pick-up voltage. Contact life is extended, and contact damage under abnormal voltage conditions is reduced, for if the voltage is sufficient to pick-up, it is also high enough to seat the armature.

If the control voltage is reduced sufficiently, the controller will open. The voltage at which this happens is called the drop-out voltage. It is somewhat lower than the seal-in voltage.

Voltage Variation

NEMA standards require that the magnetic device operate properly at varying control voltages from a high of 110 percent to a low of 85 percent of rated coil voltage. This range, established by coil design, insures that the coil will withstand given temperature rises at voltages up to 10 percent over rated voltage, and that the armature will pick up and seal in, even though the voltage may drop to 15 percent under the nominal rating.

If the voltage applied to the coil is too high, the coil will draw more than its designed current. Excessive heat will be produced and will cause early failure of the coil insulation. The magnetic pull will be too high, which will cause the armature to slam home with excessive force. The magnet faces will wear rapidly, leading to a shortened life for the controller. In addition, contact bounce may be excessive, resulting in reduced contact life.

Low control voltage produces low coil currents and reduced magnetic pull. On devices with vertical action assemblies, if the voltage is greater than pick-up voltage, but less than seal-in voltage, the controller may pick up but will not seal. With this condition, the coil current will not fall to the sealed value. As the coil is not designed to carry continuously a current greater than its sealed current, it will quickly get very hot and burn out. The armature will also chatter. In addition to the noise, wear on the magnet faces result.

In both vertical action and bell-crank construction, if the armature does not seal, the contacts will not close with adequate pressure. Excessive heat, with arcing and possible welding of the contacts, will occur as the controller attempts to carry current with insufficient contact pressure.

AC Hum

All ac devices which incorporate a magnetic effect produce a characteristic hum. This hum or noise is due mainly to the changing magnetic pull (as the flux changes) inducing mechanical vibrations. Contactors, starters and relays could become excessively noisy as a result of some of the following operating conditions:

- Broken shading coil.
- Operating voltage too low.
- Misalignment between the armature and magnet assembly — the armature is then unable to seat properly.
- Wrong coil.
- Dirt, rust, filings, etc. on the magnet faces — the armature is unable to seal in completely.
- Jamming or binding of moving parts so that full travel of the armature is prevented.
- Incorrect mounting of the controller, as on a thin piece of plywood fastened to a wall; such mounting may cause a "sounding board" effect.

Power Circuits In Motor Starters

The power circuit of a starter includes the stationary and movable contacts, and the thermal unit or heater portion of the overload relay assembly. The number of contacts (or "poles") is determined by the electrical service. In a three-phase, three-wire system, for example, a three-pole starter is required.

To be suitable for a given motor application, the magnetic starter selected should equal or exceed the motor horsepower and full-load current ratings. For example, let's assume that we want to select a motor starter for

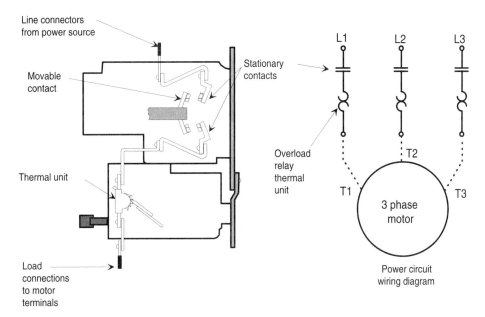

Power circuit in a typical three-pole magnetic starter

a 50-hp motor to be supplied by a 240-V, three-phase service, and the full-load current of the motor is 125 A. Refer to manufacturers' tables, available from your local electrical supplier. It can be seen that a NEMA Size 4 starter would be required for normal motor duty. If the motor were to be used for jogging or plugging duty, a NEMA Size 5 starter should be chosen.

For three-phase motors having locked-rotor kVA per horsepower in excess of that for the motor code letters in the table in following table, do not apply the controller at its maximum rating without consulting the manufacturer. In most cases, the next higher horsepower rated controller should be used.

Controller HP Rating	Maximum Allowable Motor Code Letter
1½	L
3 – 5	K
7½ and above	M

Power circuit contacts handle the motor load. The ability of the contacts to carry the full-load current without exceeding a rated temperature rise, and their isolation from adjacent contacts, corresponds to NEMA Standards established to categorize the NEMA Size of the starter. The starter must also be capable of interrupting the motor circuit under locked rotor current conditions.

Overload Protection

Overload protection for an electric motor is necessary to prevent burnout and to ensure maximum operating life. Electric motors will, if permitted, operate at an output of more than rated capacity. Conditions of motor overload may be caused by an overload on driven machinery, by a low line voltage, or by an open line in a polyphase system, which results in single-phase operation. Under any condition of overload, a motor draws excessive current that causes overheating. Since motor winding insulation deteriorates when subjected to overheating, there are established limits on motor operating temperatures. To protect a motor from overheating, overload relays are employed on a motor control to limit the amount of current drawn. This is overload protection, or running protection.

The ideal overload protection for a motor is an element with current-sensing properties very similar to the heating curve of the motor, which would act to open the motor circuit when full-load current is exceeded. The operation of the protective device should be such that the motor is allowed to carry harmless overloads, but is quickly removed from the line when an overload has persisted too long.

Fuses are not designed to provide overload protection. Their basic function is to protect against short circuits (overcurrent protection). Motors draw a high inrush current when starting and conventional single-element fuses have no way of distinguishing between this temporary and harmless inrush current and a damaging overload. Such fuses, chosen on the basis of motor full-load current, would blow every time the motor is started. On the other hand, if a fuse were chosen large enough to pass the starting or inrush current, it would not protect the motor against small, harmful overloads that might occur later.

Dual-element or time-delay fuses can provide motor overload protection, but suffer the disadvantage of being nonrenewable so they must be replaced.

The overload relay is the heart of motor protection. It has inverse trip-time characteristics, permitting it to hold in during the accelerating period (when inrush current is drawn), yet providing protection on small overloads above the full-load current when the motor is running. Unlike dual-element fuses, overload relays are renewable and can withstand repeated trip and reset cycles without need of replacement. They cannot, however, take the place of overcurrent protective equipment.

The overload relay consists of a current-sensing unit connected in the line to the motor, plus a mechanism, actuated by the sensing unit, that serves to directly or indirectly break the circuit. In a manual starter, an overload trips a mechanical latch and causes the starter contacts to open and disconnect the motor from the line. In magnetic starters, an overload opens a set of contacts within the overload relay itself. These contacts are wired in series with the starter coil in the control circuit of the magnetic starter. Breaking the coil circuit causes the starter contacts to open, disconnecting the motor from the line.

Overload relays can be classified as being either thermal or magnetic. Magnetic overload relays react only to current excesses and are not affected by temperature. As the name implies, thermal overload relays rely on the rising temperatures caused by the overload current to trip the overload mechanism. Thermal overload relays can be further subdivided into two types, melting alloy and bimetallic.

Melting Alloy Thermal Overload Relays

The melting alloy assembly of the heater element overload relay and solder pot is shown to the left. Excessive overload motor current passes through the heater element, thereby melting an eutectic alloy solder pot. The ratchet wheel will then be allowed to turn in the molten pool, and a tripping action of the starter control circuit results, stopping the motor. A cooling off period is required to allow the solder pot to "freeze" before the overload relay assembly may be reset and motor service restored.

Melting alloy thermal units are interchangeable and of a one-piece construction, which ensures a constant relationship between the heater element and solder pot and allows factory calibration, making them virtually tamper-proof in the field. These important features are not possible with any

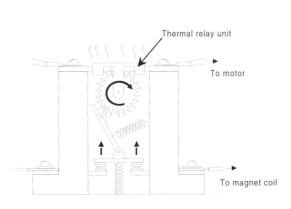

Operating characteristics of a melting-alloy overload relay

other type of overload relay construction. A wide selection of these interchangeable thermal units is available to give exact overload protection of any full-load current to a motor.

Bimetallic Thermal Overload Relays

Bimetallic overload relays are designed specifically for two general types of application: the automatic reset feature is of decided advantage when devices are mounted in locations not easily accessible for manual operation and, second, these relays can easily be adjusted to trip within a range of 85 to 115 percent of the nominal trip rating of the heater unit. This feature is useful when the recommended heater size might result in unnecessary tripping, while the next larger size would not give adequate protection. Ambient temperatures affect overload relays operating on the principle of heat.

Selecting Overload Relays

When selecting thermal overload relays, the following must be considered:

- Motor full-load current
- Type of motor
- Difference in ambient temperature between motor and controller

Motors of the same horsepower and speed do not all have the same full-load current, and the motor nameplate must always be checked to obtain the full-load amperes for a particular motor. Do not use a published table. Thermal unit selection tables are published on the basis of continuous-duty motors, with 1.15 service factor, operating under normal conditions. The tables are shown in the catalogs of manufacturers and also appear on the inside of the door or cover of the motor controller. These selections will properly protect the motor and allow the motor to develop its full horsepower, allowing for the service factor, if the ambient temperature is the same at the motor as at the controller. If the temperatures are not the same, or if the motor service factor is less than 1.15, a special procedure is required to select the proper thermal unit. Standard overload relay contacts are closed under normal conditions and open when the relay trips. An alarm signal is sometimes required to indicate when a motor has stopped due to an overload trip. Also, with some machines, particularly those associated with continuous processing, it may be required to signal an overload condition, rather than have the motor and process stop automatically. This is done by fitting the overload relay with a set of contacts that close when the relay trips, thus completing the alarm circuit. These contacts are appropriately called alarm contacts.

A magnetic overload relay has a movable magnetic core inside a coil that carries the motor current. The flux set up inside the coil pulls the core upward. When the core rises far enough, it trips a set of contacts on the top of the relay. The movement of the core is slowed by a piston working in an oil-filled dashpot mounted below the coil. This produces an inverse-time characteristic. The effective tripping current is adjusted by moving the core on a threaded rod. The tripping time is varied by uncovering oil bypass holes in the piston. Because of the time and current adjustments, the magnetic overload relay is sometimes used to protect motors having long accelerating times or unusual duty cycles.

Summary

The first of the motor-control arrangements is a plug and receptacle; next comes a fusible disconnect or circuit breaker, and then the manual and fractional-horsepower starters. The magnetic-contactor controller, however, is the type most used in electrical installations. This latter type of controller opens or closes circuits automatically when their control coil is energized. The contactors may be normally open or normally closed.

Protective devices such as overload relays, low-voltage protection devices and low-voltage release devices are an important part of a motor controller. An overload relay will open the contactors in motor circuits when current is too high; a low-voltage protective device will prevent the motor from starting as long as the full-rated voltage is not available, and manual restarting of the motor is necessary after the low-voltage protective device has operated; a low-voltage release device will disconnect the motor during a voltage dip, but the motor will start automatically when the normal voltage returns.

Controllers also contain braking arrangements, accelerators, and reversing switches which reverse the rotation of the motor.

Chapter 28

Electric Lighting

Electrical engineers and lighting designers strive to select lighting equipment that will provide the highest visual comfort and performance that is consistent with the type of area to be lighted and the budget provided. It is the electrical contractor's responsibility to see that all lighting equipment is furnished and installed exactly as selected and specified.

Lamp Classifications

In general, there are three common sources of electric light:

- Incandescent lamps
- Gaseous-discharge lamps
- Electroluminescent lamps

Despite continuous improvement, none of these light sources have a high overall efficiency. The very best light source converts only approximately ¼ of its input energy into visible light. The remaining input energy is converted to heat or invisible light. The energy distribution of a typical cool-white fluorescent lamp is shown in the illustration on the next page.

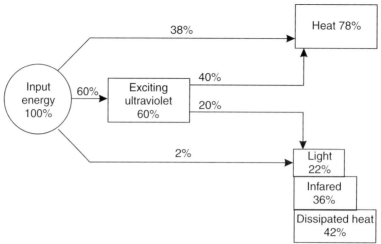

Energy distribution of a typical cool-white fluorescent lamp

Incandescent Lamps

Incandescent lamps are made in thousands of different types and colors, from a fraction of a watt to over 10,000 W each, and for practically any conceivable lighting application.

Extremely small lamps are made for instrument panels, flashlights, etc., while large incandescent lamps, over 20 in in diameter, are used for spotlights and street lighting.

Regardless of the type or size, all incandescent filament lamps consist of a sealed glass envelope containing a filament. The incandescent filament lamp produces light by means of a filament heated to incandescence (white light) by its resistance to a flow of electric current. Most of these elements are capable of producing 11 to 22 lumens per watt, and some produce as high as 33 lumens per watt.

The filaments of incandescent lamps were originally made of carbon. Now, tungsten is used for virtually all lamp filaments be-

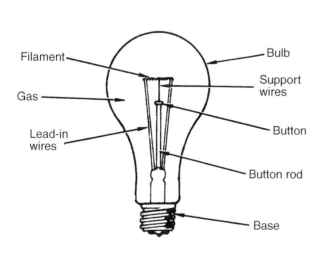

Components of a typical incandescent lamp

cause of its higher melting point, better spectral characteristics, and strength — both hot and cold.

The sealed glass envelope enclosing the filament is used to obtain a vacuum or an atmosphere of inert gas. Without such an atmosphere, the filament would rapidly disintegrate due to oxidation.

The filaments of all early incandescent lamps operated in a vacuum — all air and gas, insofar as practical, were exhausted from the space within the bulb surrounding the filament. In a vacuum lamp, the heat losses by convection and conduction are reduced, but the filament begins vaporizing at a lower temperature and therefore evaporates more rapidly than it would if pressure was applied.

The purpose of gas inside the bulb is to create pressure on the filament to retard evaporation, and this type of lamp is considered more efficient than a vacuum lamp of the same size. Since the filament can operate at a higher temperature in a gas-filled lamp, it also produces a whiter light than produced by a vacuum lamp of the same size.

Technical Descriptions of Incandescent Lamps

Incandescent lamps are made in a variety of shapes and sizes for use in many applications. Typical bulb shapes are shown below.

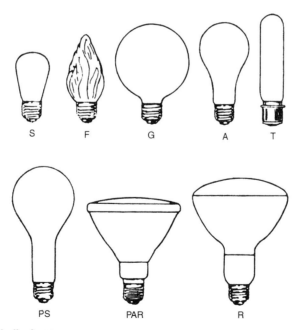

Typical incandescent bulb shapes

The sizes and shapes of lamp bulbs are designated by a letter or letters followed by a number. The letter indicates the shape of the bulb, and a few designations follow:

- S: straight side
- F: flame
- G: round or globular
- T: tubular
- A: arbitrary designation applied to lamps commonly used for general lighting of 200 W or less.

The numerals in a bulb designation indicates the maximum diameter of the bulb in eighths of an inch. For example, an A-21 bulb is 21 eighths of an inch or $2\frac{5}{8}$ in diameter at its maximum dimensions.

Bulb Finish and Color

To diffuse the light from the filament, many lamps have inside-frosted bulbs, produced by a light acid etching applied to the inner surface of the bulb. Some types of lamps are available with an inside white silica coating which provides still greater diffusion. The inside-frosted bulb absorbs no measurable amount of light, whereas the silica coating absorbs about 2 percent. With both treatments, the outer surface of the bulb is left smooth and easily cleaned. Diffusing bulbs are preferred for most general lighting purposes, but where accurate control of light is involved, as in optical systems, clear bulb lamps are necessary.

Other finishes applied to some general lighting service lamps are white bowl and silvered bowl. A white-bowl lamp has a translucent white coating on the inner surface of the bulb bowl, which serves to reduce both direct and reflected glare from open fixtures. A silvered-bowl lamp has an opaque silver coating applied to the bowl. The inner surface of this coating is a highly specula reflector which is not affected by dust or deterioration, and therefore remains efficient throughout the life of the lamp. Silvered-bowl lamps are commonly used in certain types of equipment for totally indirect lighting, and also occasionally in direct fixtures such as standard dome reflectors.

Colored light in filament lamps is produced subtractively, by means of a bulb that absorbs light colors other than that desired. Most colored bulbs are made by applying a pigmented coating to either the inner or the outer surface of a clear bulb, or by fusing an enamel into the outer surface (ceramic coating). The colors in most common use are red, blue, green, yellow, orange, ivory, flametint, and white.

Lamps with a slightly pink-colored inside silica coating (Beauty-Tone lamps) are available in the three-way design. These lamps are primarily used in residential lighting equipment. They are used where delicately tinted light is desired for a decorative effect. Ceramic coatings and inside coatings are satisfactory for either outdoor or indoor use, but most outside coatings are not permanent and are recommended for use only where they are protected from the weather.

Another type of colored lamp has a bulb of natural-colored glass, made by adding chemicals to the ingredients of the glass. Natural-colored bulbs are made in daylight blue, blue, amber, green, and ruby. They produce light of purer colors than coated bulbs and are often used for theatrical and photographic lighting purposes. Where decorative or display lighting is involved, coated lamps are preferred to natural-colored lamps because of their lower cost.

The most widely used of the natural-colored lamps is the daylight blue. The characteristics of the daylight blue bulb are such as to reduce the preponderance of red and yellow light common to incandescent lamps, with the result that the light produced more nearly approaches daylight in color. Since this is accomplished at the expense of increased lamp cost and of some 35% absorption in light, daylight blue lamps should be used only where the lighting requirements make it necessary.

Base

The base provides a means of connecting the lamp bulb to the socket. For general lighting purposes, screw-type bases are most commonly used. Most general lighting service lamps (300 W and below) have medium screw bases. The higher wattages (300 W and above) use the mogul screw base. Some of the lower-wattage lamps, particularly the sign, indicator, and decorative types, have candelabra or intermediate screw bases. See illustration on the next page for incandescent base types.

A light source (lamp filament) cannot be accurately aligned with respect to an optical system by means of a screw base. Filament orientation is provided by a number of other types of bases. The most common bases are the prefocus, bipost, bayonet, and special pin-type bases for projection lamps. A bipost base, usually used on high-wattage lamps, consists of two metal pins or posts imbedded in a glass "cup" forming the end of the lamp bulb. Most screw and prefocus bases are attached to the bulb by means of a basing cement especially designed for the purpose. Other bases used on

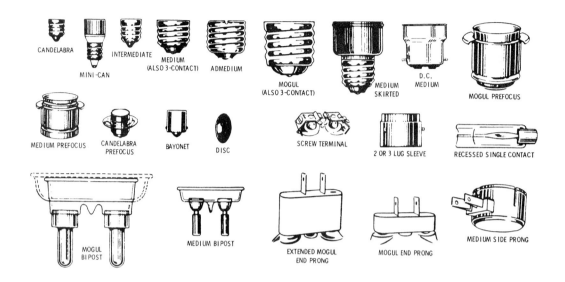

Incandescent base configurations

certain lamps include prong types, screw terminals, contact lugs, flexible leads, recessed single contact, and a number of other types for specific applications.

The filament is the light-producing element of the lamp, and the primary considerations in its design are its electrical characteristics. The wattage of a filament lamp is equal to the voltage delivered at the socket times the amperes flowing through the filament. By Ohm's law, the current is determined by the voltage and by the resistance, which in turn depends on the length and the diameter of the filament wire. The higher the wattage of a lamp of a given voltage, the higher the current, and therefore the greater the diameter of the filament wire required to carry it. The higher the voltage of a lamp of a given wattage, the lower the current and the smaller the diameter of the filament wire.

The higher the operating temperature of the filament, the greater the share of the emitted energy that lies in the visible region of the radiation spectrum. Since most filament lamps radiate as light only about 10 to 12 percent of the input energy, it is important to design a lamp for as high a filament temperature as is consistent with satisfactory lamp life. Carbon, which has a higher melting point than tungsten and was one of the early filament materials, has been almost completely replaced by tungsten because carbon at high temperatures evaporates too rapidly, whereas tungsten combines the properties of a high-melting point and slow evaporation.

Since the larger the diameter of the filament wire the higher the temperature at which it can be operated without danger of excessive evaporation, high-wattage lamps are more efficient than low-wattage lamps of the same voltage and life rating. A 150-W, 120-V general lighting service lamp, for example, produces 34 percent more light than three 50-W 120-V lamps consuming the same wattage. It also follows that low-voltage lamps, because their filament wire diameter is greater, are more efficient than higher-voltage lamps of the same wattage.

The filament forms in common use today are designated by a letter or letters indicating whether the wire is straight or coiled, a number specifying the general form of the filament, and sometimes another letter indicating

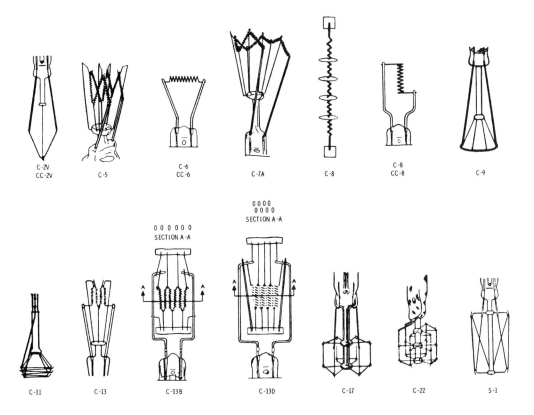

Typical incandescent filament forms

arrangement on the supports. "S" as the first letter of a filament designation means a straight (uncoiled) filament wire, "C" a coiled wire, "CC" a coiled coil, and "R" a flat or ribbon-shaped wire. The numbers and other letters assigned to the various filament forms are purely arbitrary.

Early lamps were made with straight filaments operating in a vacuum. When inert gases were introduced into the bulb, it was found that coiling the wire decreased the effective surface exposed to the circulating gas, and therefore reduced the heat lost by conduction and convection. The coils also tend to heat each other, and the coiled filament is mechanically stronger. Today, nearly all types of lamps, both vacuum and gas-filled, have coiled filaments. The single-coil filament is formed by winding the tungsten wire on a mandrel of steel or molybdenum in a continuous process. The coil with the mandrel still in place is cut into the desired lengths and immersed in an acid bath, which dissolves the mandrel but does not attack the tungsten.

Coiled-coil, or double-coiled, filaments which provide increased efficiency and reduced light source size are at present used in various general lighting service, standard-voltage lamps in the 50- to 1000-W range, also in certain types of projection lamps. The process of making coiled-coil filaments is the same as that for making single-coil-coil filaments. With the mandrel intact, the wire is wound onto another mandrel which is later "retracted," or removed mechanically. The first mandrel is then removed from the coiled coil by dissolving.

In the general lighting service type of lamp, the arrangement of the filament coil and its supports is dictated by the limiting size of the bulb neck through which it must be inserted, and by other manufacturing considerations. Mounting a filament vertically rather than horizontally (C-8, CC-8, or 2CC-8 construction), as has recently been done in general lighting service lamps, results in a higher light output because gas convection currents raise the filament temperature and because less light is absorbed by the lamp base. Further, the bulb blackening which develops as the lamp ages is localized within a smaller area, and lumen maintenance throughout the lamp life is higher. Lamps for special purposes often require certain filament forms. For projection, searchlight, spotlight, floodlight, and similar services where accurate control of light demands a small source, the filaments are concentrated into as small a space as possible. In contrast, for showcase service where a long light source is needed, the filament may be extended along almost the full length of the bulb.

Filling Gas

Incandescent lamps were first made with evacuated bulbs, the purpose being merely to keep the filament from burning up by excluding oxygen. Later it was discovered that the pressure exerted on the filament by an inert gas introduced into the bulb retarded the evaporation of tungsten, thus making it possible to design lamps for higher filament temperatures. Vacuum lamps are now designated as "type B" lamps, gas-filled lamps as "type C."

The gas removes some heat from the filament, as a result of conduction and convection losses not present in the vacuum lamp. The larger the surface of the wire in proportion to its volume or mass, the greater this cooling effect becomes, until eventually it nullifies the gain achieved by using the filling gas. Filaments with a current rating of less than $1/3$ A have a wire diameter so small that the introduction of gas is a disadvantage. For this reason, standard-voltage general lighting service lamps of less than 40 W are of the vacuum or "type B" construction, while lamps of 40 W and higher are gas-filled.

Nitrogen and argon are the gases most commonly used in lamp manufacture. Projection lamps use an atmosphere of 100 percent nitrogen. Most other types have a mixture of nitrogen and argon, the proportions varying with the lamp and the service for which it is designed. High-voltage lamps, for example, are filled with approximately 50 percent argon and 50 percent nitrogen, the higher wattage standard voltage types about 90 percent argon and 10 percent nitrogen, and the lower wattage standard-voltage types and all street series lamps about 98 percent argon and 2 percent nitrogen. Some nitrogen is necessary to prevent arcing across the lead-in wires, which would occur if pure argon were used. The greater the inherent tendency of a lamp to arc, the higher the percentage of nitrogen in its gas mixture.

Krypton is a relatively rare and expensive gas which has a higher atomic weight than either argon or nitrogen, and therefore causes less energy loss by conduction and convection. It is primarily used in certain miniature lamps such as those on miners' caps, where the limited capacity of the battery power supply makes it essential to obtain the greatest possible efficiency. Hydrogen, because of its low atomic weight, is used in certain very special types of flashing signal lamps where rapid cooling of the filament is important.

Types Of Lamps

The familiar general lighting service lamps, from the 15-W A-15 to the 1500-W PS-52, designed for multiple burning on 120-, 125-, or 130-volt circuits, are the most commonly used filament-type lamps. All standard general service lamps are equipped with screw bases. The larger wattages are manufactured in either clear or inside-frosted bulbs. Below 150 W, inside frosted and inside white silica-coated lamps are standard. The wattages most commonly used in the home are available in a straight-sided modified T-bulb shape, with the white silica coating.

High- And Low-Voltage Lamps

Lamps similar to those of the standard-voltage line are available for operation on 230 and 250 V. The low efficacy of these lamps, as compared to comparable lamps of standard-voltage rating, is a disadvantage. Other disadvantages, resulting from the smaller filament wire diameter of high-voltage lamps, are reduced mechanical strength and larger overall light-source size which makes them less satisfactory for use in floodlight and projection equipment. The only gain achieved by the industrial use of these higher voltages is the reduction in ampere load which results from doubling the voltage, and the consequence saving in wiring cost.

Projector and Reflector Lamps

PAR-bulb (projector) and R-bulb (reflector) lamps combine, in one unit, a light source and a highly efficient sealed-in reflector consisting of vaporized aluminum or silver applied to the inner surface of the bulb. The 100-W PAR38 and 150-W R-40 lamps are available in several colors. "PAR" bulbs are of hard glass. "PAR" lamps up to 160 W in size, as well as a few special service "R" lamps with heat-resistant-glass bulb, can be used outdoors without danger of breakage from rain or snow. Larger "PAR" lamps and all other "R" lamps are not recommended for outdoor use unless protected from the elements.

Higher-wattage R-52 and R-57 reflector lamps are designed for general lighting purposes. They are made in both wide and narrow distribution and are best adapted for high-ceilinged industrial areas where the atmosphere contains noncombustible dirt, smoke, or fumes. Where heat-resisting glass is required for protection against thermal shock, the R-60 lamps will perform similarly. These latter types are especially suited for outdoor floodlighting. In addition to flood and spotlight service, PAR-bulb lamps have found wide application in automotive, aviation, and other miscellaneous fields where compact lighting units of precise beam control are necessary.

Showcase and Lumiline Lamps

Low-wattage tubular-bulb lamps are used for showcase lighting and other applications where small bulb diameter is required. Some of these are designed to be used in reflectors, and others are provided with an internal reflecting surface extending over approximately half the bulb area, which concentrates the light to form a beam. The Lumiline lamp is a special type of tubular light source which has a filament extending the length of the lamp. The filament is connected at each end to a disc base which requires a special type of lamp holder. Lumiline lamps are considerably less efficient than conventional general lighting service lamps, but are useful where a linear source is necessary.

Spotlight, Floodlight and Projection Lamps

Characteristic features of all lamps designed for spotlight, floodlight, and projection applications include compact filaments accurately positioned with respect to the base, for purposes of light control; relatively short life, for high efficacy and luminance; comparatively small bulbs; and restricted burning position. Since spotlight lamps must produce narrower, more intense beams than floodlight lamps, they usually have smaller filaments and shorter lives. In projection lamps the light source is still more concentrated and lamp life is further reduced, with accompanying increased efficacy.

The objective in designing projection lamps is to fill the aperture of the projection system with a light source of high luminance and maximum uniformity. This is accomplished by arranging the filament coils in a single or double vertical plane and using a base which accurately locates the filament with respect to the optical system. The biplane (C-13D) filament, with coils arranged in two parallel rows so-placed that the coils of one row fill in the spaces between those of the other, has much greater uniformity and higher average luminance than the single-row monoplane (C-15) filament. Many projection lamps have such small bulbs and operate at such high temperatures that they cannot be burned without continuous forced ventilation, and some have designed lives as short as 10 hours. Lamps for use in certain types of projectors have an opaque coating on the top of the bulb to prevent the emission of stray light.

Halogen lamps

The halogen lamp is a relatively new concept in incandescent lamps. It uses a quartz envelope which is the basis for its many advantages, including the following:

- Compactness
- Thermal shock resistance
- High efficacy
- Almost perfect maintained light throughout the lamp life

Iodine is used in the lamp to create a chemical cycle with the sublimated tungsten to keep the bulb clean. The halogen lamp is used for floodlighting, aviation, photographic, special effects, photocopy, and other applications where its special features are desirable.

Infrared Lamps

Infrared lamps are essentially the same as lamps designed for illumination purposes; the principal difference between them is filament temperature. Since the production of light is not an objective, infrared lamps are designed to operate at a very low temperature, resulting in a low light output (about 7 or 8 lumens per watt) and a consequence reduction in glare. If only the advantage of low filament evaporation is desired, the life of infrared lamps is many thousands of hours; but because of the possibility of failure from shock, vibration, and other causes, the rated life is given merely as "in excess of 5000 hours."

Infrared lamps used in the home and for therapeutic purposes are commonly of the convenient self-contained 250-W R-10 bulb type with internal reflector and red bulb. Those used in industrial processes are of three types; reflector lamps (125-, 250-, and 375-W R-40), clear G-30 bulb lamps (125, 250, 375 and 500 W), and the more recently developed small linear sources in the T-3 quartz bulb. The latter are available in a number of sizes, and the effective heating length and the voltage rating increases with the wattage. Gold-plated or specular aluminum reflectors are most effective for use with unreflectorized infrared lamps.

Incandescent Lamp Specifications

General-Lighting Lamps: The most commonly used filament lamps are available from the 15 W, A-15 lamp to the 1500 W, PS-52 lamp. All are equipped with screw bases and are designed for use on 120-, 125-, and 130-V circuits.

High-voltage lamps are used primarily in commercial and industrial applications while soft-white lamps are popular in the home.

General-Purpose Lamps: This lamp type is inside-frosted and available for use on 120- to 130-V circuits.

Econ-O-Watt and Extended-Service Econ-O-Watt Lamps: These lamps provide similar lighting levels as standard incandescent lamps while consuming 15percent less energy. They also last two to three times longer.

INSIDE-FROSTED GENERAL-LIGHTING INCANDESCENT LAMPS – 120 V

Watts	Lamp Type	Volts	Bulb	Base	Rated Average Life (Hrs)
25	25A	120	A19	Medium	2500
40	40A15	115 – 120	A19	Medium	1000
40	40A	120	A19	Medium	1500
60	60A	120	A19	Medium	1000
75	75A	120	A19	Medium	750
100	100A	120	A19	Medium	750
150	150A	120	A21	Medium	750
200	200A	120	A23	Medium	750
300	300M IF	120	PS25	Medium	750
300	300 IF	120	PS35	Mogul	1000
500	500 IF	120	PS35	Mogul	1000

INSIDE-FROSTED GENERAL-LIGHTING LAMPS – 130 V

Watts	Lamp Type	Volts	Bulb	Base	Rated Average Life (Hrs)
15	15A15	130	A15	Medium	2500
25	25A	130	A19	Medium	2500
40	40A	130	A19	Medium	1500
60	60A	130	A19	Medium	1000
75	75A	130	A19	Medium	750
100	100A23	120 – 130	A23	Medium	750
100	100A	130	A19	Medium	750
150	150A	130	A21	Medium	750
200	200A	125 – 130	A23	Medium	750
200	200 IF	130	PS30	Medium	750

INSIDE-FROSTED ECON-O-WATT INCANDESCENT LAMPS

Watts	Lamp Type	Volts	Bulb	Base	Rated Average Life (Hrs)
34	40A-34A/EW	120	A19	Medium	1500
34	40A-34A/EW	120	A19	Medium	1500
52	60A-52A/EW	120	A19	Medium	1000
52	60A-52A/EW	120	A19	Medium	1000
67	75A-67A/EW	120	A19	Medium	750
67	75A-67A/EW	130	A19	Medium	750
90	100A-90A/EW	120	A19	Medium	750
90	100A-90A/EW	130	A19	Medium	750
135	150A-135A/EW	120	A21	Medium	750
135	150A-135A/EW	130	A21	Medium	750

INSIDE-FROSTED EXTENDED-SERVICE ECON-O-WATT INCANDESCENT LAMPS

Watts	Lamp Type	Volts	Bulb	Base	Rated Average Life (Hrs)
52	60A-52A/99/EW	120	A19	Medium	2500
52	60A-52A/99/EW	125 – 130	A19	Medium	2500
67	75A-67A/99/EW	120	A19	Medium	2500
67	75A-67A/99/EW	125 – 130	A19	Medium	2500
90	100A-90A/99/EW	120	A19	Medium	2500
90	100A-90A/99/EW	125 – 130	A19	Medium	2500
135	150A-135A/99/EW	120	A21	Medium	2500
135	150A-135A/99/EW	125 – 130	A21	Medium	2500

Clear General-Lighting Lamps: These lamps have a crystal clear bulb and a completely visible filament, so the brightness of the light is not softened. Clear general lighting can be used in a variety of appliations where brilliance and sparkle are more important than the avoidance of glare.

CLEAR GENERAL-LIGHTING LAMPS

Watts	Lamp Type	Volts	Bulb	Base	Rated Average Life (Hrs)
15	15A15/CL	130	A15	Medium	2500
25	25A/CL	120	A19	Medium	2500
25	25A/CL	130	A19	Medium	2500
40	40A/CL	120	A19	Medium	1500
40	40A/CL	130	A19	Medium	1500
60	60A/CL	120	A19	Medium	1000
60	60A/CL	130	A19	Medium	1000
75	75A/CL	120	A19	Medium	750
75	75A/CL	130	A19	Medium	750
100	100A/CL	120	A19	Medium	750

CLEAR GENERAL-LIGHTING LAMPS *(Cont.)*

Watts	Lamp Type	Volts	Bulb	Base	Rated Average Life (Hrs)
100	100A/CL	130	A19	Medium	750
150	150A/CL	120	A21	Medium	750
150	150A/CL	130	A21	Medium	750
300	300M	120	PS25	Medium	750
300	300M	130	PS25	Medium	750

Long-Life Soft-White Lamps: This lamp type provides over 33 percent longer life than standard incandescent lamps. Soft-white lamps achieve maximum diffusion of light from the filament without glare or harsh shadows.

LONG-LIFE SOFT—WHITE LAMPS

Watts	Lamp Type	Volts	Bulb	Base	Rated Average Life (Hrs)
25	25T/WL 12/2	120	T19	Medium	1350
40	40T/WL	120	T19	Medium	1350
60	60T/WL	120	T19	Medium	1350
75	75T/WL	120	T19	Medium	1000
100	100T/WL	120	T19	Medium	1000
150	150T/WL	120	T19	Medium	1000
200	200T/WL	120	T21	Medium	1000

Three-Way Soft-White Lamps: These lamps are available in both standard and extended-life types. Both types offer maximum flexibility in lighting a room or area.

Electric Lighting

THREE-WAY SOFT-WHITE LAMPS

Watts	Lamp Type	Volts	Bulb	Base	Rated Average Life (Hrs)
15	15/150T/WL 12/1	120	T19	3CT-Medium	1600
30	30/100T/SW 12/1	120	T19	3CT-Medium	1200
30	30/100T/WL	120	T19	3CT-Medium	1600
50	50/150/SW	120	T19	3CT-Medium	1200
50	50/150/WL	120	T19	3CT-Medium	1600
50	50/250/WL	120	T21	3CT-Medium	1600

TUBLUAR INCANDESCENTS

Watts	Lamp Type	Volts	Description	Bulb	Base	Rated Average Life (Hrs.)
15	15T6	115 - 125	Switchboard, clear	T6	Cand.	2000
15	15T6	140 - 150	Switchboard, clear	T6	Cand.	2000
15	15T7N	115 - 125	Appliance, clear	T7	Inter.	200 – 600
20	20T61/2/IF	120	Exit sign, frosted	T6-½	Inter.	200 – 600
20	20T61/2/DC	120	Exit sign, clear	T6-½	DCBay	200 – 600 or more
25	25T61/2/IF	115 - 125	Frosted	T6-½	Inter.	1000
25	25T61/2/IF	130	Frosted	T6-½	Inter.	1000
25	25T6/1/2	115 - 125	Clear	T6-½	Inter.	1000

TUBLUAR INCANDESCENTS

Watts	Lamp Type	Volts	Description	Bulb	Base	Rated Average Life (Hrs.)
25	25T6/1/2	130	Clear	T6-½	Inter.	1000
25	25T6/1/2DC	115 125	Clear	T6-½	DCBay	1000
25	25T6/1/2DC	130	Clear	T6-½	DCBay	1000
25	25T8DC	115 125	Appliance, clear	T8	DCBay	Varies
25	25T10	120	Clear	T10	Med.	1000
40	40T10	120	Clear	T10	Med.	1000

Rough Service and Industrial Service Lamps: These lamps feature an exceptionally rugged C-9 filament construction that is supported at five points. They are suitable for places where shocks, bumps and vibrations frequently occur. The 3500-hour lamps provide a 40 percent increase in rated average life at an increased cost of less than 25 percent over comparable wattage extended-service lamps.

General-service lamp

Industrial-service lamp

Rough-service lamp

ROUGH SERVICE, INSIDE-FROSTED LAMPS

Watts	Lamp Type	Volts	Bulb	Base	Rated Average Life (Hrs)
50	50A/RS	120	A19	Medium	1000
75	75A/RS/VS	120	A19	Medium	1000
75	75A/RS/VS	125-130	A19	Medium	1000
100	100A/RS/VS	120	A21	Medium	1000
100	100A/RS/VS	125-130	A21	Medium	1000
150	150A/RS/VS	120	A23	Medium	1000
150	150A/RS/VS	125-130	A23	Medium	1000

INDUSTRIAL SERVICE, INSIDE-FROSTED LAMPS

Watts	Lamp Type	Volts	Bulb	Base	Rated Average Life (Hrs)
60	60A19/35	120	A19	Medium	3500
100	100A21/35	125-130	A21	Medium	3500
150	150A25/35	125-130	A25	Medium	3500
200	200A25/35	125-130	A25	Medium	3500

Decorative Lamps: Decorative lamps are generally low-wattage and are available in an assortment of shapes to meet a variety of decorative needs. Lamp bases are available in candelabra (cand.) and medium.

DECORATIVE BENT-TIP LAMPS

Watts	Lamp Type	Volts	Description	Bulb	Base	Rated Average Life (Hrs.)
15	BC-15BA9C/3	120	Bent tip, clear	BA9	Cand.	1500
25	BC-25BA9C/3	120	Bent tip, clear	BA9	Cand.	1500

DECORATIVE BENT-TIP LAMPS

Watts	Lamp Type	Volts	Description	Bulb	Base	Rated Average Life (Hrs.)
25	BC-25BA91/2/3	120	Bent tip, clear	BA9-½	Med.	1500
40	BC-40BA9C/3	120	Bent tip, clear	BA9	Cand.	1500
40	BC-40BA91/2/3	120	Bent tip, clear	BA9-½	Med.	1500
60	BC-60BA9C/3	120	Bent tip, clear	BA9	Cand.	2000

DECORATIVE TORPEDO LAMPS

Watts	Lamp Type	Volts	Description	Bulb	Base	Rated Average Life (Hrs.)
25	BC-25BA91/2/3	120	Blunt tip, clear	B10	Cand.	1500
40	BC-40B10 1/2/3	120	Blunt tip, clear	B10	Cand.	1500
60	BC-60B10 1/2/3	120	Blunt tip, clear	B10	Cand.	1500

DECORATIVE FLAME LAMPS

Watts	Lamp Type	Volts	Description	Bulb	Base	Rated Average Life (Hrs.)
25	BC-25F15/3A	120	Flame, Trans. Amber	F15	Med.	1500
25	BC-25F15/3W	120	Flame, white	F15	Med.	1500
25	BC25F15/3	120	Flame, clear	F15	Med.	1500
40	BC-40F15/3	120	Flame, clear	F15	Med.	1500
40	BC-40F15/3W	120	Flame, white	F15	Med.	1500
60	BC-60F15/3	120	Flame, clear	F15	Med.	1500

DECORATIVE GLOBELAMPS

Watts	Lamp Type	Volts	Description	Bulb	Base	Rated Average Life (Hrs.)
25	BC-25G16-1/2C/3LL	120	Globe clear, LL	G16-½	Cand.	2000
25	BC-25G16-1/2C/3WLL	120	Globe white, LL	G16-½	Cand.	2000
25	BC25G16-1/2C/3W	120	Globe white	G16-½	Cand.	1500
25	25G25/3	120	Globe clear	G25	Med.	1500
25	25G25/3W	120	Globe white	G25	Med.	1500
40	40G25/3	120	Globe clear	G25	Med.	1500
40	40G25/3W	120	Globe white	G25	Med.	1500
40	BC40G16-1/2C/3W	120	Globe white	G16-½	Cand.	1500
40	BC40G16-1/2C/3	120	Globe clear	G16-½	Cand.	1500

Specialty Incandescent Lamps: Specialty incandescent lamps are used in a variety of applications. For example, the 6S6 and 7C7 are used in night lights, appliances, and in panelboards. The 11S14 is used in sign lighting applications and has a brass base.

SPECIALTY INCANDESCENT LAMPS

Watts	Lamp Type	Volts	Description	Bulb	Base	Rated Average Life (Hrs.)
6	6S6	120	Clear	S6	Cand.	1500
6	6S6	125 130	Clear	S6	Cand.	1500
7	7C7	130	Clear	C7	Cand.	3000
11	11S14	130	Clear	S14	Med.	3000

Fluorescent Lamps

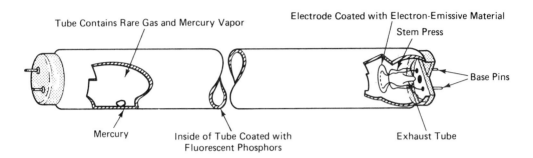

Basic components of a fluorescent lamp

Fluorescent lamps have become the major light source for general interior lighting of commercial and institutional buildings and have challenged other sources for residential, exterior, and other lighting applications.

Fluorescent lamps are available in straight, U-shaped, or circular configurations, and in various diameters.

A fluorescent lamp consists of an airtight glass tube enclosing a small drop of mercury and a small amount of argon or argon-neon gas to facilitate starting the arc. After the arc is started, the mercury vapor emits ultraviolet radiation which is invisible and does not pass through the glass. However, the inside of the glass tube is coated with a highly sensitive fluorescent powder (phosphors) which is activated by the ultraviolet radiation and in turn converts the invisible energy to visible light. By mixtures of various phosphors, a wide range of visible light colors is possible.

Colors of Fluorescent Lamps

Cool White: This lamp is often selected for offices, factories, and commercial areas where a psychologically cool working atmosphere is desirable. This is the most popular of all fluorescent lamp colors since it gives a natural outdoor lighting effect and is one of the most efficient fluorescent lamps manufactured today.

Deluxe Cool White: This lamp is used for the same general applications as the cool white, but contains more red which emphasizes pink skin tones and is therefore more flattering to the appearance of people. Deluxe cool white is also used in food display because it gives a good appearance to

Electric Lighting

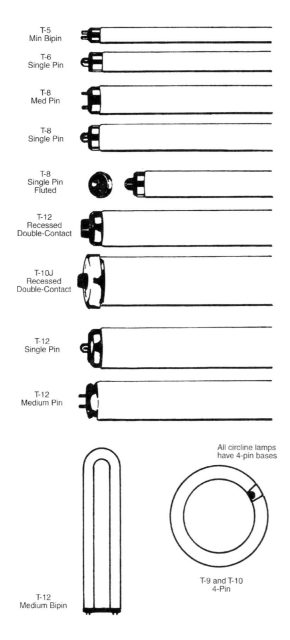

Different sizes and shapes of fluorescent lamps

lean meat; keeps fats looking white; and emphasizes fresh, crisp appearance of green vegetables. This type of lamp is generally chosen wherever very uniform color rendition is desired, although it is less efficient than cool white.

Warm White: Warm-white lamps are used whenever a warm social atmosphere is desirable in areas that are not color critical. It approaches incandescent in color and is suggested whenever a mixture of fluorescent and incandescent lamps is used. While it gives an acceptable appearance to people, it has some tendency to emphasize sallowness. Yellow, orange, and tan interior finishes are emphasizsed by this lamp, and its beige tint gives a bright warm appearance to reds; brings out the yellow in green; and adds a warm tone to blue. It imparts a yellowish white or yellowish gray appearance to neutral surfaces.

Deluxe Warm White: Deluxe warm-white lamps are more flattering to complexions than warm white and are very similar to incandescent lamps in that they impart a ruddy or tanned hue to the skin. It is generally recommended for home or social environment applications and for commercial use where flattering effects on people and merchandise are considered important. This type of lamp enhances the appearance of poultry, cheese, and baked goods. These lamps are approximately 25 percent less efficient than warm-white lamps.

White: White lamps are used for general lighting applications in offices, schools, stores, and homes where either a cool working atmosphere or warm social atmosphere is not critical. They emphasize yellow, yellow green, and orange interior finishes. This lamp, however, is seldom used in most practical applications.

Daylight: Daylight lamps are for use in industry and work areas where the blue color associated with the "north light" of actual daylight is preferred. While it makes blue and green bright and clear, it tends to tone down red, orange, and yellow.

In general, the designations "warm" and "cool" represent the differences between artificial light and natural daylight in the appearance they give to an area. Their deluxe counterparts have a greater amount of red light, supplied by a second phosphor within the tube. The red light shows colors more naturally, but at a sacrifice in efficiency.

Other colors of fluorescent lamps are available in sizes that are interchangeable with white lamps. These colored lamps are best used for flooding large areas with colored light; where a colored light of small area must be projected at a distant object, incandescent lamps using colored filters are best.

Classes Of Fluorescent Lamps

Preheat Lamps: Preheat, hot-cathode fluorescent lamps use a two-pin base and a starter which provides momentary current flow through the filament cathode in order to heat them. The radiation from the cathodes is possible only after the cathodes have been preheated. The time interval necessary for preheating is one drawback of this type of lamp, but this drawback is offset by the significant savings in ballast design and lamp life. The switch and starter connected across the lamp can be either automatic or manual.

	MINIATURE PREHEAT T-5 FLUORESCENT LAMPS				
Watts	Lamp Type	Description	Bulb	Base	Rated Average Life (Hrs.)
4	F4T5/CW	Cool white	T5	Min. Bipin	6000
6	F6T5/CW	Cool white	T5	Min. Bipin	7500
8	F8T5/CW	Cool white	T5	Min. Bipin	7500
8	F8T5/WW	Warm white	T5	Min. Bipin	7500

MINIATURE PREHEAT T-5 FLUORESCENT LAMPS *(Cont.)*

Watts	Lamp Type	Description	Bulb	Base	Rated Average Life (Hrs.)
13	F13T5/CW	Cool white	T5	Min. Bipin	7500
13	F13T5/27U	2700 Ultralume	T5	Min. Bipin	7500

PREHEAT T-8 FLUORESCENT LAMPS

Watts	Lamp Type	Description	Bulb	Base	Rated Average Life (Hrs.)
15	F15T8/CW	Cool white	T8	Med. Bipin	7500
15	F15T8/D	Daylight	T8	Med. Bipin	7500
15	F15T8/C50	Colortone 50	T8	Med. Bipin	7500
15	F15T8/WW	Warm white	T8	Med. Bipin	7500
30	F30T8/CW	Cool white	T8	Med. Bipin	7500

PREHEAT T-12 FLUORESCENT LAMPS

Watts	Lamp Type	Description	Bulb	Base	Rated Average Life (Hrs.)
14	F14T12/CW	Cool white	T12	Med. Bipin	9000
15	F15T12/WW	Warm white	T12	Med. Bipin	9000
15	F15T12/CW	Cool white	T12	Med. Bipin	9000
15	F15T12/D	Daylight	T12	Med. Bipin	9000
20	F20T12/D	Daylight	T12	Med. Bipin	9000
20	F20T12/WW	Warm white	T12	Med. Bipin	9000
20	F20T12/CW	Cool white	T12	Med. Bipin	9000
25	F25T12/CW	Cool white	T12	Med. Bipin	7500

Standard Preheat Rapid-Start Fluorescent Lamps: These lamps are designed to operate on both preheat and rapid-start ballast circuits. They are the most commonly used lamp type in the industry.

\multicolumn{6}{c}{PREHEAT RAPID-START FLUORESCENT LAMPS}					
Watts	Lamp Type	Description	Bulb	Base	Rated Average Life (Hrs.)
40	F40CW	Cool white	T12	Med. Bipin	20000
40	F40WW	Warm white	T12	Med. Bipin	20000
40	F40GO	Gold, Bug-Away	T12	Med. Bipin	20000
40	F40D	Daylight	T12	Med. Bipin	20000
40	F40/C50	Colortone 50	T12	Med. Bipin	20000
40	F410W	White	T12	Med. Bipin	20000

\multicolumn{6}{c}{PREHEAT RAPID-START EXTENDED-SERVICE FLUORESCENT LAMPS}					
Watts	Lamp Type	Description	Bulb	Base	Rated Average Life (Hrs.)
40	F40T10/CW/99	Cool white ext.	T10	Med. Bipin	24000
40	F40T10/WW/99	Warm white ext.	T10	Med. Bipin	24000

Instant Start: In order to overcome the slow starting of the preheat system and eliminate the need for a starter, the instant-start lamp was developed. Instant starting is accomplished by use of a specially designed ballast which delivers a high starting voltage and normal operating voltage once the lamps are started. Because no preheating is necessary with instant-start lamps, only a single pin on each end of the lamp is required. Hot-cathode lamps with single-pin bases are called slimline lamps.

Rapid-Start Lamps: A rapid-start fluorescent lamp retains the advantage of preheat starting, speeds up the starting interval, and eliminates the separate starter switch. The smooth, rapid start is accomplished by a built-in electrode heating coil in the ballast, and the lamp lights almost as quickly as instant-start lamps. These lamps are the most popular and important for use in fluorescent-lighting systems.

Today, the most commonly used lamp is the rapid-start type operating at 430 mA, or approximately 10 W per ft of lamp.

\multicolumn{6}{c}{RAPID-START T12 FLUORESCENT LAMPS}
Watts	Lamp Type	Description	Bulb	Base	Rated Average Life (Hrs.)
30	F30T12/WWRS	Warm white	T12	Med. Bipin	18000
30	F30T12/D/RS	Daylight	T12	Med. Bipin	18000
30	F30T12/CW/RS	Cool white	T12	Med. Bipin	18000

\multicolumn{6}{c}{ECON-O-WATT ENERGY-SAVING RAPID-START FLUORESCENT LAMPS}
Watts	Lamp Type	Description	Bulb	Base	Rated Average Life (Hrs.)
25	F30T12/WW/RS/EW-11	Warm white	T12	Med. Bipin	18000
25	F30T12/CW/RS/EW-11	Cool white	T12	Med. Bipin	18000
34	F40T12/SPEC30/RS/EW-11	3000K spec	T12	Med. Bipin	20000
34	F40T12/SPEC35/RS/EW-11	3500K spec	T12	Med. Bipin	20000
34	F40T12/SPEC41/RS/EW-11	4100K spec	T12	Med. Bipin	20000
34	F40WW/RS/EW-11	Warm white	T12	Med. Bipin	20000
34	F40CW/RS/EW-11	Cool white	T12	Med. Bipin	20000
34	F40D/RS/EW-11	Daylight	T12	Med. Bipin	20000
34	F40W/RS/EW-11	White	T12	Med. Bipin	20000
34	F40LW/RS/EW-11	Lite White	T12	Med. Bipin	20000

Typically, the use of Econ-O-Watt fluorescent lamps cuts energy costs as much as 20 percent by consuming fewer watts. Replacing standard fluorescent lamps with Econ-O-Watt lamps in a 100,000 ft^2 office building can reduce energy costs as much as 10 percent per ft^2.

Slimline Lamps: These lamps require no starters. The ballast provides sufficient voltage to instantly light the lamp. Single-pin bases can be used on this type of fluorescent lamp.

		SLIMLINE FLUORESCENT LAMPS			
Watts	Lamp Type	Description	Bulb	Base	Rated Average Life (Hrs.)
38.5	F48T12/CW	Cool white	T12	Single pin	9000
56	F72T12/CW	Cool white	T12	Single pin	12000
75	F96T12/CW	Cool white	T12	Single pin	12000
75	F96T12/D	Daylight	T12	Single pin	12000
75	F96T12/WW	Warm white	T12	Single pin	12000
75	F96T12/C50	Colortone 50	T12	Single pin	12000

		ECON-O-WATT SLIMLINE FLUORESCENT LAMPS			
Watts	Lamp Type	Description	Bulb	Base	Rated Average Life (Hrs.)
60	F96T12/CW/EW	Cool white	T12	Single pin	12000
60	F96T12/LW/EW	Lite white	T12	Single pin	12000
60	F96T12WW/EW	Warm white	T12	Single pin	12000
60	F96T12/ww/	Warm white	T12	Single pin	12000

		HIGH-OUTPUT FLUORESCENT LAMPS (800ma)			
Watts	Lamp Type	Description	Bulb	Base	Rated Average Life (Hrs.)
35	F24T12/CW/HO	Cool white (207)	T12	Recessed DC	12000
40	F30T12/CW/HO	Cool white (207)	T12	Recessed DC	12000
50	F36T12/CW/HO	Cool white (207)	T12	Recessed DC	12000

HIGH-OUTPUT FLUORESCENT LAMPS (800ma) *(Cont.)*

Watts	Lamp Type	Description	Bulb	Base	Rated Average Life (Hrs.)
55	F42T12/CW/HO	Cool white (207)	T12	Recessed DC	9000
60	F48T12/CW/HO	Cool white (207)	T12	Recessed DC	12000
75	F60T12/CW/HO	Cool white (207)	T12	Recessed DC	12000
85	F72T12/CW/HO	Cool white (207)	T12	Recessed DC	12000
85	F72T12/D/HO	Daylight (207)	T12	Recessed DC	12000
95	F84T12/CW/HO	Cool white (207)	T12	Recessed DC	12000
95	F84T12/D/HO	Daylight (207)	T12	Recessed DC	12000
110	F96T12/WW/HO	Warm white (207)	T12	Recessed DC	12000
110	F96T12/D/HO	Daylight (207)	T12	Recessed DC	12000
110	F96T12/CW/HO	Cool white (207)	T12	Recessed DC	12000
110	F96T12/C50/HO	Colortone 50 (207)	T12	Recessed DC	12000

ECON-O-WATT HIGH-OUTPUT FLUORESCENT LAMPS (800ma)

Watts	Lamp Type	Description	Bulb	Base	Rated Average Life (Hrs.)
95	F96T12/LW/HO/EW	Lite white	T12	Recessed DC	12000
95	F96T12/WW/HO/EW	Warm white	T12	Recessed DC	12000
95	F96T12/CW/HO/EW	Cool white	T12	Recessed DC	12000

All-Weather Fluorescent Lamps

Fluorescent lamps with common base sizes operate most efficiently at normal room temperatures of 70 to 80 degrees F, at which the temperature of the glass tube itself is between 100 and 120 degrees F. Where temperatures fall below this level, as in outdoor applications during winter months, a jacket placed around the outside of the lamp will maintain bulb wall temperature and will help provide reasonable light output. Rapid-start lamps with this jacket are known as all-weather fluorescent lamps.

VERY HIGH-OUTPUT FLUORESCENT LAMPS (1500ma)					
Watts	Lamp Type	Description	Bulb	Base	Rated Average Life (Hrs.)
215	F96T12/WW/HO	Warm white	T12	Recessed DC	12000
215	F96T12/CW/HO	Cool white	T12	Recessed DC	12000

LOW-TEMPERATURE JACKETED FLUORESCENT LAMPS (1500ma)					
Watts	Lamp Type	Description	Bulb	Base	Rated Average Life (Hrs.)
212	FJ96T12/CW/HO-O	Warm white	T12	Recessed DC	12000

Cold-Cathode Fluorescent Lamps

Cold-cathode-type circuits have been used for years in neon-sign tubing because it operates at relatively low current in small-diameter tubing adaptable to bending into sign letters or luminous patterns. All lamps are instant start and require special high-voltage circuits. For the same bulb size, phosphor and current loading, the lumen output and maintenance of cold-cathode lamps are identical in performance to those with a hot cathode. These types of lamps find greatest use in sign and display lighting. For some general applications, there are standardized lengths of tubing produced in T-8 glass envelopes and four, six, and eight ft in length, but few such tubes are being installed at present.

Circline Fluorescent Lamps

All circline lamps are of the rapid-start design for operation on rapid-start ballasts. They will also operate on preheat or trigger-start ballasts — making them a universal design.

T-9 4-pin Circline

Circline lamp

CIRCLINE FLUORESCENT LAMPS

Watts	Lamp Type	Description	Bulb	Base	Rated Average Life (Hrs.)
20	FC6T9/CW	Cool white	T9	4-pin	12000
22	FC8T9/CW	Cool white	T9	4-pin	12000
22	FC8T9/WW	Warm white	T9	4-pin	12000
32	FC12T9/CW	Cool white	T9	4-pin	12000
32	FC12T9/WW	Warm white	T9	4-pin	12000
32	FC12T9/D	Daylight	T9	4-pin	12000
40	FC16T9/CW	Cool white	T9	4-pin	12000

U-Bent Fluorescent Lamps

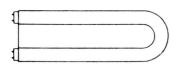

U-bent fluorescent lamp

U-bent fluorescent lamps are regular 40-W lamps bent into a U-shape. This configuration allows two or three 4-ft lamps to be used in a 2-ft square fixture. It offers the advantage of allowing wiring and lampholders to be installed at one end of the fixture.

U-BENT FLUORESCENT LAMPS

Watts	Lamp Type	Description	Bulb	Base	Rated Average Life (Hrs.)
40	FB40CW/3	Cool white (212)	T12	Med. bi-pin	12000
40	FB40WW/3	Warm white (212)	T12	Med. bi-pin	12000
40	FB40CW/6	Cool white (212)	T12	Med. bi-pin	12000
40	FB40WW/6	Warm white (212)	T12	Med. bi-pin	12000

SL and SLS Compact Fluorescent Lamps

These lamps incorporate a miniature fluorescent tube that is folded into a compact S shape. The fluorescent tube is enclosed in a lightweight polycarbonate housing and controlled by state-of-the-art electronics. Lamps produce light that closely resembles the appearance of incandescent lighting and has excellent color renditions.

SL lamps use only 18 W of electricity, compared to 60 or 75 W used by the incandescent lamps that the SL lamps are designed to replace. Additional sizes of 15, 20, and 23 W are also available. Electronic ballasts offer silent operation along with energy savings, and will not interfere with radio or television reception where the lamps are placed at least 3 ft away.

The SLS family of earthlight lamps offers the ultimate in high efficiency and compactness. SLS 15-, 20-, and 23-W lamps provide the light equivalent of 60-, 75-, and 90-W incandescent lights, respectively, while consuming as little as 25 percent the energy.

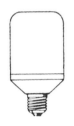
SL magnetic lamp

SL Magnetic Lamp: The short overall length of this lamp type allows them to be used as direct replacements for incandescents in many fixtures which were previously unable to accommodate compact fluorescent lamps. They may also be used in total-enclosed fixtures at temperatures as low as 32 degrees F.

Earthlight SL 18

Earthlight SL 18: This lamp type replaces 75-W standard incandescent lamps with an energy savings of 76 percent. It will fit most incandescent lighting fixtures. Applications include wall fixtures, ceiling-mounted fixtures, auxiliary lighting, and table lamps. This lamp was primarily designed for indoor use, but will also operate outdoors in an enclosed fixture at temperatures as low as 0 degrees F. The lamp has a normal service life of approximately 10,000 hours, or about 13 times longer than the equivalent incandescent lamp.

SL 18/R40 reflector lamp

SL 18/R40 Reflector Lamp: This lamp contains the same components as the SL18 except it has an R-40 reflector designed specifically for high-hat, recessed down-lighting. However, this lamp is not designed for use on circuits with dimmers. A light output of 800 lumens is achieved using only 18 W of energy.

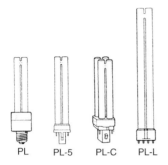

PL and PLC Compact Fluorescent Lamps: These low-energy, long-life lamps are available in many wattages and sizes. They offer high color rendition and energy savings in a variety of applications. These lamps are designed to replace 25 to 100-W incandescent lamps, but use only about 25 percent of the energy. Furthermore, they last up to 13 times longer than the incandescent equivalent.

PL and PLC compact fluorescent lamps

High-Intensity Discharge (HID) Lamps

High-intensity discharge lamps are available in a variety of sizes, shapes and colors and offer the following important benefits:

- High efficiency
- High lumen output
- Not affected by ambient temperatures
- Long useful life
- Compact size

HID lamps are divided into three families:

- High-pressure sodium (HPS)
- Metal Halide (MH)
- Mercury vapor (MV)

HID lamps are most often used in industrial, roadway, sports, and some commercial applications. They should be used only with ballasts that match the lamp.

High-Pressure Sodium Lamps

The high-pressure sodium lamp utilizes an arc tube to enclose gases through which an electric current passes. The unique light-transmitting ceramic tube enables sodium to be operated at higher temperatures and pressure than other types of HID lamps. The result is a warm yellow light at nearly maximum theoretical efficiency — 100 to 115 lumens per watt.

This lamp type is excellent for street lighting and general outdoor lighting. Since some of all colors are present in this type of lamp, it has application for virtually all general lighting under most conditions.

Metal-Halide Lamps

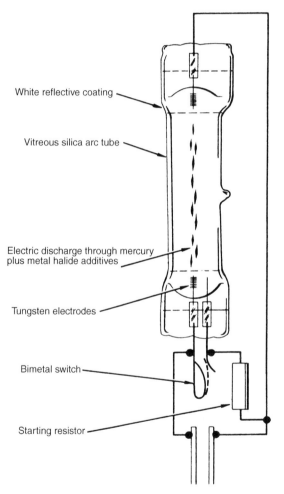

Basic components of a metal-halide lamp

The metal-halide lamps closely resemble a regular clear mercury lamp, but the inner arc tube contains additional halide chemical compounds to increase the light output and improve lamp color.

Since the color produced by metal-halide lamps is much "warmer" than regular mercury lamps, it is suitable for many indoor applications including food displays. It has found more use, however, in outdoor floodlighting, sports-lighting, and certain general street-lighting applications.

The efficiency of the metal-halide lamp is approximately twice that of conventional mercury lamps — producing from 75 to 105 lumens per watt. However, the life of this lamp type is shorter than regular mercury lamps, which average from 6,000 to 10,000 hrs. All other operating characteristics are similar to those of the regular mercury lamp, a description of which follows.

Mercury-Vapor Lamps

The mercury-vapor lamp produces light directly as a result of a current passed through gas or vapor under pressure.

While a lighted mercury-vapor lamp appears to emit white light, it actually produces light with a predominance of yellow and green rays and a small percentage of violet and blue rays. Red is absent in the basic lamp, and therefore red objects appear black or dark brown under mercury-vapor lamps. This color distortion initially prevented its use for many applications, but this disadvantage was overcome by the use of red-light-generating chemicals with the bulb.

Typical mercury-vapor lamp components include the following:

- Arc tube made to withstand the high temperatures generated as the lamp builds up to normal wattage
- Two main operating electrodes, located on opposite ends of the tube
- A starting electrode connected in series with a starting resistor and connected to the lead wire of the lower operating electrode
- Tube leads and supports
- An outer phosphor-coated glass bulb that helps stabilize the lamp operation and prevents oxidation of metal parts

The American National Standards Institute (ANSI) has developed a system of codes for mercury lamp types. This system designates a letter "H" followed by a number and two letters. The letter "H" stands for the chemical symbol "Hg" for mercury and indicates the lamp is a mercury type. The number represents the ballast type, and the two letters to follow define the physical lamp characteristics. Additional letters are used to identify the type of phosphor coating on the inside of the bulb. They are as follows:

- C — Color-improved phosphor
- W — High-efficiency phosphor
- DX — Deluxe
- Y — Yellow

Ballasts

A ballast is a current and voltage regulating device acting as an electrical brain for an electrcal-discharge lamp. It performs these main functions:

- Transforms line voltage to proper open circuit voltage necessary for the particular lamp it will operate

- Provides a specific amount of electrical energy to preheat lamp electrodes, either temporarily as in a starter-type lamp, or permanently as in a rapid start lamp

- Supplies a controlled surge of high voltage to initiate the arc throughout a starter-type lamp

- Controls lamp current by reducing open circuit voltage to safe operating voltage within the limits prescribed by the lamp manufacturer

Any coil of wire wound on an iron core has two peculiarities: When connected to an alternating current circuit, it tends to resist any change of current flowing through it. When the current flowing through it is cut off, it delivers momentarily a voltage much higher than the voltage applied to it. The ballast for a fluorescent lamp is just such a coil.

Fluorescent Starters

The starter is designed so it is ordinarily closed when the lamp is turned off. When the lamp is turned on, the starter opens a second or so after the current starts to flow and then stays open until the lamp is turned off.

When a fluorescent lamp is turned on, current flows through the ballast, through one filament or "cathode", as it is called in the case of the fluorescent lamp, through the starter, through the other filament or cathode, and back to the line. During this period, the lamp glows at each end but it does not light. Then the starter opens and the ballast does its work—it delivers a high voltage, as mentioned previously, a voltage considerably above 115 V and high enough to start the lamp. The current can no longer flow through the starter because it is open. It then flows through the tube, jumping gap and forming an arc inside the glass tube. The current flows first in one direction, then in the opposite direction, because the current involved is alternating current. The ballast then performs its other function;

it limits the current flowing through the lamp (current flow) to a predetermined safe value.

General Ballast Information

There are five manufacturers of general ballasts for fluorescent lighting in the United States:

- Advance
- General Electric
- Jefferson
- Universal
- Westinghouse

Each of these ballast manufacturers provides a cross reference guide which has been prepared to show the various equivalent numbers of each manufacturer for interchange when necessary.

Each manufacturer codes and dates the ballast in various ways, either stamped on the back or the front and most provide a one or a two year replacement warranty on their product. The beginning date of the warranty is always indicated on the ballast.

For most purposes 120- or 277-V ballasts are the most commonly used. There are variations to the voltages for specific applications such as 236 or 480 V.

A representative listing of available ballasts is as follows:

RAPID-START BALLASTS			
Number of Lamps	Lamp Description	Circuit (Volts)	Catalog Number
1	F17T8	120	R-1P817-TP
3	—	120	REL-3P32-TP Electronic
3	F17T8	277	VEL-3P32-TP Electronic
4	—	120	REL-4P32-TP Electronic
4	F17T8	277	VEL-4P32-TP Electronic

RAPID-START BALLASTS *(Cont.)*

Number of Lamps	Lamp Description	Circuit (Volts)	Catalog Number
1	F25T8	120	R-IP825-TP
3	—	120	REL3p32-TP Electronic
3	F25T8	277	VEL3P32-TP Electronic
4	—	120	REL4P32-TP Electronic
4	F25T8	277	VEL4P32-TP Electronic
1	F30T12 F40T12	120	RL-140-TP
1	—	120	HM-IP30-TP Electronic
1	—	120	REL-1540-TP Electronic
1	F30T12	277	VM-1P30-TP
1	—	277	VEL-1540-TP Electronic
2	—	120	RM-23P30-TP
2	—	120	R-25P30-TP Markill
2	—	120	REL-2540-TP Electronic
2	F30T12	277	VM-25P30-TP
2	—	277	V-25P30-TP
2	—	277	VEL-2S4O-TP Electronic
3	—	120	REL-3540-TP Electronic
3	F30T12	277	VEL-3540-TP Electronic
1	F40T8	120	R-IP840-TP
2	F40T8	120	R-2P840-TP
3	—	120	REL-3P32-TP Electronic

RAPID-START BALLASTS *(Cont.)*			
Number of Lamps	Lamp Description	Circuit (Volts)	Catalog Number
3	F40T8	277	VEL-3P32-TP Electronic
1	FC8T9	118	RLQ-122-TP
1	FCl2T10	120	RL-140-TP
1	FC16T12	120	RLCS-140-TP
1	FC8T9	120	R-22-32-TP
1	FC12T10 (1 each)	120	RS-22-32-TP
1	FCl2T10	120	R-32-40-TP
1	FCl6Tl2 (1 each)	120	RS-32-40-TP

Core and Coil Ballasts

Core and coil ballasts feature small size, light weight, low cost, and permit a wide variety of mounting options allowing for flexibility in fixture design. For consistent performance, MagneTek core and coil ballasts feature coils precision wound on bobbins, and are impregnated with a protective high temperature varnish, insulation Class 180.

Core and coil ballasts can be used in ambient temperatures of considerable variance, indoors and out.

All constant wattage autotransformer (CWA) type core and coil ballasts are manufactured with dual common leads. This permits quicker and easier installation in new or existing fixtures.

Core and coil ballasts can be mounted by using the mounting holes pre-drilled through the laminations, by optional welded mounting brackets, or by adjustable mounting brackets which are routinely supplied with replacement kits.

Core and coil ballasts for many lamp types are offered in dual voltage, tri-voltage and multi-voltage designs for added versatility and to lower inventory requirements. On most, the 120 V terminal can be used as a tap for standby lighting when the ballast is utilized for any of the higher voltages.

All multi-voltage ballasts can be utilized for any one of four single voltages-120, 208, 240 or 277 V/60 Hz. The core and coil is manufactured with four male quick-connect blade terminals permanently mounted to the coil bobbin. Line wires identified for each voltage and equipped with female quick-connect blade terminals are connected onto the corresponding individual male blade terminals. For wiring ease, each wire is imprinted for its purpose — "line", "common", "lamp", "capacitor." Lead wires not being used can be disconnected and discarded. No cutting or insulating of the unused leads is necessary.

The core and coil ballasts are manufactured for use on high-pressure sodium, metal halide and mercury lamps.

35 W S76 HIGH-PRESSURE SODIUM LAMP BALLAST		
Line Voltage	Catalog Number	Circuit Type
120	1233-251U	R-NPF
50 W S76 HIGH-PRESSURE SODIUM LAMP BALLAST		
120	1233-35U	R-NPF
120	—	R-HPF
120 208 240 277	12310-95	HX-HPF
70 WS62 HIGH-PRESSURE SODIUM LAMP BALLAST		
120	1233-142U	R-NPF R-HPF
480	12310-148R	HX-HPF
120 208 240 277	12310-153	HX-NPF
400 W S51 HIGH-PRESSURE SODIUM LAMP BALLAST		
120 208 240 277	1230-93S	CWA
120 208 240 277	123-93U	CWA

Electric Lighting

\multicolumn{3}{c}{1000 W S52 HIGH-PRESSURE SODIUM LAMP BALLAST}
Line Voltage	Catalog Number	Circuit Type
120 208 240 277	1230-97S	CWA
\multicolumn{3}{c}{250 W M58 OR M103 METAL HALIDE LAMP BALLAST 250 W M37 MERCURY LAMP BALLAST}
480	1130-32	CWA
120 208 240 277	1130-92	CWA
\multicolumn{3}{c}{400 W M59 METAL HALIDE OR H33 MERCURY LAMP BALLAST}
480	1130-32R	CWA
120 208 240 277	1120-93R	CWA
\multicolumn{3}{c}{1000 W M47 METAL HALIDE OR H36 MERCURY LAMP BALLAST}
480	1130-57	CWA
120 208 240 277	1130-97	CWA
\multicolumn{3}{c}{1500 W M48 METAL HALIDE LAMP}
480	1130-69R	CWA
120 208 240 277	1130-99R	CWA

Lighting Fixtures

Surface-Mounted Corridor Fixture: This fixture type is available in the 120- and 277-V styles. It uses a rapid-start ballast. The housing is die-formed code gauge prime cold-rolled steel with baked-on white finish. It has a low brightness, extruded acrylic refractor. It requires one or two F40 lamps. The fixture is 48 in x $6\frac{1}{8}$ in x $4\frac{5}{8}$ in.

Surface-Mounted Wraparound Fixture: This fixture is 120 V or 277 V, 60 Hz. It uses an energy-saving rapid-start ballast. The housing is die-formed code gauge prime cold-rolled steel. It has a baked-on white finish. It has a clear acrylic prismatic refractor/low brightness. It uses 2 or 4 F 40 lamps. Dimensions of the fixture are 48 in x $8\frac{1}{8}$ in x $2\frac{5}{8}$ in for the 2 lamp configuration and 48 in x 14 in x $2\frac{5}{8}$ in for the 4 lamp configuration.

Recessed GP Static: This fixture has a 60 Hz energy-saving rapid-start ballast and a $3\frac{3}{4}$-in deep recessed housing. It is one-piece, die-formed code gauge, prime cold-rolled steel. It has a baked white finish, reinforced heavy gauge, flat white steel door, and mitered corner with cam latches/safety integral hinges. It has an acrylic prismatic No. 12 pattern shielding.

Commercial Narrow Strip Fixture: This fixture type is made of channel die-formed code gauge prime cold-rolled steel. It has pressure lock lampholders and a baked white finish. The end plate converts to channel connector for continuous row mounting.

Commercial Wide Strip Fixture: This fixture is channel die-formed code gauge prime cold-rolled steel. It has pressure lock lampholders for rapid-start lamps and spring-loaded lampholders for slimline lamps. It has a baked white finish. The end plate converts to channel connector for continuous row mounting.

Industrial Fixture: This fixture has an energy-saving rapid-start and slimline 800 ma ballast. The channel is code gauge prime cold-rolled steel. It has a heavy-gauge aperture (IA) and closed-top (IC) reflector and baked white finish.

Industrial Vapor-Tight Fixture: This fixture type has a 60 Hz energy-saving rapid-start ballast. It has a reinforced fiberglass housing and is waterproof and chemical resistant. It has a low brightness acrylic shield with high impact additive and has a baked white finish liner.

1 x 4 and 2 x 2 Grid Static Fixture: This fixture has a flush steel door and hinges and latches from either side. It has a baked polyester white finish. It uses 2 lamps and an integral grid clip (RA). It is UL damp location listed.

Premium Specification Grade Parabolic Fixture: This fixture has a 3-in deep parabolic louver and has uniform cell brightness. The louver is protected during shipment and installation by polyethylene sheet full black reveal. The louver is finished in semi-specular silver. It has integral grid clips and is UL 1570 damp location listed.

2-Ft x 2-Ft Lay-In Troffer Fixture: This fixture has an acrylic lens, a full door and uses a U-bent F40 lamp.

Parabolic 2-Ft x 4-Ft Lay-In: This is an 18 cell and 3-lamp high efficiency fixture.

Parabolic 2-Ft x 2-Ft Lay-In: This fixture is a 9-cell and uses 2 F40 U-bent lamps.

Incandescent Recessed Fixtures

Standard Recessed Housing Number H7T: This is versatile, reliable and easy to install. It is the most widely used housing in the lighting industry. The integral thermal protector guards against over-lamping and misuse of ceiling insulation. It has a feed-through junction box, adjustable hanger bars and adjustable socket bracket for different size lamps. It is $7\frac{1}{2}$ tall and is UL listed for damp locations and feed-through branch circuit wiring.

Standard Remodel Housing Number H7RT: This fixture is designed for use in remodelling installations. It slips through a $6\frac{1}{2}$-in ceiling opening. The integral thermal protector guards against over lamping and misuse of ceiling insulation. Its height is $7\frac{1}{2}$ in and uses a $6\frac{1}{2}$-in ceiling opening. It has an adjustable socket bracket for different size lamps. It is UL listed for damp locations.

Slope Ceiling Housing Number H47T: This is designed for use in a sloped ceiling construction (10 to 27-degree slope). The adjustable socket plate maintains proper lamp aiming. The integral thermal protector guards against over lamping and misuse of ceiling insulation. The housing adjusts vertically for different joist construction. It is $7\frac{3}{8}$ in to $9\frac{3}{8}$ in high, and uses a $7\frac{1}{4}$-in ceiling opening. It has a feed-through junction box and adjustable hanger bars. It is UL listed for damp location and feed-through branch circuit wiring.

Small Aperture Housing Number H99T: The small aperture housing is ideal for accent lighting. The integral thermal protector guards against over lamping and misuse of ceiling insulation. It is $5\frac{1}{2}$-in high and needs a $4\frac{1}{4}$ in ceiling opening. It has a feed-through junction box and adjustable hanger bars. It is UL listed for damp locations and feed-through branch circuit wiring.

Small Aperture Remodel Housing Number H99RT: The small aperture remodel housing is ideal for accent lighting. The integral thermal protector guards against over lamping and misuse of ceiling insulation. The unit slips through a $4\frac{1}{4}$-in ceiling opening. It is $5\frac{1}{2}$ high and is UL listed for damp locations.

Small Aperture Insulated Ceiling Housing Number H99ICT: This small aperture housing is for insulated ceiling installations. It maintains unbroken insulation barrier to prevent heating an air conditioning losses. The integral thermal protector guards against over lamping. It is $5\frac{1}{2}$ high and uses a $4\frac{1}{4}$-in ceiling opening. It has a feed-through junction box and adjustable hanger bars. It is UL listed for damp locations and feed-through branch circuit wiring.

Square Housing Number H1T: This housing is for use with lens trims. The integral thermal protector guards against over lamping and misuse of ceiling insulation. The integral magnetic transformer lowers line voltage (120 V) down to lamp operating voltage (12 V). It is $5\frac{3}{8}$ in high and uses an $8\frac{1}{4}$-in square ceiling opening. It has a feed-through junction box and adjustable hanger bars. It is UL listed for damp locations and feed-through branch circuit wiring.

Recessed High-Intensity Discharge Lamp Fixtures

HID Down Lighting: This housing is for low-wattage, energy-efficient, long life HID lamps. Order the housing, choice of ballast and choice of trim for a complete unit. It has a separate ballast with quick mount plug-in connectors and dual tap (120/277 V) ballast. It is $9\frac{3}{8}$ high and uses a 7-in ceiling opening. It has a feed-through junction box and adjustable hanger bars. It is UL listed for damp locations and feed-through branch circuit wiring.

Chapter 29

Overcurrent and Disconnecting Devices

Service Switches

To disconnect a building from its source of electrical supply, a device must be provided at or near the point where the service-entrance wires enter the building. In addition, it is also necessary to have a device or devices which will disconnect single circuits from the source of supply. This constitutes a safety factor, because there are times when it is absolutely necessary to disconnect or "kill" parts, or all of the wiring, such as in the case of fire or when working on components within the system.

Equally important, is providing overcurrent protection that protects the installation as a whole, or parts of individual circuits, against short circuits or overloads. Often the disconnecting means and the overcurrent protection come as a single unit housing all the necessary components. The disconnecting device or devices must be listed by UL and bear the service equipment label. The National Electrical Code (NEC) states that the disconnect device must be "externally operable," which fulfills the condition if the switch can be operated without the operator being exposed to live parts.

Service switches, or main breakers as they are often referred to, are rated at 30, 60, 100, 200, 400, 600, 800, and 1200 A with no in-between ratings. Due to the increased usage of electrical equipment and appliances, the 100-A size is the minimum size specified by the NEC in most cases. Even the 200-A size is becoming fairly common and is already specified as a minimum in some local codes.

Section 230-70(a) of the NEC states that a means must be provided for disconnecting all conductors in the building from the service-entrance conductors.

Service switches are rated either 250 or 600 V in either fused or unfused styles. Switches suitable for indoor use only are rated NEMA 1; those suitable for outdoor use are rated NEMA 2.

The disconnecting means must be located at a readily accessible point nearest to the entrance of the conductors, either inside or outside the building or structure. Sufficient access and working space must be provided about the disconnecting device.

Service switches are also rated as either single-phase or three-phase and with or without a solid neutral. The term *solid neutral* means that the neutral wire that connects to the switch is not interrupted by a switch or fuse. Instead, it runs to the neutral busbar in the switch. This neutral busbar is merely a copper strip with the number of terminal lugs to which the neutral of the incoming wires, the neutral wires of all the 120-V branch circuits, and also the equipment grounding wire are connected.

All switches must be marked with the current, the voltage, and if horse-power rated, the maximum rating for which they are designed. Fused switches must not have fuses connected in parallel. All 600-V knife switches, auxiliary contacts of the removal or quick-break type — or the equivalent — must be provided on all switches designed for use in breaking currents over 200 A.

Safety switches are also available with the NEMA 12 rating enclosure for several industrial and commercial installations, where it is desired to exclude such materials as dust, lint, fibers or filings, oil or coolant seepage, etc. The dusttight and oil-tight construction of these types of enclosures assures safety, performance, and continuity of service within such difficult atmospheres.

In an apartment block or multiple occupancy building, each occupant must have access to the disconnecting means. A multiple occupancy building having individual occupancies above the second floor must have service equipment grouped in a common accessible place. Furthermore, the disconnecting means must consist of not more than six switches or six circuit breakers.

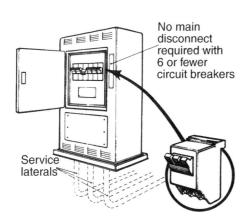

Not more than six disconnects are allowed at the main service location

Permitted Disconnects

The following are types of disconnects that are permitted by the National Electrical Code:

- A manually operable switch or circuit breaker equipped with a handle or other suitable operating means, positively identified and marked for mechanical operation by hand.
- An electrically operated switch or circuit breaker, provided the switch or breaker can be opened by hand in the event of failure of power supply, and the open and closed positions are clearly indicated to the operator.

Whatever disconnecting means are provided, they must be plainly marked to indicate whether it is in an open or closed position; that is, OFF or ON, respectively. Where two to six service disconnects are installed at one service location, each such disconnecting device must be permanently marked to identify it as a service disconnecting means. The service conductors must be attached to the disconnecting means by pressure connectors or other approved means. Connections that depend on solder must not be used.

Panelboards

Two basic types of panelboards are defined in the National Electrical Code.

1. Lighting and appliance branch circuit panelboards.
2. All others.

A lighting and appliance branch-circuit panelboard is defined as a panelboard having more than 10 percent of its overcurrent devices rated at 30 A or less, and for which neutral connections are provided. It is not permitted to contain more than 42 overcurrent devices in addition to any main overcurrent device. A two-pole circuit breaker is considered as two devices and a three-pole circuit breaker is counted as three. The panelboard is also required to be provided with some physical means that will effectively prevent the installation of a greater number of overcurrent devices than that for which the panelboard is designed.

Panelboards may be equipped with circuit breakers or fuses; the latter with or without snap switches. Snap switches are required to have a rating not less than that of the feeder ampacity as determined by the connected load. All panelboard overcurrent devices must not be loaded to more than 80 percent of their rating when loads are continuous for more than three hours, unless the panelboard, and all overcurrent devices, are approved for continuous duty at 100-percent capacity. Overcurrent protection for panelboards having snap switches rated at 30 A or less is limited to 200 A. Circuit breakers are not considered to be snap switches.

Each lighting and appliance branch-circuit panelboard is required to be provided with individual overcurrent protection, which may consist of not more than two main circuit breakers or sets of fuses, with a combined rating not in excess of the rating of the panelboard except when:

1. The feeder overcurrent device has a rating not greater that that of the panelboard, or

2. A lighting an appliance branch circuit is used as the service equipment for an individual residential occupancy, such as an apartment, provided that each set of fuses supplying 15 or 20 A branch circuits is protected on the supply side with a main overcurrent device.

Location In Premises

The NEC requires that overcurrent devices must be located where they will be:

- Readily accessible except as provided in NEC Sections 230-91 and 230-92 for service equipment, and Section 364-11 for busway.

- Not exposed to physical damage.

- Not in the vicinity of easily ignitable material.

- All overcurrent devices protecting the conductors supplying their occupancy must be readily accessible to each occupant.

Exception: In a multiple occupancy building, where electric service and electrical maintenance are provided by the building management and where these are under continuous building management supervision, the service overcurrent devices and feeder overcurrent devices supplying more than one occupancy may be accessible to authorized management personnel only.

Enclosure for Overcurrent Devices

The NEC specifies three types of enclosures for overcurrent devices.

- *General:* Overcurrent devices shall be enclosed in cutout boxes or cabinets, unless a part of a specially approved assembly which affords equivalent protection, or unless mounted on switch boards, panelboards or controllers located in rooms or enclosures free from easily ignitable materials and dampness. The operating handle of a circuit breaker may be accessible without opening a door or cover.
- *Damp or Wet Locations:* Enclosure for overcurrent devices in damp or wet locations shall be of a type approved for such locations and shall be mounted so there is at least ¼ in of air space between the enclosure and the wall, or other supporting surface.
- *Vertical Position:* Enclosures for overcurrent devices shall be mounted in a vertical position unless in individual instances that is shown to be impractical.

Overcurrent Protection

Current in excess of the normal rating for conductors or equipment will cause excessive heating, and unless properly and rapidly interrupted, may cause fire. Overcurrent protection is provided to prevent damage to conductors or equipment. Overcurrent protection for conductors usually will not provide adequate protection for equipment. Therefore, additional protective devices for most equipment are needed.

Protective devices are actuated either thermally or magnetically. The most common types of overcurrent protective devices for conductors and some types of equipment are fuses and circuit breakers. For motors, thermal overload devices are used.

In larger installations, protective relays are specified. Standard ampere ratings for fuses and nonadjustable trip circuit breakers are from 15 to 6000 A. In general, overcurrent devices must be located where they are readily accessible but not exposed to physical damage or to material that is readily combustible.

Simple calculations of circuit loads and conductor ampacities may not always determine the proper overcurrent protection. In a circuit having a high impedance, a serious delay may occur in the operation of an overcurrent protective device. Section 110-10 of the NEC requires that the overcurrent protective devices along with the total impedance and other characteristics of the circuit be so selected and coordinated that the circuit and its components will be protected against damage that could result from either a phase-to-phase or phase-to-ground fault.

Overcurrent protective devices must have sufficient interrupting capacity to open the circuit under both simple overloads and maximum fault conditions. Without such interrupting capacity, an overcurrent protective device may fail physically at a time when its operation is essential.

Any device that limits the current in a wire to a predetermined number of A is called an overcurrent device in the NEC.

History of Fuses

A fuse acts as a "safety valve" for electrical systems. It is to the electrical circuit what the safety valve is to the steam boiler. It is designed to open the circuit when too much current is flowing, or when there is a sudden rush of current of dangerous proportions, as from a short circuit.

If the fuse did not open the circuit, such excess current would burn up the wiring or cause fire or injury, and result in costly repair bills.

The earliest form of fuses consisted of a piece of copper wire, smaller in size than the line wires. However, such fuses would reach very high temperature before they melted, and were soon found to be quite hazardous. This led to the development of a lead-alloy wire for use in fuses.

Initially, poor contact of the fuse terminals frequently caused premature blowing. Therefore, the next development was the soldering of slotted terminals to the fuse wire.

Various forms of fuse blocks with removable covers were designed to enclose such fuse links which lessened the hazard of such fuses on light overloads but these fuse blocks would, on short circuits, be frequently shattered.

The lead allow wire was then enclosed in a tube of fiber, glass, porcelain or other insulating material. Doing so was an advantage on small overloads but on heavy short circuits, these tubes would also shatter.

As the art progressed, a zinc strip or wire enclosed in a fiber tube filled with an insulating, fire-resisting powder evolved. It was found that the lesser tendency of zinc to hold and arc, combined with the arc quenching properties of the powder, allowed a cartridge to be made which would not explode or emit flame or molten metal.

In 1904, a standard on fuses was adopted which is still in effect. By this standard, two different forms of cartridges were adopted and dimensions were standardized. Thus, we have our standard 250 and 600 V fuses that are used to this day.

How A Fuse Operates

If a current of more than the rated load is continued sufficiently long, the fuse link becomes overheated, causing the center portion of the link to melt. The melted portion drops away, but due to the short gap, the circuit is not immediately broken and an arc continues, burning back the metal at each end until the arc is stopped because of the very high increase in resistance, and because the material surrounding the link tends to mechanically break the arc. The center portion melts first because it is furthermost removed from the terminals which have the highest heat conductivity.

On a short circuit, the entire center section is instantly heated to an extremely high temperature, causing it to volatilize (turn to vapor). This vapor has a high current-carrying capacity which permits the arc to continue but the arc-extinguishing filler cools, condenses the vapor, and stops the arc. The degree or extent of the volatilization of metal is dependent upon the design of the fuse.

General Constriction of Fuses

Ordinary fuses, Fusetron and Low-Peak fuses for 250 or 600 V typically are made with fiber tubes. The fiber withstands a high internal pressure generated when a fuse is blown under short circuit conditions, and yet will not carbonize from the high temperature of the arc occurring when a fuse blows.

Ordinary plug fuses, Fusetron dual-element plug fuses and Fustat fuses are made with a porcelain body in ratings up to 30 A for circuits of 125 V or less. Porcelain is used because it is an excellent electrical insulator. It can be readily molded into the shape required and is strong enough to withstand the high pressure developed in these fuses under short-circuit conditions.

Glass tubes are used on fuses of 250 V circuits in low ratings and on circuits up to 30 A. Glass can be used on these fuses because the amount of metal that can be volatilized when the fuse blows is very limited. However, they would explode on heavier or high-voltage circuits.

Fusible Elements

The fusible element in all ordinary fuses, except on very small sizes is made of zinc. Zinc is a fairly good conductor. Other metals of lower melting points would require a much larger volume of metal, and in view of the fact that vapors and pressure developed on a short-circuit blow are in proportion to the amount of metal. Consequently, the less metal that can be used, the less likelihood there is of the fuse exploding. Lead and other materials are used on very low-amperage circuits because with zinc, the element might be so small as to lack mechanical strength. Zinc has a reasonably low melting point. Silver, copper and aluminum are better conductors than zinc and would therefore require less volume of metal. However, the temperatures at which such metals melt are very much higher than zinc, and would create too high a temperature to be safe in an ordinary fuse.

In Fusetron and Fustat fuses, the link is made of copper. In these devices, however, the link is made heavier than in ordinary fuses of the same rating and overloads will not cause the link to melt, since the thermal cut-out opens the circuit before this could occur. In the Low-Peak High-Cap and Limitron fuses, the fusible element is made of silver, contained in a special arc-quenching filler.

Open fuses are usually made of an alloy consisting mostly of lead because the melting point of lead is much lower than that of zinc, and the lead alloy does not have a tendency to glow red for a long time when blowing on light overloads, as other metals do.

With the tremendous development in the use of electricity came an instant demand for a renewable fuse. By the use of a properly designed link, by making the cartridge much longer, and providing proper venting on the larger sizes, the powder filler was omitted and fuses made so that when blown, they can be restored to original condition by simply replacing the blown link.

The next development was a renewable fuse with a link that retained all the advantages of the old type, but gave a greater time lag when subjected to harmless overloads than had ever been possible before.

Overcurrent and Disconnecting Devices

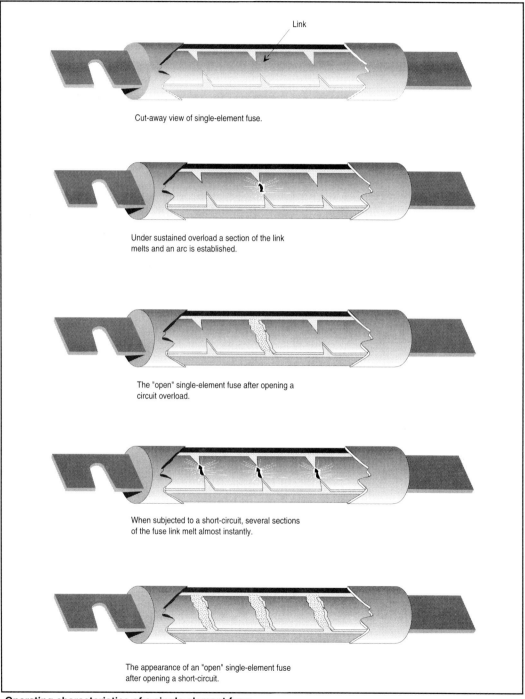

Operating characteristics of a single-element fuse

How A Renewable Fuse Works

On a heavy overload or short circuit, the weak spots of the link are suddenly heated to a very high temperature, causing these portions of the strip to volatilize. Due to the short length of these weak spots, an arc continues, but as the arcs are in series and voltage across each is low, the arc is stopped and the circuit is open with but very little metal having been volatilized, destructive pressure generation within the fuse has been avoided.

On overloads, only one weak spot may melt. On light overloads, where blowing time is long continued, its use may melt it in the center because weak spots are near the heavy metal contacts and the heat generated is conducted away almost as fast as generated. As the center portion has very little means of having heat conducted away, it may melt before the end weak spots melt.

Dual-Element Fuses

The increased use of small motors on ordinary branch circuits in a household or business caused the blowing of plug fuses unnecessarily from the starting current demand of the motors. The Fusetron dual-element idea was incorporated into plug fuses, one with a time-lag so great that it withstands starting currents and can actually be used not only alone for ordinary circuits, but also in mains and feeder protection, and for the actual protection of motors against overloads.

The Fusetron dual-element fuse has a fuse link element and a thermal cutout element. On overloads, the circuit is opened by the thermal cutout. This cutout has a very long time lag, so the Fusetron fuse will not open on harmless overloads or ordinary motor starting currents. The fuse link is made heavier than that used in an ordinary fuse of the same rating. It protects only against short circuits and can therefore be made of copper, which permits the use of smaller amounts of metal which in turn means that the Fusetron dual-element fuse will open on short circuits as safely as an ordinary fuse.

This construction gives the fuse a lesser electrical resistance than ordinary fuses, so it will not cause as much heating as ordinary fuses.

How a Dual-Element Fuse or Low-Peak Fuse Operates

On a short circuit, the fuse link blows just as in the case of other types of fuses. On an overload, the fuse link remains intact. The heat generated by an overload is fed to the center mass of copper on which is mounted a

Overcurrent and Disconnecting Devices 447

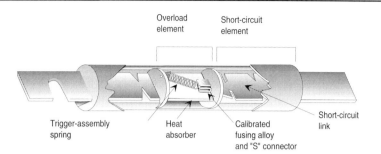

Overload element | Short-circuit element

Trigger-assembly spring | Heat absorber | Calibrated fusing alloy and "S" connector | Short-circuit link

The true dual-element fuse has distinct and separate overload and short-circuit elements.

Under sustained overload conditions, the trigger spring fractures the calibrated fusing alloy and releases the "connector."

The "open" dual-element fuse after opening under an overload.

Like the single-element fuse, a short-circuit current causes the restricted portions of the short-circuit elements to melt and arcing to burn back the resulting gaps until the arcs are suppressed by the arc-quenching material and increased arc resistance.

The "open" dual-element fuse after opening under a short-circuit condition.

Operating characteristics of a dual-element fuse

spring with a short connector. This connector is held in place by low melting solder, and connects the center mass of copper to the fuse link. When the temperature is increased by the overload to such an extent that the solder melts, this connector is pulled out of place, thereby opening the circuit.

A later development in dual-element fuses was a Low-Peak, incorporating the thermal cutout element of the Fusetron fuse with a high capacity current-limiting fuse element. The Low-Peak fuse affords all the advantages of the Fusetron, plus much higher interrupting capacity and great current limitation on short circuits.

The Limitron fuse was then designed to combine high-interrupting capacity with great opening speed and limitation of let-through current for use as short circuit protection for circuits carrying loads from 1000 to 6000 A at 600 V or less. They have an interrupting rating of 200,000 A and limit the peak of let-through current to a very small fraction of the current available.

Edison-base plug fuse

Plug Fuses

Until 1935, all plug fuses were made with the Edison base. Inspectors found that many people, who were ignorant of the importance of proper fuse protection used fuses that were too large or else used pennies or other materials to bridge the fuse, thereby wiping out all protection.

Fustats

To guard against the use of pennies or other materials to bridge a fuse, the Fustat fuse was developed. This type fuse is made to prevent the use of oversized fuses on a circuit, and is so designed to prevent the use of bridging materials, making it virtually impossible to alter the fuse protection. The Fustat is constructed with the Fusetron dual-element feature, so unnecessary blows from motor-starting currents are eliminated.

An ordinary fuse cannot protect a motor. An ordinary fuse has only one part, the link that must handle all overloads. This link cannot be made to do two jobs. An ordinary fuse, small enough to protect a motor against continued excess current, will blow every time on the

An Edison-base fuse (left) with a Fustat and base (right)

heavy-starting current. An ordinary fuse large enough to hold a starting current will not blow on small excess currents that can burn out the motor.

A Fustat has two parts: a fuse link element which handles overloads only, and a thermal cutout element which handles light overloads. A fuse link will only open very heavy overloads, but easily holds the motor starting current. The thermal cutout must first melt the solder before it can open; the heavy starting current does not last long enough to do this. Yet a small excess of current, long continued, will melt solder. A Fustat fuse is small enough to protect a motor but will not open on overloads until they have been continued long enough to endanger the motor.

The Fustat fuse is made to fit into a regular Edison-base fuse holder by means of a simple inexpensive adapter which locks in place. The Fustat fuse can then be removed or inserted in the same manner as the ordinary fuse.

Standard Fuse Requirements

UL standards on fuses require that fuses be of certain dimensions within specified tolerances, and that the fuses meet the following tests:

- It must withstand heavy short circuits. Failure to do this will result in a very great fire hazard and such failure would mean the fuse would explode or belch fire.

- It must operate at a reasonably low temperature. Failure to do so would make a hazard of the fuse.

- It must carry rated current. Failure to do so would result in premature opening of the fuse, causing unnecessary expense and annoyance. All fuses are required to carry a 10 percent overload indefinitely, when tested in the open. When fuses or switches are installed in panelboards, as is usually the case, they may open on less currents. The 10 percent overload requirement assures that the fuse will carry the rated current in actual practice.

- The fuse must open within prescribed time limits when tested on overloads. Failure to do so might allow an excessive amount of current to damage wiring or equipment before this fuse would open. When tested in open air on a 35-percent overload, fuses rated from 0 to 60 A are required to open within an hour, and larger fuses within two hours.

When An Ordinary Fuse Blows

When a fuse blows, one of the following conditions should be the cause:

1. A short circuit. If a short circuit blew the fuse, make sure the line has been cleared and the cause of the short circuit removed; then install a new fuse and proceed.

2. An overload has occurred. If the circuit is overloaded, obviously the overload will have to be relieved. If it was a temporary overload, install a new fuse and continue.

3. The fuse did not have sufficient time-lag. If a motor starting current or other harmless overload blew the fuse, the fuse did not have sufficient time-lag. A larger size fuse would reduce the protection and would not comply with the NEC.

4. Poor contact. Poor contact on, near or in the fuse causes useless blowing of fuses and very often causes destruction of the switch or panelboard. No matter how careful maintenance personnel may be, poor contact conditions might develop due to vibration, heating, cooling or other operating conditions. Very often, with poor contact, the fuse will indicate this. For example, if the surfaces that make contact with the fuse clips are discolored, the fuse has been making poor contact in the clips. If the contact has been tight, there will be very little, if any, air getting to the portions that make contact and contact surfaces will remain bright and clean. If only one end of the fuse has the contact surface badly oxidized, this is a positive indication that poor contact exists at the end where discoloration has occurred.

5. Overheating of fuses. Where a number of fuses are installed in an enclosure, the heating of the fuses reduces their carrying capacity. Where such trouble causes blowing a fuse, the only answer is to use Fusetron dual-element fuses or Low-Peak fuses, which have materially less resistance and therefore create much less heat.

6. Wrong size fuse. The size of the fuse may have been too small to begin with, which would cause blowing of fuses.

Avoid such useless blowing by being sure you use the right size fuse.

7. Fuse located in high ambient temperature. If the fuse is installed at a point where the surrounding air is of a very high temperature, such as near a heater of some kind, it cannot carry its stated load. Fuses should be installed where they will not be subjected to high temperatures.

8. Vibration. Many makes of fuses are seriously affected by vibration causing the link to break or causing contact points on renewable fuses to loosen.

The Right Kind of Fuse

For Circuits Not Over 150 V To Ground: Plug fuses (0 to 30 A) are usually used on such circuits. There are three distinctly different types of plug fuses available. The ordinary plug fuse, Fusetron dual-element plug fuse and the Fustat fuse. The ordinary plug fuse has very little time-lag and if any small motor is connected to the circuit, the starting current of the motor will very often blow the fuse, thereby causing an entirely unnecessary interruption of service.

The Fusetron dual-element fuse eliminates such unnecessary blows. It has a tremendously long time lag that permits it to hold when the motor is being started. In spite of the long lag-time of Fusetron dual-element fuses, their operation on short circuits is just as fast as with ordinary plug fuses.

Fustat fuses have the same feature as Fusetron fuses, but in addition, are made so too large a size cannot be used, making bridging practically impossible. Fustat fuses are listed by UL as Type S fuses. On new installations, fuse holders are required to be of Type S.

Fuses For 250- and 600-V Circuits: These fuses come in two distinctly different shapes, depending on their rating. They are available from 0 to 60 A in the cartridge type, and 70 to 600 A in the knife-blade type.

Ordinary one-time fuses and ordinary renewable fuses have comparatively little time-lag, but will very often blow on harmless overloads, or from the starting current of motors that might be connected. In addition, in sizes above 30 A, such fuses, when well loaded, will cause considerable heating of the switch or panelboard. The cartridge fuse from 0-60 A is also known as a ferrule type fuse. The ferrule construction is used only on fuses rated 60 A or less, and the knife-blade construction type cartridge fuse is used on fuses rated over 60 A.

SC or Class G Fuse

This is a comparatively new type fuse, called Class G by UL and 5C by the manufacturer. It was developed primarily for commercial and industrial buildings with lighting circuits operating at 277 V. In such installations, the overall size of the panelboards is greatly reduced by use of this type fuse in place of the large 600 V type formerly required. The Class G fuse may also be used for residential and farm installations in panelboards or fuse cabinets designed for it. The Class G is the time delay type.

Panelboards and fuse cabinets for the Class G are furnished empty, except for an assembly of terminals and busbars. When installing, individual plug-in units of appropriate type and size are plugged in to complete assembly. The plug-in unit comes in two types: Type 1 contains a switch for turning the circuit on and off, a fuse-holding block that can be removed for replacement of fuses, and a signal light glows only when the the fuse has blown. The plug-in unit fits only panelboards designed for it.

In the second type (Type 2), the switch is omitted. Each plug-in unit of the Type 2 unit contains two removable fuse-holding blocks for two 115 V circuits, or two fuses for one 230 V circuit. The fuse-holding block opens the circuit; insert the block upside down turns the circuit off. The signal light glows only when the fuse has blown. This type fits in only fuse cabinets designed for it.

In both Type 1 and Type 2 units, the plug-in units carry ampere ratings so it is impossible to use fuses for a higher amperage rating.

Circuit Breakers

The NEC defines a circuit breaker as "a device designed to open and close a circuit by non-automatic means, and to open the circuit automatically on a predetermined overload of current without injury to itself when properly applied within its rating." All switches break circuits but are not circuit breakers, and the NEC definition refers only to the type that opens a circuit when amperage greater than that for which it was designed flows through it.

Q-type circuit breakers that are used in residential construction looks like a somewhat overgrown toggle switch. Essentially, it consists of a carefully calibrated bimetallic strip similar to that used in a thermostat. As the current flows through this strip, it bends enough to release the trip that opens the contacts, interrupting the circuit just as it is interrupted with a fuse blowing, or when the switch is opened. In addition to the bimetallic strip that operates on heat, most breakers have a magnetic arrangement that opens the breaker instantly in case of a short circuit. A circuit breaker is, in fact, a switch that opens itself in case of an overload.

Circuit breakers are rated in amperes, just as fuses are rated. Breakers carry their rated loads indefinitely and will carry a 50-percent overload for perhaps a minute, a 100-percent overload for about 20 seconds, and even a 200 percent overload for about 5 seconds — long enough to carry the heavy current required to start a motor.

The trend is away from fuses to circuit breakers, for they have many advantages. When a fuse blows, spare fuses may or may not be on hand. When a circuit breaker trips, reset it and restore the circuit to normal. The circuit breaker provides protection on all significant overloads, but also have a time-delay feature that prevents them from tripping on small, temporary overloads. Most modern residential homes are equipped with circuit breakers.

The NEC general specifications for circuit breakers are:

- *Method of operation:* In general, circuit breakers shall be capable of being closed and opened by hand, without employing any other source of power, although normal operation may be other power, such as electrical, pneumatic, etc. Large circuit breakers which are to be closed and opened by electrical means, pneumatic or other power, shall be capable of being closed by hand for maintenance purposes and shall also be capable of being tripped by hand under load, without the use of power.

- *Injury to operator:* Circuit breakers shall be arranged and mounted so operation is not likely to injure the operator.

- *Indication:* Circuit breakers shall indicate whether they are in the open or closed position.

- *Nonoperable:* An air circuit breaker used for branch circuits described in Article 210 of the NEC, shall be of such design that any alteration of its trip point (calibration) or in the time required for its operation, will be difficult.

- *Marking:* Circuit breakers shall be marked with their rating in such a manner that the marking will be durable and visible after reinstallation, except that it may be necessary to remove a trim or cover. The ampere rating of circuit breakers rated 100 A or less and 600 V or less shall be molded, stamped, etched, or similarly marked into the handle or escutcheon area of the circuit breaker. Each circuit breaker rated 240 V or less and 100 A or less and having an interrupting rating other than 5000 A, shall have its interrupting rating shown on the circuit breaker or on its label. Each circuit breaker rated more than 240 V or more than 100 A, and having an interrupting rating other than 10,000 A shall have its interrupting rating shown on the circuit breaker or on its label.

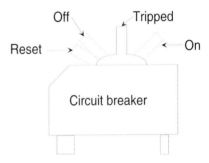

Operating characteristics of a circuit breaker

Chapter 30

Electric Motors

Alternating Current Induction Motors

Motor Identification and Selection

To order correct motors for new applications, it is necessary to have certain information. The selection information described below is written in general terms and is intended only as an aid. If further assistance is required in selecting the proper motor, you should contact a motor supplier or factory representative.

Power Supply: Is the power supply three-phase or single-phase? Most home and farm applications require single-phase motors, while most factories and large commercial and industrial users require three-phase motors. Single-phase motors can be used on three-phase systems. Three-phase motors, however, cannot be operated from single-phase systems unless a phase converter is used.

Enclosure: Motors are available in drip-proof, totally enclosed and explosionproof enclosures. Drip-proof motors have open enclosures and are suitable for indoor use and in relatively clean atmospheres. Totally-enclosed motors are suitable for use in humid environments or dusty, contaminated atmospheres. Explosionproof motors are also totally enclosed but designed for applications in hazardous atmospheres which contain explosive gases or dusts. Consult the National Electrical Code for more information regarding explosionproof regulations.

Starting Methods: Most motors are either capacitor-start, split-phase, permanent-split-capacitor or three-phase. Capacitor-start motors have high starting torque, high breakdown torque and relatively low starting current. Split-phase motors have medium starting torque and medium starting current. Permanent-split-capacitor motors have low starting torque, extra high breakdown torque and typically very low starting current.

Mounting: There are several mounting configurations available: rigid base or resilient base for belted drive applications (the resilient base also minimizes vibration and noise), C-base, D-flange, P-base, through-bolt, resilient ring (only) and belly band mount also for direct drive applications.

Service Factors: Motors are available in either NEMA, or 1.0 service factors. A 1.0 service factor motor will perform to its nameplate rating. A NEMA service factor motor can exceed its horsepower rating periodically. (*see* the table below)

NEMA Service-Factor Ratings: The table to follow lists the NEMA service factors for single-phase, drip-proof motors. Totally enclosed and explosionproof motors have 1.0 service factors, except where noted.

Single Phase

	SERVICE FACTOR AT THE LISTED RPMs			
HP	3600	1800	1200	900
1/12	1.4	1.4	1.4	1.4
1/8	1.4	1.4	1.4	1.4
1/6	1.35	1.35	1.35	1.35
1/4	1.35	1.35	1.35	1.35
1/3	1.35	1.35	1.35	1.35
1/2	1.25	1.25	1.15	1.15
3/4	1.25	1.25	1.15	1.15
1	1.25	1.15	1.15	1.15
1.5 up	1.15	1.15	1.15	1.15

Agencies: Where applicable, many motors have UL recognized construction for explosionproof motors, 56 through 445T frame, under File No. E 12044, and non-explosionproof motors, 48 through 365T frame, under File No. E49747. Many motors also are CSA certified for explosionproof motors, 56 through 445T frame, under File Nos. LR21839 and LR47504, and non-explosionproof motors, 4S through 500 frame, under File Nos. LR2025, LR37479 and LR45148.

Horsepower: Motors are available in the $\frac{1}{12}$ to 500 horsepower range. Speed or RPM: 3600, 1800, and 1200 are the most common speeds. Some motors are available with more than one speed.

Rotation: Most motors are reversible by electrical recondition or by physical orientation.

Voltage: Standard voltages are 115V, 115/230V, 230V and 230/460V. Some newer ratings are 115/208-230 or 208-230/460. Dual voltage motors are easily reconnectable from information shown on the nameplate of the motor.

Frame Size and Dimensions: Most of the motors described are built to NEMA dimensions. Check the dimension diagrams discussed later in this chapter for more specific information.

Bearings: Motors come with either sleeve bearings, ball bearings or roller bearings. Sleeve bearings are more economical and quieter than ball. Ball bearings carry heavier loading and can withstand more severe applications. Roller bearings are used in large motors for belt loads.

Overload Protection: There are four choices in protection: manual (inherent type), automatic (inherent type), thermostats and none. A manual overload must be physically reset to restart the motor. An automatic thermal overload will stop the motor when it is overloaded or over heated and restart it after the motor has cooled down. None means the motor has no protection. Thermostats are embedded in the winding and connected to the motor starter control circuit.

CAUTION

A motor with an automatic reset protector must not be used where automatic restarting (after motor cool-down) would endanger personnel or equipment. Such applications should use a manual reset protector.

General-Purpose, Definite-Purpose and Special-Purpose Motors: There are three groups of motors as defined by the National Electric Manufacturers Association (NEMA). They are general-purpose, definite-purpose and special-purpose. General-purpose motors can handle a wide variety of applications. They have NEMA torques, NEMA limited current, conservative temperatures and rugged construction. Special service motors are more economical than general purpose. They have moderately high torques and are designed for specific loads. Definite-purpose motors are designed for a specific application. Often they are restricted in use by their physical and electrical characteristics. Good examples would be oil burner, condenser fan and double shaft fan and blower motors. Special-purpose motors are custom designs to fit specific customer applications.

Typical Motor-Starting Characteristics

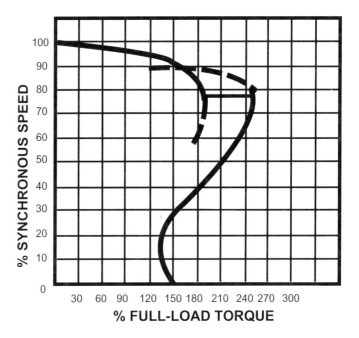

The split-phase motor start winding is disconnected at a predetermined speed by a centrifugal switch. While the start winding is in use, pull-up torque of motor will greatly exceed full-load torque figures. As motor speed causes the switch to operate, torque drops to within full load torque range, and continues to drop slowly until motor reaches 100 percent synchronous speed. Desirable characteristic include good efficiency and power factor at continuous full load.

Capacitor-Start Motor

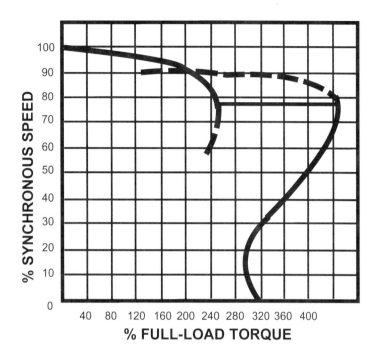

The capacitor-start motor has additional starting torque with considerably less starting current in comparison with split-phase type motors which makes the capacitor-start motor highly desirable for applications requiring multiple start-stop cycles and high starting torque. A centrifugal switch, as in the split-phase motor, removes the starting capacitor from the circuit as motor speed reaches a predetermined level.

Permanent Split-Capacitor Motor

Low starting-torque characteristics of permanent split-capacitor motors limit their applications to light starting loads such as direct connected fans and blowers. PSC type motors run extremely quietly as a starting switch is not required for operation The capacitor is permanently wired in series with start winding and in parallel with run winding.

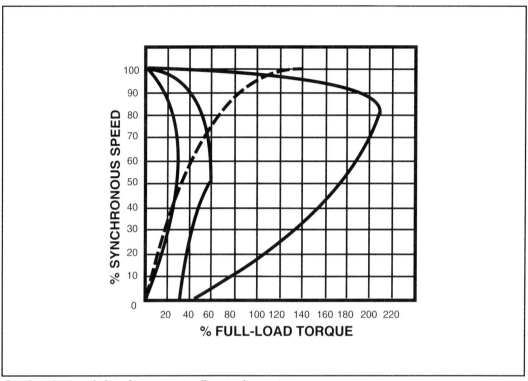

Starting characteristics of permanent split-capacitor motors

Dimensions of Integral Horsepower, General-Purpose Motors

Single- and Three-Phase: Characteristics and types available include: rigid base, drip-proof, TFC, explosionproof, from 143T and larger frames.

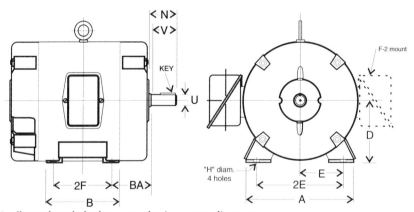

Approximate dimensions in inches; see chart on opposite page

Electric Motors

Frame	A (Max)	B (Max)	D	2E	2F	H	N	U	V	BA
143T	7.00	6.50	3.50	5.50	4.00	.340	2.31	.87	2.25	2.25
145T	7.00	6.50	3.50	5.50	5.00	.34	2.31	.87	2.25	2.25
182T	8.75	5.62	4.50	7.50	4.50	.41	2.78	1.12	2.62	2.75
184T	8 75	6.62	4.50	7.50	5.50	.410	2.78	1.12	2.62	2.75
213T	10 38	7.00	5.25	8.50	5.50	.41	3.56	1.38	3.25	3.50
215T	10.38	8.50	5.25	8.50	7.00	.41	3.56	1.38	3.25	3.50
254T	12.00	10.25	6.25	10.00	8.25	.53	4.15	1.62	3.88	4.25
256T	12.00	12.00	6.25	10.00	10.00	.53	4.15	1.62	3.88	4.25
284T	12.75	11.50	7.00	11.00	9.50	.53	4.81	1.88	4.50	4.75
284TS	12.75	11.50	7.00	11.00	9.50	.53	3.44	1.62	3.12	4.75
286T	12.75	13.00	7.00	11.00	11.00	.53	4.81	1.88	4.50	4.75
286TS	12.75	13.00	7.00	11.00	11.00	.53	3.44	1.62	3.12	4.75
324T	14.50	12.25	8.00	12.50	10.50	.66	5.50	2.12	5.12	5.25
324TS	14.50	12.25	8.00	12.50	10.50	.66	4.00	1.88	3.62	5.25
326T	14.50	13.75	8.00	12.50	12.00	.66	5.50	2.12	5.12	5.25
326TS	14.50	13.75	8.00	12.50	12.00	.66	4.00	1.88	3.62	5.25
364T	17.75	13.25	9.00	14.00	11.25	.66	6.12	2.38	5.62	5.88
364TS	17.75	13.25	9.00	14.00	11.25	.66	4.00	1.88	3.50	5.88
365T	17.75	14.25	9.00	14.00	12.25	.66	6.12	2.38	5.62	5.88
365TS	17.75	13.25	9.00	14.00	12.25	.66	4.00	1.88	3.50	5.88
404T	19.75	15.00	10.00	16.00	i2.25	.81	7.50	2.88	7.00	6.62
404TS	19.75	15.00	10.00	16.00	12.25	.81	4.50	2.12	4.00	6.62
405T	19.75	16.50	10.00	16.00	13.75	.81	7.50	2.88	7.00	6.62
405TS	19.75	15.00	10.00	16.00	13.75	.81	4.50	2.12	4.00	6.62
444T	21.75	17.00	11.00	18.00	14.50	.81	8.75	3.38	8.25	7.50
444TS	21.75	17.00	11.00	18.00	14.50	.81	5.00	2.38	4.50	7.50
445T	21.75	19.00	11.00	18.00	16.50	.81	8.75	3.38	8.25	7.50
445TS	21.75	17.00	11.00	18.00	16.50	.81	5.00	2.38	4.50	7.50
505U	25 00	20.50	12.50	20.00	18.00	.94	10.38	3.88	9.88	8.50
505US '	25 00	20.50	12.50	20.00	18.00	.94	5.00	2.38	4.50	8.50
508U	25.00	27.50	12.50	20.00	25.00	.94	11.88	4.12	11.38	8.50
508US	25.00	27.50	12.50	20.00	25.00	.94	7.00	3.38	6.50	8.50
510US	25.00	34.50	12.50	20.00	32.00	.94	7.00	3.38	6.50	8.50
447T	21.75	22.50	11.00	18.00	20.00	.81	8.75	3.38	8.25	7.50
447TS	21 75	22.50	11.00	18.00	20.00	.81	5.00	2.38	4.50	7.50
449T	21 50	28.25	11.00	18.00	25.00	.81	8.75	3.38	8.25	7.50
449TS	21.50	28.25	11.00	18.00	25.00	.81	5.00	2.38	4.50	7.50

Industrial Motors

General-Purpose, Single-Phase, Drip-Proof, Rigid-Base, Capacitor-Start Motor: This type of motor is rated for continuous duty at 40 degrees C ambient temperature and is available from ¼ to 10 hp. This motor type is double-shielded or has double-sealed ball bearings. Its insulation is rated as Class B. NEMA design L (integral frames). F-1 assembly. Reversible by electrical connection. High-starting torque and low-starting current. Heavy gauge rolled-steel construction. UL recognized and CSA certified. This type of motor is used on loads that require high-starting torque and include:

- Compressors
- Pumps
- Conveyors
- Milking machines
- Blowers

General-purpose, single-phase motor

General-Purpose, Single-Phase, Totally-Enclosed, Rigid Base, Continuous-Duty, 60 Hz Motor: 40-degree C ambient temperature, double-shielded or double-sealed ball bearings. NEMA design L-1 hp ratings. Class B insulation. Reversible by electrical reconnection. Power switch (629 design and integral frames). UL recognized and CSA certified. Available from ¼ to 10 hp. Applications include:

- Compressors
- Pumps
- Machine tools
- Fans
- Blowers
- Farm equipment
- Conveyors

General-purpose, single-phase, totally-enclosed motor

Electric Motors 463

General-purpose, single-phase explosionproof motor

General-Purpose, Single-Phase, Explosionproof, Capacitor-Start Motor: This motor type has automatic thermal protection, high-starting torque, low-starting current, and is rated for continuous duty at 40 degrees C ambient temperature. It utilizes ball bearings and has a 1.0 service factor. The motor is reversed by reconnecting its motor leads.

This motor type is UL listed under File No. E-12044 and CSA certified under File No. LR21839. It is available from ¼ to 10 hp.

Applications include:

- For use in chemical/petrol refineries
- Grain elevators
- Paint booths

General-Purpose, Three-Phase, Dripproof, Rigid-Base Motor: This motor type is for use in environments where dirt and moisture are minimal. Rated at 60 Hertz, continuous duty, 40 degrees C ambient. Ball bearings. NEMA design B, class B insulation. F-1 assembly. Rolled-steel frame construction. Suitable for 230 volt PWS. UL recognized and CSA certified. Available ¼ to 3 hp. Typical applications include:

- Pumps
- Fans
- Blowers
- Compressors
- Conveyors

General-purpose, totally-enclosed rigid-base motor

General-Purpose, Totally-Enclosed, Rigid-Base, Three-Phase Motor: Service Factor of 1.15. 60/50 Hertz (50 Hertz, 190/380 V). Next lower hp on 208 – 230/460-V motors). Continuous duty. Ball bearings. NEMA design B. Class F insulation on 180 frame and above. Class B insulation on 56 – 140 frame. All 230/460-V motors are suitable for use on 230-V PWS. Available ¼ to 10 horsepower. Applications: For use in environments which are damp, dusty or dirty. Typical applications include:

- Pumps
- Fans
- Blowers
- Compressors
- Conveyors

High-efficiency, severe-duty, three-phase motor

General-Purpose, Premium Ultra-High Efficiency, Severe-Duty, Rigid-Base Motor: 1.15 Service Factor NEMA design B. All cast-iron construction. Oversized, gasketed threaded cast-iron conduit box. Internal surfaces treated with corrosion-resistant epoxy paint. Brass drain and breather in shaft end bracket. Bearing caps, (front and rear) frame 254T and larger. Lead separator. Low-loss electrical grade lamination steel. External epoxy paint. Double-shielded ball bearing. Regreasable without disassembly. Bi-direction, nonsparking, polypropylene fans. Stainless-steel nameplate. zinc dichromate hardware. Molded neoprene shaft slinger. 1200 percent copper windings. UL recognized, non-hygroscopic Class F insulation system. CSA certified. These high-efficiency motors offer substantial energy savings over standard designs. Rugged construction and premium features make these motors ideal for the process industries. Available $^3/_4$ to 100 hp.

AC and DC Electric Motors, 1 to 35,000 Horsepower

The Optim line of motors is manufactured by the Westinghouse Motor Co., who has four lines of electric motors.

1. The Optim T-Frame motors, SE, HE mill and chemical, and XP in sizes 1 to 250 HP, 1200 to 3600 RPM, 230/460 V.

2. Induction motors and generators in the World Series (LhC): 250 to 8000 HP, 300 to 3600 RPM, 380 to 6900 V, in the M-Series: 250 to 1250 HP, 300 to 3600 RPM, 380 to 4000 V, in the H-Series (MAC) & PAM; 4000 to 25,000 HP, 300 to 3600 RPM, 2300 to 13800 V.

3. Synchronous motors and generators in slow speed and high speed models: 2000 to 30,000 HP, 180 to 1800 RPM. 2300 to 13800 V.

4. DC motors and generators in mill, marine, and industrial types 250 to 35000 hp in the AISE Mill: MC600, 700, 800 Series.

T-Frame, AC, Three-Phase Motors: TEFC Designs Optim Motors are a high quality line of AC induction motors that consists of four T-frame products: Optim SE (standard efficiency), Optim HE (high efficiency, Optim Mill and Chemical (mill & chemical duty and high efficiency), and XP (explosionproof). Design features include: a rugged cast-iron frame and end brackets. Class F insulation system. High performance bearings, 1.5 service factor, 90 degrees rotatable conduit box. High-grade steel laminations, corrosion-resistant external fan, reinforced end-turn lacing, copper windings with advanced varnish process, neoprene lead separator, rust-proof base coat on all surfaces. 1 to 250 hp size range.

Single-phase, rigid-base motor

Single-Phase, 1.0 Service Factor, Rigid-Base Motor: 60 Hertz continuous duty, 40 degrees C ambient temperature. Double-shielded ball or all angle-sleeve bearings. Class B insulation. Reversible by reversing motor leads. Moderate starting torque. Power switch (629 design). UL recognized and C5A certified. Available $\frac{1}{4}$ to 1 horsepower. Suitable for use on:

- Saws
- Tools
- Fans blowers
- Coolers
- Other applications where power does not exceed nameplate rating.

Single-phase, resilient-base motor

Single-Phase, 1 0 Service Factor, Resilient-Base Motor: 60 Hertz, continuous duty, 40 degrees C ambient temperature. All angle-sleeve or double-sealed (629 design) ball bearings. Class B insulation.

Reversible by reversing motor leads. UL recognized and CSA certified. Available in fractional-hp sizes from $\frac{1}{4}$ to 1 hp. Suitable for use with:

- Fans
- Blowers
- Tools
- Milking machines
- Evaporator coolers
- Pumps
- Other applications requiring power not exceeding the nameplate rating.

Capacitor-Start TEFC Electric Motor: Reversible by changing motor leads. 60 Hertz tri-V auto-thermal protection. Suitable for 50 Hz at 1.0 service factor at 110/220 V. Rigid-base mounting. Ball bearing. 40 degrees C ambient temperature, continuous duty. Used for:

- Pumps
- Compressors
- Air handlers
- Machine tools
- Blowers in dusty noncombustible areas

Capacitor-start, TEFC motor

Belt-drive, split-phase motor

Belt-Drive Split-Phase Motors: Suitable for a wide range of commercial and industrial application. 115V and 230V, 115/230V 60 Hz (50 Hz at 1.0 service factor) Class B insulation. Reversible. Continuous duty. Sleeve and ball bearings. Open and TEAO. Single and two speed models available. Horsepower rating $\frac{1}{6}$ to $\frac{1}{2}$.

Commercial face-mounted motor

Commercial Face-Mounted Motors: Open Drip-Proof; both single-and three-phase models are available. Suitable for fan, gear and pump applications. Single and dual V. 60 Hz 1 Speed. 4 and 6 pole reversible. NEMA "56C" mount. $\frac{5}{8}$-inch keyed shaft. Open drip proof. Ball bearing and 40 degree C ambient temperature rated. Split Phase, reversible model available $\frac{1}{6}$ - $\frac{1}{2}$ hp. Capacitor-start, reversible models are available in $\frac{1}{3}$ - $1\frac{1}{2}$ hp. Three Phase, reversible models are available in $\frac{1}{3}$ - 2 hp.

HVAC Fan/Blower Motors

Double-shaft, 5-in diameter motor

Double-Shaft, 42 Frame, 5-In Diameter Motors on Cradle Base: 115, 208 - 230 V, 60 Hz only. 40 degrese C ambient temperature. Auto protector, continuous air-over cooling. CCWLE rotation. Sleeve bearing. $\frac{1}{2}$-in diameter flatted shafts. Open construction, $3\frac{1}{2}$ shaft height. Available in $\frac{1}{30}$ to $\frac{1}{4}$ hp.

Double-shaft, 5-in diameter motor without base

Double-Shaft 42 Frame, 5-Inch Diameter Motors Without Base: 115, 208-230,277 V 60 Hz only. 4O degree C ambient temperature. Auto protector. CCWLE Rotation, sleeve bearing. $\frac{1}{2}$-in diameter - flatted shafts. Open construction. $2\frac{1}{2}$ -in diameter rings. Available $\frac{1}{30}$ to $\frac{1}{5}$ hp.

Double-shaft, high-efficiency motor

Double-Shaft High-Efficiency Motor: Available for connection to 115, 230, 277 V 60/50 Hz. 40 degree C ambient temperature. CCWLE rotation, continuous air over. Auto protector. Energy efficient. Sleeve bearing. Semi-enclosed construction. $\frac{1}{2}$-in diameter, double-shaft with length adapter. $2\frac{1}{4}$-in to $2\frac{1}{2}$-in resilient rings. Available $\frac{1}{8}$ to $\frac{3}{4}$ hp.

Three-Phase Commercial Deluxe Fan Canopy Motor: "Patented" fan canopy. 200-230/460 V and 575 V 60 Hz only. Class B insulation. Heavy-duty ball bearing. 60 degree C ambient temperature. Auto protector. High-thrust capacity. Reversible. Available in sizes from $\frac{1}{2}$ to 1 hp.

Commercial Condenser Motor: 200-230/460V. 60/50 Hz. 60 degree C ambient temperature. Class 8 insulation and auto protector. Reversible. Quick-connect terminal board. Double sealed ball bearings. $6\frac{1}{2}$-in diameter with $\frac{5}{8}$-in key/flatted shaft. Moisture-protecting rain shield. Threaded conduit hole-shaft end. Locating screws for ease of assembly. Hybrid-epoxy corrosion -resistant paint finish. Available 1 and $1\frac{1}{2}$ hp.

Indoor blower, high-efficiency 48-frame motor

HVAC PSC Indoor Blower, Special High-Efficiency 40 Frame Motor: $2\frac{1}{4}$ in - $2\frac{1}{2}$ in resilient rings. 60/50 Hertz 40 degree C ambient temperature. Auto protector. Continuous air over. High efficiency. llS, 208/230 and 277 V. Reversing plug. Multiple mount. Sleeve bearing. $\frac{1}{2}$ in diameter flatted shaft. 26 inch long leads (minimum). Open construction. Available $\frac{1}{6}$ to $\frac{3}{4}$ horsepower.

Outdoor fan, energy efficient motor

HVAC PSC Outdoor Fan Motor: Semi-enclosed 208-230V and 460V 60/S0 Hz. Class B insulation (4-, 6-, or 8-pole). 60 degree C ambient. 26-inch leads and lead reversible. Energy efficient. Ball bearing. Vertical (shaft up) or horizontal mount. Closed except lead end. $\frac{1}{2}$-in diameter shaft, double-flat. Shaft slinger. Available $\frac{1}{6}$ to $\frac{3}{4}$ hp. Used for heat pumps, and refrigeration — specially suited to cold weather.

Single-phase, explosionproof split-phase motor.

HVAC Fan/Blower, Totally-Enclosed, Explosionproof, Split-Phase Motor: Resilient-base, single-phase motor: Residential and commercial fans and blowers for air conditioning, centrifugal pumps, roof ventilators, exhaust fans, and other applications where required starting torque is low to moderate. 60 hertz, continuous duty, 40 degree C ambient. Double sealed ball bearing. 1.0 service factor. Class B insulation. Reversible by electrical reconnection.

Extended through-bolts. UL recognized and CSA certified. Available $\frac{1}{4}$, $\frac{1}{3}$ and $\frac{1}{2}$ hp.

Definite-purpose, direct-drive. three-speed motor

Definite-Purpose HVAC Direct-Drive Fan and Blower Motor: Three-speed, electrical reversible motors that are replacement motors for: Bard, Chrysler, York, Armstrong, Johnson, Janitrol, Bonn, Lennox, Carrier, American Standard Westinghouse and other brands of HVAC equipment. Permanent split capacitor. Energy efficient. Sleeve and ball bearings. Continuous duty. Air over. Open construction 1075 RPM. Available $\frac{1}{4}$ to $\frac{3}{4}$ hp.

Air-compressor motor

Definite-Purpose Air-Compressor Open Drip-Proof Motor: Continuous duty, 40 degree C ambient. 1.0 service factor. Reversible (connected CWLE). 3450 RPM. Rigid base. Manual thermal overload. Ball bearing. Available $\frac{3}{4}$ to 5 horsepower.

Pool-filter pump motor.

Definite-Purpose Pool Filter Pump Motor: 115/230 V, 60 Hz, 52 Degrees C ambient. NEMA C-face mount. Sealed ball bearing. All copper windings. Open drip proof. Sealed switch design. 304 bearing, shaft end. 303 stainless steel shaft on threaded (J)shaft. Capacitor start. Capacitor run. Available $\frac{1}{2}$ to 3 horsepower.

Definite-Purpose Easy-Clean Wash-Down Motor: Totally enclosed, three-phase. These totally-enclosed motors have been specifically developed to provide long, reliable service life in the most severe wash-down processing applications. Each one has been designed and manufactured to resist the highly corrosive effects of cleaning in place with steam, high-pressure water, detergents and sanitizing materials. Each motor includes these easy-clean-plus features: L-Base,heavy-duty, zinc-plated. Bearings are double-shielded. Gasketed top-mounted saddle box; F-l/f-2 mounting is standard. End shields and rugged cast-iron frame. Corrosion-resistant polypropylene. Fan Cover is feavy-duty corrosion-protected smooth steel. Finish is USDA accepted gloss white epoxy. Frame is heavy-duty rolled steel, corrosion-protected with plating. insulation-treated for moisture and vibration resistance. moisture drains. Drilled and unplugged. Nameplate- Special easy clean stainless steel. Rotor-Cast-iron aluminum with epoxy coating. Shaft-300 series stainless steel. Shaft is double-lip seals . . . both ends. 60 Hz, 40 degree C ambient temperature, continuous duty. Class F insulation. 1.15 service factor. Available in $\frac{1}{2}$ to 3 horsepower.

Definite-purpose Inverter-duty motor

Definite-Purpose, Inverter-Duty Motors: Variable torque and rated at both standard and severe duty. Standard Features- Class F thermostats, continuous duty, 40 degree C ambient. Cast-iron frame construction. 1.0 service factor on VFD power. Multiple dips and bakes. Shaft slinger. 230/460 V through 100 hp; 460 V only 125 through 350 hp. Severe-duty features. Internal surfaces treated with corrosion-resistant epoxy paint. Lead separator. All cast-iron construction. Brass drain and breather, shaft end. Available 1 through 350 horsepower.

Index

A

ac hum, 386

ac induction motors
    identification and selection, 455-464

ac and dc motors, 464

Acoustical T-bar clips, 162

Adapters, twist lock, 73

Adjustable bar hangers, 196, 197

Air
    compressors, 121
    handling equipment, 121

Allen wrenches, 126

Allen wrench socket set, 126

Aluminum scaffolding planks, 99

Ampacity of wire, 238

Analog/digital multimeter, 112

Anti-oxidant, Noalux, 284

Appliance carts, 87

Arc welders
    gasoline powered, 64
    electric, 65

B

Ballasts
    definition, 424
    general information, 425-429

Band saws, 50

Bar hangers, adjustable, 196, 197

Bars
    digging, 76
    Johnson, 86
    wonder/pry, 128
    wrecking/crow, 128

Basket, wire pulling, 34

Battery operated lift, 103

Beam clamp, 163

Beam hanger clamps, 162

Bellhanger bit, 46

Benders, conduit
    cable, 57
    Chicago-type, 7
    electric, 3, 9
    hand, 8
    hickey, 8
    hydraulic, 4
    manual, 5
    mechanical, 7
    offset, 8
    portable, 9
    ratchet, 7
    rigid, 5
    table, 6

Bits
    bellhanger, 46
    core, 45
    drill, 42
    ship auger, 46
    woodboring, 45, 46

Bolt cutter, 137

Box covers, 198-201

Box-end wrench, 123

Boxes
    conductor support, 222
    flanged cover, 230
    gang, 87
    gasketed screw-cover, 226
    hinged cover, 230, 231
    hinged-cover pull, 226
    pull and junction, 221
    screw cover pull, 225
    unflanged, 229

Boxes, floor
    see Floor boxes

Boxes, outlet
    see Outlet boxes

Boxes, switch
    see Switch boxes

Brass bound level, 128

Bushed elbow, 170

Builder's level, 128

Bushed elbow, 170

Bushings
    bonding, 168
    capped, 168
    male enlarger, 169
    metal, 168
    metallic insulated, 168
    plastic, 168
    reducing, 169
    threadless, 169

C

Cabinet, transformer, 226

Cable
    benders, 57, 58
    core bits, 45
    coupling, PVC, 145
    cutters, 59
    gripping gloves, 30
    hickey, 57
    hoist, 89
    locator/ground fault finder, 113
    paralleling and coiling machine, 25
    penciling tool, 58
    splicer's knife, 134
    strippers, 58
    transporting reels, 26
    types of, 240-242

Cadweld tool, 61

Canadian Standards Association
    see CSA

Carpenter's square, 130

Carrier, reel, 24

471

472 Handbook of Electrical Construction Tools and Materials

Carts
 appliance, 87
 wire, 33

Chain
 hoist, 89
 saw, 50

Chase nipple, 166

Chisels
 cold, 132
 flooring/electrician's, 133
 wood, 133

Circuit breakers, 452-454

Circuits, isolated grounding
 see Isolated grounding circuits

Clamps
 beam, 163
 beam hanger, 162
 GRC-EC, 157
 GRC-PC, 157
 GRC-RC, 157
 rigid conduit, 158

Class ratings, 294-296

Classified location enclosures, 271

Clips
 acoustical T-bar, 162
 EMT, 158, 159, 160
 EMT and GRC malleable, 154
 EMT, one-hole, medium, 153
 flexible cable, 161
 multifunction, 162, 163
 series "M" flange mount, 159
 series "P" flange mount, 159
 standard MC/AC cable, 164
 through-stud support, 164
 "Z" purlin, 161

Circular saw, 54

CO_2 fish tape system, 19

Code, National Electrical
 see NEC

Combination
 coupling, 172
 square, 129
 wrench, 123

Compacting equipment
 ground pounder, 79
 whackers, 79

Compression
 connector, 166, 171, 172

Compression *(Cont.)*
 coupling, 172
 tools, 60

Compression lugs
 one-hole, dual-rated, 284
 two-hole, dual-rated, 285

Compression splicers
 dual-rated, 286, 287
 long barrel copper, 290
 short barrel copper, 291, 292

Compression terminals
 one-hole, long barrel copper, 289
 two-hole, long barrel copper, 288

Compressor, air, 121

Concrete cutting saw, 52

Conductors
 sizes and types of, 233

Conduit
 benders, 3-10
 box support, 160
 clip, 160
 EMT, 149
 fittings, 149
 flexible, 150, 151
 measuring tape, 35
 PVC, 145
 rigid aluminum, 146
 rigid, boxes for, 181-184
 rigid galvanized, 147

Conduit bodies for EMT
 Type C, 180
 Type LB, 180
 Type LL, 180
 Type LR, 180
 Type T, 180

Conduit bodies
 covers for, 180-181
 gaskets for, 181

Conduit bodies for rigid and IMC
 Type C, 178
 Type E, 178
 Type LB, 178
 Type LL, 179
 Type LR, 179
 Type T, 179
 Type TA, 179
 Type TB, 179
 Type X, 179

Connectors
 dual-rated transformer, 283

Connectors *(Cont.)*
 EMT compression, 171
 EMT set-screw, 171
 flex metal conduit, 174, 175
 gutter, 166
 insulated steel compression, 172
 liquidtight flex, 175, 176
 rigid compression, 166
 rigid set-screw, 166
 steel set screw, 173
 two- and split-bolt, 277-280
 wire-nut, 259-263

Contactors
 lighting, 368
 magnetic, 366
 mechanically held, 371

Controls, motor
 see Motor controls

Coping saw, 55

Cords, extension, 72

Corner adapter elbow, 170

Corner elbow, 170

Corner pull-in elbow, 170

Corrosion resistant devices
 connectors, 307, 310, 311, 317
 explanation of, 303
 plugs, 307, 310, 311, 314, 318
 receptacles, 304-306, 308, 310-315

Coupling
 EMT combination, 172, 174
 EMT compression, 172
 EMT set-screw, 170
 rigid Erickson, 165
 rigid split, 165
 steel compression, 173
 steel set-screw, 173

Covers
 box, 198-201
 floor box, 207
 weatherproof, 217-219

Covers for conduit bodies
 Type BS, 180

CSA, 337, 350

Current master tester, 111

Cutters
 bolt, 137
 cable, 59
 pipe, 13

Cutting oil, use of, 12, 17
Cutting torches, 66

D

Data communication cables, 256-258

Devices, corrosion resistant
   see Corrosion resistant devices

Devices, hospital grade
   see Hospital grade devices

D-handle, 38
   extension for, 41
   hammer, 39
   motors, 38
   press, 43
   right-angle, 40
   roto-hammer, 42
   stands for, 43
   variable speed, 39

Digging tools
   bars, 76
   hand post hole, 77
   manual, 76
   posthole spoons, 77
   sharpshooter, 77
   shovels, 76

Digital meters, 109

Dimmers, 363

Disconnects, crimp
   see Crimp disconnects

Disconnects, permitted
   see Permitted disconnects

Dispenser, wire, 33

Dollies, four wheel, 86

Drill
   concrete-core, 44
   cordless, 38

Drives
   extensions, 126
   hinge-handle, 125
   nut, 127
   ratchet, 125
   T-handle, 125
   universal joint, 125

Dual-element fuses, 446

E

Elbows
   EMT 90° short, 170
   GRC, 148
   GRC 90° bushed, 170
   GRC 90° corner, 170
   GRC 90° corner adapter, 170
   GRC 90° corner pull-in, 170
   GRC 90° short, 169
   GRC 90° threadless, 169
   PVC, 146

Electrical contractors, 72

EMT
   benders, 8
   clips, 153, 154, 158, 159, 160
   conduit, 149
   conduit fittings, 149
   GRC Naylon straps, 155
   GRC shoot up straps, 155
   GRC two-hole straps, 155
   one-hole jiffy straps, 154

Enclosures
   classified location, 225
   motor starter, 231
   nonclassified location, 223, 224
   oil-tight pushbutton, 227-229

Engraver, 137

Equipment
   air handling, 121
   paint spray guns and, 122

Erickson coupling, 165

Extension cords, 72

E-Z check plus GFI circuit
   tester, 110

E-Z grip hangers, 158

F

Fastener, flex conduit stud, 161

Fault finder, cable locater, 113

Feeding sheave, 32

Fiber optic cable puller, 30

Fiberglas fish tapes, 21

Fittings, EMT steel, 172

Fittings, service
   see Service fittings

Fixtures
   incandescent recessed, 435
   lighting, 434
   recessed HID lamp, 436

Flange mount clip
   series "M", 159
   series "P", 159

Flanged cover boxes, 230

Flex conduit stud fastener, 161

Flexible
   cable, 245-248
   cable clip, 161
   cable support, 164
   conduit, 150, 151

Floor boxes
   covers for, 207
   types of, 205-207

Flooring/electrician's chisel, 133

Fluorescent lamps
   all weather, 421
   circline, 422
   classes of, 420-425
   cold-cathode, 422
   colors, 414-416
   SL and SLS compact, 424
   U-bent, 423

Fluorescent starters, 428

Folding wood rules, 131

Footcandle meter, 119

Four-way switches, 361

Four wheel dolly, 86

Frequency meter, 115

Front-end loader, 79

Fuses
   construction, 443
   dual-element, 446
   history, 442
   operation, 443, 444
   plug, 448
   requirements, 449
   right kind of, 451
   SC or Class G, 452
   when one blows, 450

Fusible elements, 444

Fustats, 448

G

Galvanized rigid conduit (GRC)
  clamps, 157
  couplings, 148
  elbows, 148
  use of, 147

Gang boxes, 87

Gasketed screw-cover boxes, 226

Generators
  portable, 69, 70
  standby, 71
  trailer mounted, 70

Gloves, cable gripping, 30

Grip, wire, 34

Ground hog digger, 75

Ground pounder, 79

Grounding
  connectors, 275, 276
  devices, 197, 198

Gutter
  definition, 222
  weatherproof rain-tight, 227

H

Hanger
  E-Z grip, 157
  minerallac conduit, 156
  snap clip flex, 161
  trapeze, 163
  U-bolt, 158
  unistrut, 156
  unistrut flex, 160

Hand hack saw, 55

Hand truck, 86

Heating blanket, PVC, 10

Hernia, prevention of, 81

Hickey, cable, 57

HID lamps
  benefits of, 425
  high-pressure sodium, 425
  mercury-vapor, 427
  metal-halide, 426

High-clearance scaffolding, 98

High-intensity discharge lamps
  see HID lamps

High-voltage power cable, 248-251

Hilti gun, 135

Hinged cover box, 230, 231

Hoe, mortar, 78

Hoist
  cable, 89
  chain, 89
  tug-it, 88

Hole saws, 53

Hospital grade devices
  receptacles, 300-302, 304,
    308, 309, 313-315

HVAC fan/blower motors, 467-470

Hy-jacker lift, 102

I

Impact wrench, electric or air, 127

Incandescent fixtures
  recessed, 435

Incandescent lamps
  base, 397-400
  bulb finish and color, 396
  filling gas, 401
  halogen, 404
  high- and low-voltage, 402
  infrared, 404
  projector and reflector, 402
  showcase, 403
  specifications, 405-414
  spotlight and floodlight, 403
  types of, 402

Industrial wagons, 87

Infrared thermometer, 113

Insulation
  classes of, 239
  resistance testing, 114
  types of, 239

Isolated grounding circuits
  explanation of, 303

J

Jacks
  pallet, 84
  reel, 32
Jig saw, 53

Johnson bars, 86

K

Knives
  cable splicer's, 134
  pocket, 134
  putty, 133
  skinning, 134
  slitting-blade, 134
  utility, 134

Knockout punch driver, 49

Knockout punches, 48, 49

L

Ladders, 106

Lamp classifications
  fluorescent, 414-425
  high-intensity discharge, 425
  high-pressure sodium, 425
  incandescent, 394-413
  mercury-vapor, 427
  metal-halide, 426

Lampholders
  covers, 219, 220
  outdoor, 219, 220

Lamps, trouble, 73

Levels
  brass bound, 129
  builders, 128
  lighted torpedo, 129

Lifting, correct, 82

Lifts
  battery operated, 103
  Hy-jacker, 102
  Selma, 102
  Tel-Hi Scoper, 104
  Tiger, 105

Lighted torpedo level, 129

Lighting
  contactors, 368
  control, 355
  fixtures, 434
  temporary, 73

Lights, portable flood, 74

Liquidtight adapter, 176

Liquidtight flex connectors, 175, 176

Loader, front end, 79

Locknuts, rigid, 167

Index

Long handle scraper, 78
Low-voltage remote-control switching, 374
Lubricants
  cutting oil, 12, 17
  wire pulling, 27-29
Lugs
  compression, 284, 286

M

Machine
  cable paralleling and coiling, 25
  taps, 17
Magnetic contactors, 366
Male enlarger, 169
Manhole sheaves, 31
Manual digging tools, 76
Mechanical lug connectors
  L-style lug, 264
  panelboard, 266
  parallel-tap, 272, 273
  splicer reducers, 267
  switchgear, 269
  tee-tap, 274, 275
  two-barrel, 269
Mechanical solderless lugs
  short-tang solderless, 282
  single-hole offset, 281
  single-hole solderless, 281, 282
Mechanically held contacts, 371
Megohmmeter, 114
Metal shear, 138
Meters
  digital, 109
  footcandle, 119
  frequency, 115
  power factor, 116
Minerallac conduit hangar, 156
Mining cable, 252
Mortar hoe, 78
Motor controls
  attributes of, 375
  fractional horsepower manual starters, 379
  integral horsepower manual starters, 380
  magnetic, 381
  magnetic coils, 384

Motor controls *(Cont.)*
  manual motor-starting switches, 380
  manual starters, 378
  plug-and-receptacle, 376
Motor identification and selection
  ac and dc motors, 464
  capacitor-start motor, 459
  general-purpose, 460
  industrial motors, 462
  motor-starting characteristics, 458
  permanent split-capacitor, 459
  single-phase, 456
Motor starters, power circuits in, 386
Motorized electric scaffolding, 101
Motors, drill, 38
Multifunction clips, 162
Multimeter, analog/digital, 112

N

National Electrical Code
  see NEC
National Electrical Manufacturer's Association
  see NEMA
NEC, 144, 147, 153, 168, 185, 236, 303, 376, 437, 439, 440, 441, 452
NEC requirements for switches, 354
NEMA
  configurations for locking plugs and receptacles, 299, 300
  configurations for nonlocking plugs and receptacles, 298
  definition, 293
  ratings, 293, 294, 368, 385,
Nipples
  chase, 166
  close, 167
  offset connector, 167
  steel offset, 174
Nonclassified location enclosure, 223, 234
Nylon fish tapes, 20
Nylon slings, 85

O

Occupational Safety and Health Act
  see OSHA
Oil-tight pushbutton enclosures, 227-229
Oil bucket and pump, 17
Open-end wrench, 123
OSHA, 37, 63, 72
Outlet boxes
  gang, 195
  most used, 186
  octagonal, 186-189
  specifications for, 185
  square, 190-193
  tile wall, 196
  utility, 194, 195
Overcurrent devices, enclosure for, 441
Overcurrent protection, 441
Overload protection
  bimetallic thermal overload relays, 390
  definition of, 388
  melting alloy thermal overload relays, 389
  selecting overload relays, 390

P

Paint spray guns, 122
Pallet jack, 84
Panelboards, 439
Penciling tool, cable, 58
Permitted disconnects, 439
Phase-sequence indicator, 119
Photoelectric switches, 362
Pipe
  and bolt threading machine, 14
  cutters, 13
  taps, 17
  wrench, 16
Pipe threaders
  geared, 12
  pipe and bolt, 14
  pony, 14
  power head, 15
  ratchet, 11
  series, 12
  tri-stand chain vise, 16

Plastic bushing, 168
Plastic pipe hacksaw, 55
Plug fuses, 448
Plugs
    definition of, 281
Pocket knife, 134
Pocket-size circuit tester, 110
Pop riveting tool, 136
Portable
    band saw, 49
    flood lights, 74
    generator, 69
Post hole digger
    hand, 77
    spoons, 77
    two-man gas-powered, 75
Power circuits in motor starters, 386
Power factor meter, 116
Power-Glo pocket tester, 111
Press, motor driven drill, 43
Propane torch, 67
Pulling socks, wire, 34
Pump and oil bucket, 17
Punches
    center, 132
    drift, 132
    drive pin, 132
    knockout, 48
    Whitney, 48
Putty knife, 133
PVC
    benders, 9
    conduit, 145
    coupling, 145
    elbows, 146
    female adapter, 145
    heating blankets, 10
    male adapter, 146

Q

Quick square layout tool, 130

R

Ratchet drives, 125
Receptacles

Receptacles *(Cont.)*
    see also Corrosion resistant devices
    combination voltage, 327-334
    definitions, 297
    125 V, 319-321
    250 V, 321-323
    277 V, 323, 324
    480 V, 324-326
    600 V, 326, 327
Reducing bushing, 169
Reel
    carrier, 24
    jack, 32
Relays
    definition of, 362, 372
    solid-state, 373
    timers and timing, 372
Remote-control switching, low-voltage, 374
Rigid
    aluminum conduit, 146
    aluminum fittings, 147
    bonding bushing, 168
    capped bushing, 168
    close nipple, 167
    conduit clamp, 158
    locknut, 167
    Meyers hub, 166
    metal bushing, 168
    metallic insulated bushing, 168
    offset connector, 167
    threadless bushing, 169
Rigid conduit
    outlet boxes for, 181-184
Riveting tool, 136
Rollers, 85
Rules
    folding wood, 131
    tapes, 131

S

Safety, 90-92, 96, 97, 107
Saws
    band, 50
    bench or table, 50
    chain, 80
    concrete cutting, 52
    coping, 55
    circular, 54

Saws *(Cont.)*
    jig, 53
    hand hacksaw, 55
    hole, 53
    plastic pipe, 55
    portable band, 55
Sawzall, 52
Scaffolding
    high-clearance, 98
    motorized electric, 101
    planks, 99
    safety, 91, 92, 96, 97, 107
    span, 92
    steel, 100
    VX folding, 97
    walk-up, 96
Scraper, long handle, 78
Screw-cover pull boxes, 225
Screw gun, 47
Selma lift, 102
Service
    fittings, 207, 208
    temporary, 72
Service switches
    kinds of, 437-440
    location, 440
Set-screw
    connector, 166, 171, 173
    coupling, 171, 173
Shackles, 85
Sharpshooter digger, 77
Sheaves
    feeding, 32
    manhole, 31
    right-angle, twin-yoke, 31
    tray-type, 32
Short elbow, 170
Shovels, 76
Skids, roller, 85
Skinning and slitting knife, 134
Sleeve, nylon, 173
Slings
    nylon, 85
    steel, 85
Snap clip flex hanger, 161
Socket wrenches, 124

Index 477

Sockets and accessories
　Allen wrench set, 126
　drive extensions, 126
　hinge-handle, 125
　individual, 124
　nut driver, 127
　ratchet drives, 125
　speed handle drives, 125
　T-handle drives, 125
　universal joint drive, 125

Solid-state relays, 373

Span scaffolding, 92-96

Special cable fabrications, 254

Specifications, switch, 351

Split coupling, 165

Squares
　carpenter's, 130
　combination, 129
　quick layout, 130
　try, 130

Stand, drill press, 43

Standard MC/AC cable clip, 164

Standby generator, 71

Staple gun, 136

Steel
　compression coupling, 173
　fish tapes, 20
　fittings, 172
　scaffolding, 100
　slings, 85
　snap-in blanks, 173

Straps
　EMT and GRC Naylon, 155
　EMT and GRC shoot-up, 155
　EMT and GRC two-hole, 155
　one-hole EMT Jiffy, 154

Strippers, cable, 58

Support
　flexible cable, 164
　through stud conduit, 164

Switch boxes
　nonmetallic, 210-212
　types of, 202-205, 208-210
　weatherproof, 213-217

Switches
　dimmer, 363
　four-way, 361
　identification, 349

NEC requirements, 342
photoelectric, 362
specifications, 351
terms, 348
three-way, 357
types of, 347

Switches, service
　see Service switches

Synchroscope, 117

T

Table saws, 51

Tachometer, 118

Tapes, fish
　CO_2, 19
　conduit measuring, 35
　fiberglas, 21
　measuring, 131
　nylon, 20
　steel, 20
　vacuum, 21

Taps
　machine, 17
　pipe, 17

Tel-Hi Scoper lift, 104

Temporary
　electric services, 72
　lighting, 73

Test equipment, 109

Testers
　analog/digital multimeter, 112
　cable locator/fault finder, 113
　current master, 111
　digital voltage/continuity, 111
　E-Z check plus GFI circuit, 110
　footcandle meter, 119
　frequency meter, 115
　infrared comparative
　　thermometer, 113
　Lite voltage/continuity, 111
　megohmmeter, 114
　phase-sequence indicator, 119
　pocket-size circuit, 110
　power factor meter, 116
　power-glo pocket, 111
　synchroscope, 117
　tachometer, 118
　voltage continuity, 110

Testing, insulation resistance, 114

Thermometer, infrared
　comparative, 113

Threaders, 12

Three-way switches, 357

Through-stud conduit support, 164

Through-stud flex conduit
　support clip, 164

Tiger lift, 105

Time and material management,
　139-141

Timers and timing relays, 372

Tool pouch and bag, 135

Torches
　cutting, 66
　propane, 67

Torque wrench, 127

Trailer mounted generator, 70

Transformer cabinets, 226

Trapeze hangers, 163

Tray cable, 245

Tray-type sheave, 32

Tripod, 16

Tri-stand vise, 16

Trouble lamps, 73

Try square, 130

Tug-it hoist, 88

Twist lock cord adapters, 73

U

U-bolt hanger, 158

Underwriters' Laboratories, 143,
　284, 300, 350

Unflanged pull boxes, 229

Unistrut
　flex hanger, 160
　hanger, 156

Uni-clip, 156

Using time wisely, 139-141

Utility cables, 254-256

V

Vacuum fish tapes, 21

Vol-Con

Vol-Con *(Cont.)*
    digital voltage/continuity tester, 111
    Lite voltage/continuity tester, 110
    voltage/continuity tester, 110

Voltage drop, 236

Voltage variation, 385

VX folding scaffolding, 97

W

Wagons, industrial, 87

Walk-up scaffolding, 96

Weatherproof
    boxes, 213-217
    covers, 217-219
    rain-tight gutter, 227

Welders
    Cadwell tool, 61, 62
    electric, 65
    gasoline powered arc, 64
    industrial, 66

Welders
    spot, 65

Whackers, 79

Wheelbarrow, 87

Whitney punch, 48

Wire
    aluminum, 237
    ampacity of, 238
    cart, 33
    dispenser, 33
    grips, 34
    hook-up, 251
    identification of, 238
    pulling sock, 34
    stranded, 237
    types of, 236

Wire-nut connectors, 259-263

Wire pulling systems
    Easy Tugger, 23
    fiber optic cable, 30
    hand powered, 22
    motor driven, 22
    portable CO_2, 19

Wire pulling systems *(Cont.)*
    self-propelled, 24

Wireway, 222, 223

Wonder Bar pry bar, 128

Woodboring bit, 45

Wood chisels, 133

Wrecking/crow bar, 128

Wrenches
    Allen, 126
    box-end, 123
    combination, 123
    impact, 127
    open-end, 123
    socket, 124
    speed, 122
    torque, 127

Z

"Z" purlin clips, 161

ABOUT THE AUTHOR

Gene Whitson is a highly skilled consultant in Scottsdale, Arizona, with more than 40 years of experience in commercial and industrial contracting, including 27 years in the electrical contracting industry. He is a former vice president of Cannon & Wendt Electric Company in Phoenix, where he was in charge of finance, tools, material, safety, contract administration, insurance, and claims adjustment. Whitson is the coauthor (with Ralph Johnson) of *Successful Business Operations of Electrical Contractors* and holds a degree in accounting from the International Accounting Society.